ANIMAL SCIENCE, ISSUES AND PROFESSIONS

NEMATODES

MORPHOLOGY, FUNCTIONS AND MANAGEMENT STRATEGIES

Animal Science, Issues and Professions

Additional books in this series can be found on Nova's website under the Series tab.

Additional e-books in this series can be found on Nova's website under the E-book tab.

ANIMAL SCIENCE, ISSUES AND PROFESSIONS

NEMATODES

MORPHOLOGY, FUNCTIONS AND MANAGEMENT STRATEGIES

FAUSTO BOERI
AND
JORDAN A. CHUNG
EDITORS

Nova Science Publishers, Inc.
New York

For permission to use material from this book please contact us:
Telephone 631-231-7269; Fax 631-231-8175
Web Site: http://www.novapublishers.com

Additional color graphics may be available in the e-book version of this book.

Library of Congress Cataloging-in-Publication Data

Nematodes : morphology, functions, and management strategies / editors, Fausto Boeri and Jordan A. Chung.
p. cm.
Includes bibliographical references and index.
ISBN 978-1-61470-784-4 (hardcover)
1. Nematodes--Morphology. 2. Nematodes--Physiology. 3. Nematodes--Control. I. Boeri, Fausto. II. Chung, Jordan A.
QL391.N4N3787 2011
592'.57176--dc23

2011026732

Published by Nova Science Publishers, Inc. † New York

Contents

Preface

Nematodes are the most numerous multicellular animals in the biosphere encountered in terrestrial or aquatic ecosystems among multispecies communities and are integrated in the food chain at five trophic levels according to their feeding habits. In this book, the authors present topical research in the study of nematodes, including the biological control of animal-parasitic nematodes and plant-parasitic nematodes; resistance and virulence in plant-nematode interactions; free-living nematodes in the marine ecosystem; management and control of the Pine Wood Nematode and slug pest management through nematodes in agricultural and horticultural crops.

Chapter I - Nematodes are the most numerous multicellular animals in the biosphere encountered in terrestrial or aquatic ecosystems among multispecies communities and are integrated in the food chain at five trophic levels according to their feeding habits: bacterivorous, fungivorous, herbivorous, carnivorous and omnivorous. The majority are not parasitic and play an important role in soil biology by cycling most nutrients. However, some species are animal- or plant-parasitic and may limit agriculture and livestock productivity around the world, leading to significant economic losses. Moreover, some nematodes have a great impact on human health because of the severe diseases they can cause. It is estimated that the number of nematodes in agricultural soil can reach twenty million per square meter and that yearly worldwide agriculture losses exceed US$100 billion.

Chapter II - All cultivated plant species are subjected to biotic stresses that can endanger crop yields and cause relevant economic losses. Genetically-based resistance may be a simple and efficient solution to protect plants against pathogens and pests. Resistance to plant parasitic nematodes has been proved to be an effective management tool that improves crop yields, lowers pest population densities, and favors the developments of effective rotation systems. Most of the studies on resistance and virulence in plant-nematode interactions have been focused on those nematode families that are the most diffused and cause the major economic losses to worldwide agriculture, i.e. the sedentary endoparasites belonging to the cyst (*Globodera* and *Heterodera* spp.) and root-knot (*Meloidogyne* spp.) nematodes. Sedentary endoparasitic nematodes enter the roots as motile second-stage juveniles (J2) that do not kill parasitized cells and induce very specialized and complex relationships with their hosts. Cyst-forming J2 induce the formation of feeding sites, called syncytia, in which few cells merge by dissolving their cell walls, while root-knot nematodes induce the formation of few discrete giant or nurse cells; both types of feeding sites have the role of actively transferring solutes and nutrients toward the developing nematode. Once their feeding sites have been arranged, J2 become sedentary and begin to develop into enlarged adult females through subsequent molts to J3 and J4. Several genes conferring resistance to sedentary endoparasitic nematodes (*R*-genes) have been isolated from plants. They are generally expressed constitutively in proteins that are assigned to the recognition of the presence of nematode avirulent factors. The interaction between *R*-proteins and avirulent factors induces conformational changes in *R*-proteins that lead to signaling of defense response. Avirulence to specific *R*-genes is inherited as a single dominant trait in some plant parasitic nematode species. It seems that the loss of this trait determines the failure of pest recognition by resistant plants and produce the development of virulent populations. Virulent populations are able to develop on resistant crop plants and can be selected in laboratory and in the field by repeated exposure to *R*-genes. This chance to break resistance by field populations can endanger resistance durability and the use of resistance as a valid management against nematodes for farmers. However, *R*-genes exert high selection pressures on nematode populations, thus leading to fitness costs, such as loss of competitiveness and lower reproduction potentials on susceptible hosts with respect to wild field populations. Moreover, virulent populations specifically develop only on plants that carry the gene on which selection occurred. In this chapter, the genetic bases of resistance and virulence, along with the most up-to-date studies on the molecular events and interactions leading to incompatible/compatible responses of the plants to nematodes, will be described, with particular attention to the role of some important hormones in plants, such as salicylic acid, and genetic variability of nematode populations.

Chapter III - The phylum Nematoda (roundworms), encompasses up to 500,000 species. Nematode infections in humans include ascariasis, trichuriasis, hookworm, enterobiasis, strongyloidiasis, filariasis, and trichinosis, etc. Recent data has demonstrated that approximately 60 species of roundworms parasitize humans. According to World Health Organization (WHO) report 2005, approximately 0.807-1.221 billion humans have ascariasis, 604-795 million have trichuriasis, 576-740 million have hookworm infections, and more than 120 million have lymphatic filariasis, worldwide. Lymphatic filariasis is one of the only six infectious diseases considered by WHO as eradicable with the currently available tools by the year 2020. The main problem regarding the chemotherapy of filariasis is that no safe and effective drug is available to combat the adult human filarial worms. For a rational approach to antifilarial chemotherapy, knowledge of the biochemical composition and metabolic pathways operative in filarial parasite are of paramount importance, so that more potent antifilarial agents based on specific drug targets can be identified. In fact, even the most widely employed drug in the treatment of lymphatic filariasis, diethylcarbamazine (DEC), was discovered by a random screening of compounds fifty years ago and still does not have a defined mechanism of action. Further advancement of drug treatment and vaccine development would benefit greatly from an improved understanding of basic filarial biology. The drugs used for lymphatic filariasis are albendazole, diethylcarbamazine (DEC) and ivermectin. None of these is effective in killing the long-lived adult worms (macrofilariae) and the treatments are therefore aimed at reducing transmission and pathology. Although it is impractical, prolonged treatment with doxycycline significantly reduces the numbers of microfilaria in circulation, which is an important strategy to control transmission of filariasis by mosquito vectors. The ~90 megabase genome of the human filarial parasite *Brugia malayi* consists of 5 chromosomes, including an XY sex-determination pair and predicts ~11,500 protein coding genes in 71 Mb of robustly assembled sequence. The filarial parasites are capable of assimilating oxygen under appropriate conditions, but none can completely oxidize substrates to carbon dioxide and water. All accumulate organic end products, indicating little or no terminal respiration. Energy yielding biochemical pathways, nerve transmission and neuromuscular conduction regulating parasitic motility represent the targets for action of many antifilarials. Folate, nucleic acid, polyamine metabolism and other biochemical pathways are being explored as alternate target for chemotherapy. Microfilariae have proved to be the most vulnerable target so far. However, microfilaricides yield only transient cure. Vulnerable targets of adult parasites are badly needed to design macrofilaricidal drugs for any lasting solution to filariasis. The following review summarizes the biochemistry of filarial parasites and the metabolic difference with the respective hosts in the gross pathway as well as in molecular and biochemical properties. This information in turn can be meaningfully utilized for understanding the host invader interaction and managing it in favour of mammalian host.

Chapter IV - Nematodes are the most abundant metazoans on Earth. Estimates of global nematode species diversity vary widely, but currently seem to agree that there are about one million species, a number that is comparable to what is reported for other very diverse animal phyla. Nematodes are the dominant taxon in marine sediments in terms of both abundance and biomass. However, the great importance of this group is not only related to these characteristics, but also to its close relationship with the other organisms playing a key role in trophic webs. Theeffects of nematodes on sediment stability are similarly crucial. The relative stability of populations, the short turnover rate, and the generallyhigh tolerance to ecosystem alterations make Nematoda particularly suitable as bioindicators, a benefit that is enhanced because collection methods and sample treatment for these species are relatively simple. For this reason this phylum has been used as an indicator for assessing the ecological quality of marine ecosystems according to the Water Framework Directive (WFD, Directive 2000/60/EC). In addition, the high taxonomic diversity of nematodes is also related to their large trophic spectrum and the variety of their buccal structures. The latter, and other traits such as body and tail shape and life strategy, are easily recognizable, meaning that nematodes can be distinguished on both a morphological and functional basis. Functional diversity is an important component of biodiversity, but if compared to taxonomic diversity, it is still poorly applied. Nevertheless it could be very promising when it comes to improving the authors' knowledge of ecosystems and relationship between the diversity patterns of nematodes and the abiotic environment. Recent studies have suggested that biodiversity loss might impair the functioning and sustainability of ecosystems. Accordingly, monitoring the quality of the environment appears to be essential for devising effective protection strategies and appropriate forms of management. In this respect, the benthic environment is the best domain and nematodes the ideal candidates. The aim of this paper, therefore, is to review past and present advances in the use of nematodes as bioindicators for the monitoring of marine ecosystems, and to highlight future perspectives.

Chapter V - *Caenorhabditis elegans* is a small, free-living soil nematode that survives by feeding on microbes, primarily bacteria. *C. elegans* has been used as a model organism for a variety of reasons, including economy, and the ease of maintaining a worm population in the laboratory. A useful feature of *C. elegans* is that it is relatively straightforward to test pathogenesis. As the worm feeds on bacteria, when feeding bacteria are replaced by pathogenic bacteria, the worm will suffer an infection. The number of surviving worms and other pathogenic traits can be easily monitored, helping in the deciphering the underlying pathogenic traits. In this chapter, the authors will present and discuss some examples on the use of the nematode *C. elegans* as an infection model to study the pathogenic traits of several Gram-positive and Gram-negative pathogens. Results from the authors' own experience in using the nematode as an infection model to unveil virulence determinants of bacteria of the *Burkholderia cepacia* complex are also described. The major advantages and pitfalls of the *C. elegans* infection model are also discussed.

Chapter VI - The Pine Wood Nematode (PWN) *Bursaphelenchus xylophilus* (Steiner and Buhrer) Nickle is the causal agent of the pine wilt disease, and a quarantine organism within the European Union. This pathogenic organism was detected in Portugal in 1999 associated to dead maritime pines (*Pinus pinaster*), with its vector among trees being the native insect *Monochamus galloprovincialis* (Oliv.). Over the years, management and control of both the nematode and the vector has been implemented by the Portuguese National Programme to Control the Pine Wood Nematode (PROLUNP). This programme set the guidelines for phytosanitary measures and procedures incorporating the latest scientific knowledge available. In the period from 1999 to 2008 the nematode was restricted to Setúbal Peninsula. Subsequently new pine wilt focuses were detected in Central Continental Portugal and Madeira Island. The current management strategy is focused on three core points: (i) identification of nematode-affected areas; (ii) prevention of dispersal of wilt disease to new locations, (iii) and reduction of wilt incidence in affected areas. Identification of affected areas (i) is based on periodic surveys in pine forest during the autumn and winter periods (when wilting symptoms are conspicuous) and on the collection of wood samples from symptomatic trees for PWN lab identification. The detection of nematode in new locations implies the delimitation of an Intervention Zone encompassing the infected area and a bordering buffer zone (nematode-free). The spread of wilt disease (ii) occurs by the flight of the insect vector and by transportation of wood infected with the PWN and the vector. Scientific evidence suggests that *M. galloprovincialis* makes short-distance flights of only a few hundred meters if adequate hosts are available. Dispersal of wilt disease can be avoided by trunk-injection with insecticides in healthy pines, along with an intensive monitoring system in the buffer zone. To prevent anthropogenic dispersion, transport of wood between infected and non-infected areas is officially regulated, and technological procedures such as heat-treatment of wood are also applied. Reduction of pine wilt incidence in affected areas (iii) is achieved by the harvesting of wilted trees and biomass (including branches of all dimensions) from November to April, when the insect vector is inactive. In the period May-October an active complementary management approach against other declining agents such as bark beetles is implemented, consisting on the immediate cut of affected trees and the installation of a trap network baited with pheromones and host kairomones to monitor and diminishing both the bark beetle and the *Monochamus* populations. The authors' case-study in Tróia consisting in a 10-year management program in a 400ha maritime pine forest has successfully diminished wilt incidence and *M. galloprovincialis* populations to residual levels. This is clearly demonstrative that the aforementioned control and management strategies are successful, when implemented with efficiency and within the correct time schedules.

Chapter VII - Gastro-intestinal nematode infections are highly prevalent in herbivores worldwide and cause serious economical losses in ruminant breeding. Up to 11 classes of anthelmintic drugs exist to control gastro-intestinal nematode infections. Each class of anthelmintic drugs is characterized by a specific mode of action although multidrug resistance mechanisms may be involved. Two main groups can be defined according to their action on membrane ion channels, or on their biochemical action on enzyme inhibition. Anthelmintic resistance has been recorded extensively for only 3 or 4 classes of anthelmintic drugs, namely, benzimidazoles, macrocyclic lactones and tetrahydropyrimidines or levamisole. These anthelmintic drugs classes are efficient against gastro-intestinal and pulmonary nematodes and represent the main portion of anthelmintic drugs used in the world. Newly developed drugs include monepantel and derquantel Despite a remarkably diverse choice of anthelmintic drugs, gastro-intestinal nematode infections are poorly controlled in the field. Anthelmintic resistance has become so frequent worldwide that even if anthelmintic drug classes act in a different manner on nematode biology, it is unlikely they can reverse the increasing reports of multi-resistant populations of nematodes. Mathematical models have been elaborated to address the importance of genetic mechanisms (monogenic or multigenic) and to evaluate the sustainability of breeding management systems (number of treatment, management of pastures…). This theoretical approach allows for the simulation, *in silico*, of very low initial resistance allelic frequencies and over successive years which is not tractable via experimentations. In this chapter, the authors propose an exhaustive presentation of the available classes of anthelmintic drugs, their mode of action, and if known, the identified drug resistance mechanisms (benzimidazoles), or putative ones (macrocyclic lactones, tetrahydropyrimidines, levamisole). After a presentation of factors associated with the increase in anthelmintic resistance, the available strategies to slow down the development of anthelmintic drugs resistance such as: Targeted Selective Treatment (TST), Targeted-Treatment (TT), and the use of drug combinations (combos) will be evaluated. The authors also discuss the agricultural practices implemented to control the dynamics of gastro-intestinal nematode infections in field conditions.

Chapter VIII - Slugs are considered a persistent problem in a variety of agricultural and horticultural crops. They are widely dispersed, prolific and damage the crops reducing their yield and quality, thus making them unmarketable. One individual slug can kill up to a 50 wheat seeds in the first week of sowing. An yield loss of 50-90% in corn and soybean has been reported due to slugs. The high cost of chemical pesticides and their hazardous effect on the environment demand an alternative approach for slug pest management that should be eco-friendly and cost effective. The nematode belonging to the species *Phasmarhabditis* serve as a biological control agent of slugs. The infective juveniles of *Phasmarhabditis hermaphrodita* move through the soil, locate and infect the slug by penetrating through a natural opening at the posterior end of the mantle. Once inside the host the nematode releases a bacteria *Moraxella osloensis*, the main killing agents which multiply profusely and produce endotoxin which is tolerant to heat and protease treatments. Thus, the nematode just acts as a vector in transporting the bacterium into the shell cavity of the slug. The nematode-bacterium complex is a lethal combination which leads to the death of slug within 7-21 days. Although this bioagent has been commercialized in some countries under different trade names, however, there is still a need to find out its vital optimum application rates, timing and proper application technology to make it cost effective and include it in the pest management programme.

Chapter IX - The communities of nematodes in six orographic units of the Slovak Republic and in three different types of grassland were investigated. The types of grassland included new meadow established on arable soil, permanent meadow and permanent pasture.

Out of a total nematode specimens isolated from investigated grassland, 90 species and 66 genera were identified. Majority of identified species belonged to the orders Tylenchida, Dorylaimida and Rhabditida. The highest diversity of nematode species and genera was recorded at localities with lover altitude and higher average annual temperature. Cluster analysis showed that the structure of nematode communities and diversity depended more on meadows age and on their utilization than on geographical, climatic and soil conditions of sites. Plant parasitic nematodes were dominant trophic group in all types of grassland. Bacterial feeders were subdominant trophic group only in new meadows in arable soil, where soil was characterised by higher resources of organic matter. On the other hand, omnivores characterised by a long life cycle, and low abundance especially in cultivated soil, were subdominant trophic group in permanent meadows and pastures. The Maturity Index (MI) values were lover in new meadows than in permanent meadows and pastures. Low proportion of Plant Parasitic Index and Maturity Index (PPI/MI) in permanent meadows and pastures is characteristic for natural ecosystems, without environmental destruction. Using the Structure Index and Enrichment Index, all localities of new meadows on arable soil were characterised as moderately disturbed with matured food web and N-enriched conditions, and many localities with permanent pastures were characterised as undisturbed with structure food web conditions.

Chapter X - Anthropogenic habitat loss and global climate change are factors that have been demonstrated to directly affect host parasite systems because they increase the possibility of contact among potential hosts (animals, human). Zoonosis, the transfer of parasites from domestic or wild animals to man, is a problem of medical and economic importance in the Americas. Several studies exist concerning zoonosis in Latin America, principally reported transfer of parasites from domestic animals. However, there are virtually no studies of zoonosis with wild mammals, even those that are relatively tolerant of human presence and frequently found in suburban and urban habitats. This is of special concern in Mexico, which ranks second in total mammal diversity, and the study of parasite communities in Mexican wild mammals remains relatively unexplored. This chapter presents a revision of the information available concerning the parasitic nematodes of wild animals of Mexico with emphasis on the most abundant and ecological generalist wild canids: coyotes (*Canislatrans*) and the gray fox (*Urocyoncinereoargenteus*). These canids are of special interest because of their flexible and generalist feeding behavior and because urban encroachment of their natural habitat has increased the probability of parasite transfer to humans (eg. *Ancylostomacaninum*, *Toxocara* sp., *Strongyloidesstercoralis*) and the encounter chances with domestic and feral dogs. Management strategies, based on basic knowledge of the parasitofauna of these two canid species are needed to monitor, prevent and control of zoonosis including identification of high risk zones, animal treatment and relocation and increase of public awareness. Such strategies may help to avoid negative consequences for wild mammals, and parasite monitoring should be part of an integrated conservation and public health program for Mexico and others countries of Latin America.

Chapter XI - The authors are investigating alien rodents that occur within Japanese Islands for their parasitic nematodes, because some alien nematodes derived from such alien hosts could become highly pathogenic agents to human and livestock, and because the parasites potentially could switch their hosts to and having a negative impact on native host-parasite relationships that have evolved as a result of adaptive radiation or co-speciation between the native rodents and parasitic nematodes. The authors are generalizing the patterns of host-parasite associations, and providing 4 types: alien rodents and alien nematodes such as *Callosciurus erythraeus* and *Brevistriata callosciuri*, and *Myocastor coypus* and *Strongyloides myopotami*; alien rodents and native nematodes such as *Rattus rattus* and *Heligmosomoides kurilensis* shifted from *Apodemus speciosus*; native rodents and alien nematodes such as *Apodemus argenteus* and *Heligmosomoides polygyrus* shifted from *Mus musculus*; native rodents and native nematodes such as *Apodemus* spp. and *Heligmosomoides* spp. Hence, aberrant host-parasite relationships due to the presence of alien rodents and/or nematodes comprise the 1st, 2nd and 3rd groups. These relationships should be discriminated from the native host-parasite relationships within Japanese natural ecosystem. Some potential strategies including monitoring survey for reduction of risk affected by the aberrant relationships were discussed.

Chapter XII - Nematode parasites of marine fishes have been recognized as an important component of global biodiversity, but also some species represent a potential risk for human health. The aim of the present study was to assess the changes with time in abundance and prevalence of anisakid nematodes in two commercially important fish species (the red rockfish *Sebastes miniatus*, and the ocean whitefish *Caulolatilus princeps*) from Baja California, Mexico. A total of 301 fish were analyzed (210 of *S. miniatus* and 91 of *C. princeps*) from February to December 2005. The nematode species were identified by the traditional morphological methods. Three nematode species belonging to the genera *Anisakis*, *Hysterothylacium* and *Pseudoterranova* (Anisakidae) were found, which showed certain specificity for host organ. The most abundant species was *Hysterothylacium* sp. (*S. miniatus* = 37 ind/host, and *C. princeps* = 15.5 ind/host), and the parasitic prevalence was higher in *Anisakis* sp. (*S. miniatus* = 91.8% and *C. princeps* = 95%). The group of nematodes showed a significant change in abundance with the time in *S. miniatus* (KW, $H = 13.05$, $p = 0.023$), and *C. princeps* (KW, $H = 23.37$, $p = 0.000$), with a marked decrease during winter. Larval stages of *Anisakis* and *Hysterothylacium* were particularly important due to their high abundance and prevalence and because they represent a human health risk (anisakiasis). In this respect, this chapter also provides an illustration of the species found, its location within the host and some recommendations to prevent infection in people.

Chapter XIII - *Lagochilascaris minor* is a nematode only rarely encountered in animals and man that causes purulent abscesses in the subcutaneous tissue of the ear, neck, jaw, orbit, mastoid and retropharyngeal tissue of humans. Sometimes, *L. minor* invades the pulmonary tissue and central nervous system in fatal cases. The extraordinary capacity of *L. minor* to migrate across different human tissues can also be observed in animal models of the disease such as mice and cats. In mice orally inoculated with *L. minor* infective eggs, hatched larvae can soon be observed migrating in the intestinal tract. After the process of hatching, third stage larvae (L3) migrate through intestinal mucosa reaching vessels and hepatic parenchyma and disseminate to other tissues such as lungs, skeletal muscles, subcutaneous tissues, renal parenchyma and pancreas. In cats inoculated with skinned carcasses of infected mice, L3 migrate through the esophagus, pharynx, trachea, rhino and oropharynx, and cervical lymph nodes.

Chapter XIV - The nematophagous fungus *Pochonia chlamydosporia* (Clavicipitaceae) can parasitize eggs of nematodes including cyst (*Globodera* spp., *Heterodera* spp.), root-knot (*Meloidogyne* spp.), false-root knot (*Nacobbus* spp.) and reniform (*Rotylenchulus reniformis*) nematodes. Its potential as a biological control agent has been the subject of numerous studies aimed at understanding the micro-ecological conditions, including the tritrophic (*e.g.*, plant, fungus, nematode) and host-parasite relationships that allow the fungus to thrive in the soil and rhizosphere environment. In order to effectively exploit the fungus in regulating plant endo-parasitic nematodes, a careful selection of fungal isolates (biotypes) appropriate for both host plant and nematode is essential. *Pochonia* survives in soil in the absence of a nematode host and, although it behaves as a saprophyte, research evidence points to a physiological 'switch' from the saprophytic to the parasitic stage that is triggered by nutrition. Fungal chlamydospore-based formulations for application to soil as an inoculum have been shown to be viable commercial formulations. The basic biology and physiology of the fungus is herein reviewed to provide new insights into the host-parasite interaction and commercial production methodology.

Chapter XV - Nematodes play important role in terrestrial and marine ecosystems; nevertheless nematode species diversity in ecology has been poorly studied. Taxonomic intractability is one of the main reasons that has resulted in poor understanding of biodiversity of nematodes from different ecological niches. The era of genomics is playing a pivotal role in exploring and understanding nematode biodiversity in different environments ranging from terrestrial to deep-sea ecosystems. In this review paper the authors will discuss how DNA barcoding and massively parallel sequencing approaches have helped the authors gain a better understanding of nematode diversity from different environments. Also the authors will discuss wider applicability of massively parallel sequencing technique at the transcriptome level so as to understand the role of this key metazoan group facilitating major ecosystem processes.

In: Nematodes
Editors: F. Boeri and J. A. Chung
ISBN: 978-1-61470-784-4

Chapter I

Biological Control of Animal-Parasitic Nematodes and Plant-Parasitic Nematodes

Alessandro Spalenza Maciel, Hugo Agripino de Medeiros, Everaldo Antônio Lopes and Paulo Afonso Ferreira
Federal University of Viçosa, Brazil.

1. Introduction

Nematodes are the most numerous multicellular animals in the biosphere encountered in terrestrial or aquatic ecosystems among multispecies communities and are integrated in the food chain at five trophic levels according to their feeding habits: bacterivorous, fungivorous, herbivorous, carnivorous and omnivorous (Yeates et al., 1993; Kimpinski and Sturz, 2003). The majority are not parasitic and play an important role in soil biology by cycling most nutrients (Bilgrami and Gaugler, 2004). However, some species are animal- or plant-parasitic and may limit agriculture and livestock productivity around the world, leading to significant economic losses (Bekal and Becker, 2000; Ahman et al., 2002; Jansson and Lopez-Llorca, 2004; Chauhan et al., 2005). Moreover, some nematodes have a great impact on human health because of the severe diseases they can cause (Bundy 1994; Crompton, 2000;). It is estimated that the number of nematodes in agricultural soil can reach twenty million per square meter (Barron, 2003) and that yearly worldwide agriculture losses exceed US$100 billion (Bird and Kaloshian, 2003).

Currently, the expression "sustainable agriculture" has being highlighted and is generally defined as the production of foods for human consumption with maintenance of ecosystem productivity, preservation of environmental quality. and conservation of nonrenewable resources and the biological diversity (Ferraz et al., 2010). There is increasing pressure by customers for animal and plant products free of chemical residues as well as a substantial interest by agriculturists for production systems with low use of chemical inputs. Therefore, for development of sustainable alternatives for plant and animal nematode control, it is necessary to understand traditional methods, which are expensive and temporary efficiency, based on the use of nematicides and anthelmintics, and associated with environmental pollution and harmful effects to humans, the fauna and the flora (Kerry, 2000; Ahman et al., 2002). Furthermore, the effectiveness of these chemicals has being threatened by resistant parasites which requires a continuous supply of new molecules (Ahman et al., 2002; Waghorn et al., 2002; Chandrawathani et al., 2004, Jackson and Miller, 2006). In this context, there arises integrated control which promotes the rational use of a set of biological, biotechnological and chemical measures, along with cultural practics and production systems of agricultural and livestock activity with the intent of decreasing the use of chemical control agents while maintaining productivity levels.

The soil microbiological community is represented by a wide and diversified population in a dynamic equilibrium state. Thus, the existence of a microorganism in a certain time and place is a result of its evolution in that place, the physical and chemical factors favorable to its development, and the existence of associated microorganisms and competitors, antagonists and predators. All organisms in an ecosystem are subject to a constant natural balance, and individual species tend to be limited by abiotic factors and controlled by biotic factors which with they coexist (Sayre and Walter, 1991; Stirling, 1991; Chen and Dickson, 2004). Such phenomenon may also be inducted by artificial introductions of antagonists or by stimulating the activity of native antagonists in the soil (Jatala, 1986; Stirling, 1991).

Animal- or plant-parasitic nematodes may also be controlled by other organisms which feed on them since part of their lifecycle occurs in the soil where their pre-parasitic stages are exposed to the action of such agents (Soulsby, 1982; Stirling, 1991). This natural control is sufficient to avoid an indiscriminate increase of nematode populations in the soil, however, it is not enough to protect plants and animals from the damages caused by these pathogens (Mendoza-de-Gives et al., 1994). Human manipulation of natural enemies (predators, parasites or pathogens) with the objective of regulating the population of a target-organism at a non-harmful level to the soil ecosystem is called "applied biological control" (Stirling, 1991; Larsen, 1999; Chen and Dickinson, 2004). Ideologically, such control is an alternative or complementary approach to traditional methods of chemical control, having no negative effects on the environment (Gronvold et al., 1996a). The objective of biological control of nematodes is to impact the pre-parasitic population of these plant and animal pathogens, decreasing their number to levels which are not harmful to the host and do not cause economic losses.

A wide variety of soil microorganisms are described as antagonists of nematodes: protozoa (Canning, 1973; Sayre and Wergin, 1979), turbellarians (Sayre and Powers, 1966; Sayre and Wergin, 1979), tardigrada (Doncaster and Hooper, 1961; Sayre, 1971; Sayre and Wergin, 1979), copepods (Lehman and Reid, 1993), collembola (Lysek, 1963; Decker, 1989; Lee and Widden, 1996), mites (Imbriani and Mankau, 1983; Inserra and Davis, 1983; Bilgram, 1997), predatory nematodes (Small and Grootaert, 1983; Stirling, 1991; Eisenback, 1993; Yeates and Wardle, 1996), earthworms (Gronvold et al., 1996a; Waghorn et al., 2002), viruses (Mankau, 1981; Stirling, 1991), insects (Gronvold et al., 1996a; Vlassoff et al. 2001; Waghorn et al., 2002), bacteria (Stirling, 1988b; Larsen, 1999; Kerry, 2000; Tian et al., 2007), and fungi (Barron, 1977; Nordbring-Hertz, 1988). Beside these organisms, some plants also are antagonists of nematodes (Pandey et al., 2003; Ferraz and Freitas, 2004; Deyn et al. 2004).

Despite the diversity of antagonists, not all present potential for use in nematode biological control programs (Ferraz et al., 2010). For a microorganism to be considered as a good agent of nematode biological control, it must present several attributes (Jatala, 1986; Kerry, 1989; Kim and Riggs, 1992), including: a) be highly pathogenic to several species of animals- or plants-parasitic nematodes; b) not be pathogenic to cultivated plants, humans and animals; c) be identifiable and have a known action mode; d) be efficient in the reduction and suppression of a high-density nematode population; e) survive in extreme soil conditions in the absence of the target host; f) present a high capability for dissemination and establishment in the soil; g) grow in artificial environment with appropriate pH and within a temperature range; h) be easily produced in mass and be economically viable; i) be produced in a usable formulation; j) remain infective after long storage periods; k) be compatible with fertilizers, chemical inputs and cultural practices; l) not modify the habitual cultural practices for application; m) have little or no impact on soil microbiota. In the control of animal-parasitic nematodes the ability to survive in the gastrointestinal tract and the ability to grow and be effective in the feces are other necessary attributes of a biocontroller (Larsen, 2006).

Fungi and bacteria possess many of these characteristics and therefore are the most studied biocontrollers of nematodes (Jatala, 1986; Stirling, 1991; Waller and Faedo, 1996; Chen and Dickson, 2004), and for this reason they are emphasized in this chapter.

2. Nematophagous Fungi

Fungi are eukaryotic organisms whose nuclei are dispersed in continuous or septate hyphae. They have a cell walls similar to that of plants regarding chemical and structural composition, but because fungi are not photosynthetic, their nutrition is obtained by absorption and therefore they remain restricted to a saprophytic or parasitical existence (Alexopoulos et al., 1996). Nematophagous fungi are one of the potential microorganisms which may be used in the control of nematodes. For several decades fundamental studies have been performed on their morphology, nutrition, evolution, distribution, ecology, systematics, and their role as potential agents in the biological control against animal- or plant-parasitic nematodes (Kerry, 2000; Larsen, 2000; Li et al., 2000; Chen and Dickson, 2004).

These ubiquitous hyphomycetes of the soil have suppressive effects on multiplication of nematodes due to parasitism, predation or antagonism in all stages of its lifecycle; and its principal species can be found in the majority of soils around the world (Gray, 1987; Stirling, 1991; Siddiqui and Mahmood, 1996). Some nematophagous fungi are obligate parasites which require the nematode for survival, others are facultative or opportunistic parasites that live saprophytically and others have intermediate characteristics between these two categories (Gray, 1987). The parasitism of nematodes results in its complete digestion and this activity supplies the fungus with nutrients and energy for its continued growth (Lopez-Llorca et al., 2008).

There are around 200 species of taxonomically diversified fungi with ability for attacking eggs, juveniles and adult female nematodes and use them as a source of nutrients (Li et al., 2000; Nordbring-Hertz et al., 2006) but few are adequate to be biocontrollers due to their low efficacy (Siddiqui and Mahmood, 1996). The nematophagous fungi are widely distributed geographically across the world in several climates and live in all types of soil, especially those rich in organic matter (Barron, 1977; Gray, 1987). Studies have shown that they are more frequently found in organic production systems than in any other (Jansson and Lopez-Llorca, 2004).

It is possible that the antagonism between fungi and nematodes has occurred for millions of years since it has been found in southwest France in the form of an amber fossil of a predator fungus, possessing hyphae rings as capturing structures, as well as its potential to prey on nematodes dated back 100 million years to the Cretaceous period of the Mesozoic Era which marks the height of the dinosaurs era (Schmidt, 2007). In Mexico a fossil of the nematode *Oligaphelenchoides atrebora* parasitized by a predator fungus dated back 22.5-26 million years between the Paleogene-Neogene periods of the Cenozoic Era, although the capturing structures of the fungus are not clearly identifiable (Jansson and Poinar, 1986). However, the ecologic relation between these microorganisms was first scientifically reported in the 19th century. According to Drechsler (1941) the first description of a nematophagous lifestyle was made by Lohde in 1874 for the endoparasitic fungus *Harposporium anguillulae* which was isolated from a dead *Anguillula nematode*. Previously, the nematophague fungi *Arthrobotrys superba* and *A. oligospora* had been described in 1839 by Corda and in 1852 by Fresenius, but at that time the ability to capture nematodes was not perceived and they were considered saprophyte fungi, as reported by Duddington (1956). In 1870, Woronin observed that conidia of *A. oligospora* were able to germinate and form hyphal networks with unknown functions; but only in 1888, for first time in history, Zopf verified that the structure of this fungus was involved in capture, penetration and consumption of the active nematodes (Barron, 1977). In 1877, Kühn was the first to observe the parasitism of the *Heterodera schachti* females nematode by a fungus identified in 1881 as *Tarichum auxiliare*, renamed one century later as *Catenaria auxiliaris* (Kerry, 1980; Siddiqui and Mahmood, 1996). In 1933, Drechsler showed that the networks contained an adhesive substance which held the nematode (Barron, 1977). Furthermore, transmission electron micrographs and chemical methods revealed that such adhesive substance present in traps of the *A. oligospora* fungus were polymers of proteins and carbohydrates which accumulated in the linking site between the fungus and the nematode and were involved in the adhesion process (Tunlid et al., 1991).

The understanding of antagonist fungi was further emphasized in the extensive work of Drechsler (1937), containing detailed observations on 21 species. This classic of micologic literature has generated numerous investigations which are still being performed by researchers today. Among the researchers who contributed most to the understanding of nematophagous fungi, specially referring to isolation and description of new species, Drescher, Duddington and Barron standout. Other important researchers were Cooke and Godfrey (1964) who compiled information with respect to the identification of nematophagous fungi which were found scattered and disorganized.

According to Lopez-Llorca et al. (2008), nematophagous fungi are found in the phyla Ascomycota (predatory fungi, endoparasitic, egg- and female-parasites), Basidiomycota (endoparasitic fungi, toxine-makers and predators), Zygomycota (predatory fungi), Chytridiomycota (endorapasitic fungi) and Oomycetes (endoparasitic fungi, egg- and female-parasitic). In the past the predator fungi were considered mitosporic, classified as Deuteromycetes, class Hyphomycetes, order Hyphomycetales and family Moliniaceae, showing septate and well developed mycelium, reproducing itself agametically by exogenous spores formed on the hyphae branches (Barnett and Hunter, 1998). However, sexual reproduction stages of these fungi were observed for some species which began to be recognized as belonging to the phylum Ascomycota (Griffin, 1994; Pfister, 1997). In the modern classification the most common predator fungi belong to the family Orbiliaceae of the phylum Ascomycota (Yang et al., 2007; Liu et al., 2009) and their taxonomy has been organized according to the type of capturing structure (Ahrén et al., 1998).

In 1937, Linford pondered the possibility that nematophagous fungi control plant-parasitic diseases caused by nematodes since he observed in *in vitro* studies that the addition of organic matter to nematodes-infested potting soil stimulated the development of nematophagous fungi which resulted in a drastic reduction of free-living and plant-parasitic nematode populations (Linford, 1937; Linford et al., 1938). The first experiment of nematode control under field conditions using nematophagous fungi was performed in Hawaii by Linford and Yap (1939). The researchers tested the efficiency of *Arthrobotrys oligospora*, *A. musiformis, Monacrosporium ellipsosporum, M. thaumasium*, *Dactylaria candida* species to control *Meloidogyne* sp. in pineapple plants (*Ananas comosus*). During 15 months of plant cultivation in infested soil, only *M. ellipsosporum* restricted the injuries caused by the nematode, decreasing the number of galls and increasing the weight of pineapple plants. Inspired by these experiments, a group of scientists from the Pasteur Institute of France began pioneer experiments between the end of 1930 and beginning of 1940 using nametophagous fungi to control animals-parasitic nematodes in *in vitro* tests in agar media as well as tests with a small number of animals (Descazeaux, 1939ab; Descazeaux and Capelle, 1939; Deschiens, 1939ab; Roubaud and Descazeaux, 1939, 1941ab; Roubaud and Deschiens, 1939).

Progress in biocontrol using nematophagous fungi in recent years has been remarkable (Akhtar and Malik, 2000) and evidences show that the use of fungic antagonists to control nematodes is promising (Chen and Dickson, 2004; Stear et al., 2007). The using of plant-parasitic nematodes biocontrollers began with the nematicide crisis in the 1970's in which drugs no longer decreased nematode populations, presenting high cost, caused environmental problems, caused problems to soil microbiota, contaminated groundwater and intoxicated plants and animals (Mankau, 1981; Thomason, 1987). However, studies involving sustainable alternative strategies to control animal-parasitic nematodes using nematophagous fungi were the consequence of worrying about resistance to the anthelmintic drugs, environment degradation and residues in the foods (Waller and Larsen, 1993).

With the interest for this particular group of fungi and research progress, scientists were able to verify that hundreds of fungi species feed on nematodes, adopting different action strategies (Gray, 1987; Nordbring-Hertz, 1988; Stirling, 1991; Chen and Dickson, 2004) which allow them to be classified as: a) endoparasites, that use adhesive or non-adhesive spores (adhere to the nematode cuticle or are ingested by them); b) nematode-trapping fungi, that use specialized mycelial structures, adhesive or non-adhesive (adhesive hyphae, adhesive knobs, adhesive branches, adhesive networks, constricting or non-constricting rings, and stephanocysts), to capture juveniles nematodes; c) parasites of eggs, cysts or females, that use vegetative hyphae with or without an appressorium-like swelling; and d) producers of nematode toxins by special structures that paralyze nematodes before penetration. Recently, Luo et al. (2006) reported an efficient method for killing of nematodes *Panagrellus redivivus* and *Bursaphelenchus xylophilus* by the fungus *Stropharia rugosoannulata* as a result of severe mechanical injuries to their cuticles carried out by acanthocytes (large cells with a distinct and easily recognizable spiky shape), which might be a nematode-attacking device representative of a fifth group (Liu et al., 2009). Other novel fungal structures called spiny balls produced by the fungus *Coprinus comatus* can immobilize the nematodes *P. redivivus* and *Meloidogyne incognita* by nematotoxins (seven toxins identified) and kill them with mechanical forces which are somewhat similar to acanthocyte. However, the long projections (acanthae) of acanthocyte can pierce nematode bodies and cause more serious damage than spiny balls (Luo et al., 2007).

2.1. Endoparasitic Fungi

Some endoparasitic fungi that attack vermiform nematodes are facultative parasites and may live saprophytically without them in the environment. However, the majority of endoparasitic fungi are obligate parasites, completing their lifecycle inside the infected nematodes and not developing hyphae outside the body of the host, being encountered in the environment in the form of conidia which are liberated from nematodes disintegrated by the parasitism (Gray, 1987; Lopes-Llorca et al., 2008). These obligate parasites are grouped according to their infection mechanism and they infect nematodes through several types of specialized structures (Chen and Dickson, 2004): a) encysting zoospores (*Catenaria anguillulae, C. auxiliaris, Haptoglossa dickii, Myzocytium humicola* e *Nematophthora gynophila*); b) adhesive conidia (*Acrostalagmus obovatus, Cephalosporium balanoides, Meristacrum asterospermum, Nematoctonus leiosporus, N. concurrens, Hirsutella rhossiliensis, H. minnesotensis, Drechmeria coniospora* e *Dilophospora agrostis*); c) ingestible conidia (*Harposporium anguillulae* e *H. oxycoracum*); d) or special infective cells known as gun cells (*Haptoglossa heteromorpha*).

The ability to attract nematodes appears to be increased with increase of the dependency of the fungi on the nematode as source of nutrients (Jansson, 1982a). Due to the fact that endoparasitic fungi are obligate parasites and have low ability of saprophytic competition, they present a great ability to attract nematodes (Jansson and Nordbring-Hertz, 1979). This attraction is the result of chemotaxis by compounds released from the conidia (Nordbring-Hertz et al. 2006), being exercised more strongly by adhesive conidia than by those non-adhesive (Jansson, 1982a). Such ability is considered ecologically significant for these fungi (Siddiqui and Mahmood, 1996). This attraction mechanism also occur in a manner contrary to the case of zoospores because they are mobile and may be attracted by exsudates released from orifices on the body of nematodes such as sialic acid-like carbohydrates (Jansson and Nordbring-Hertz, 1984). After adhered to the cuticle of the nematodes, the spores germinate and emit hyphae which penetrate the host and proliferate inside it, using the pseudocoelomic content for their nutrition. In few days the conidia are produced and released to the external environment beginning a new infection cycle (Lackey et al., 1992).

Hirsutella rhossiliensis, Haptoglossa dickii, Catenaria auxiliaris, C. anguillulae, Nematophthora gynophila, Drechmeria coniospora, Nematoctonus spp., Acrostalagmus spp., Harposporium spp., Myzocytium spp. and Haptoglossa spp. are examples of endoparasitic fungi studied for the biological control of nematodes (Stirling, 1991; Siddiqui and Mahmood, 1996; Chen and Dickson, 2004; Nordbring-Hertz et al., 2006). However, only four species can be cultivated in vitro: Nematoctonus concurrens, N. haptocladus, Drechmeria coniospora and Hirsutella rhossiliensis (Kerry, 1987). According to Lopez-Llorca et al. (2008), the genres of endoparasitic fungi of nematodes may occupy different taxonomic positions and be found in Basidiomycetes (Nematoctonus), Ascomycestes (Haptocillium, Harposporium and Drechmeria), Chytridiomycetes (Catenaria) or Oomycetes (Stylopage and Cystopage).

Hirsutella rhossiliensis behaves as an ecological endoparasite by infecting nematodes via adhesive conidia and can be grown in artificial media in the lab. The potential of this fungus as a biological control agent has been controversial. Although it has suppressed the population of nematodes such as *Pratylenchus penetrans* (Timper and Brodie, 1994), *Globodera pallida* (Velvis and Kamp, 1996), *Heterodera schachtii* (Jaffee and Muldoon, 1989) and H. glycines (Chen and Liu, 2005; Zhang et al., 2008) in greenhouse tests, it was not effective on cyst and gall nematodes in some field tests (Tedford et al., 1993; Jaffee et al., 1996; Chen and Reese, 1999). However, the fungus was able to supress *Mesocriconema xenoplax* in peach orchards in the United States (Zehr, 1985; Jafee et al., 1988) and *H. schachtii* in some beet fields in Germany (Müller, 1982). It appears to profusely colonize the rhizoplane, where it produces viable conidiophores (Lopez-Llorca et al., 2006). Because it is an obligate parasite dependent on the nematode for nutrition in nature, the percentage of parasitized nematodes is positively correlated with the host density (Jaffee et al., 1992; Jaffee et al., 1993; Lackey et al., 1994).

Nematoctonus (anamorph of *Hohenbuehelia*) is a peculiar genus of nematophagous fungi due to the fact that while some species are classified as predatory, other are clearly endoparasitic (Barron, 1977; Koziak et al., 2007). Moreover, there are 15 described species, where 10 are predatory which produce an extended mycelium and capture nematodes by means of adhesive hourglass knobs formed over the hyphae, and 5 are endoparasitic which infect these hosts through adhesive conidia (Thorn and Barron, 1986), similar to the adhesive knobs of the predatory species (Chen and Dickson, 2004). These germinated spores secrete a nematoxin which causes rapid immobilization and death on the nematodes before penetrating them (Giuma et al., 1973). Giuma and Cooke (1974) evaluated the potential of the *N. concurrens* and *N. haptocladus* species and observed that these fungi decreased the number of plant-parasitic nematodes only in sterile soil, suggesting that failure in natural soil was due to micostatic and lytic effects. In *in vitro* tests the *N. robustus* species was able to capture animal-parasitic nematodes such as *Ancylostoma* spp. (Maciel et al., 2009) and *Strongyloides papilosus* (Campos, 2006).

Drechmeria coniospora is a nematophagous fungus which infects nematodes through conidia which possess an adhesive knob in the apical region (Dijksterhuis et al., 1993). As a prerequisite, infection is dependent of the presence of this structure formed in the maturation phase of conidia which assists in adhesion to the surface of the host (Boogert et al., 1992). Conidia of *D. coniospora* present different patterns of adhesion to the cuticle of the various nematode species and may be restricted to the head or tail regions or occur alongside the cuticle, as well as be completely absent of adherence (Jansson et al., 1985b). Adhesion preferentially occurs in chemosensory organs located in the head and tail regions (Jansson and Nordbring-Hertz, 1983). In susceptible nematodes, adhesion leads to penetration of the cuticle with the invasion of the victim and production of new conidia from the decomposed nematode, where production is mediated by proteins in the adhesive portion of the conidia which are linked to proteins excreted from the nematode sensory organs (Jansson, 1993). Conidia of this fungus adhere to the cuticle of several nematode species, be they may be free-living such as *Panagrellus redivivus, Caenorhabiditis elegans* and *Acrobeloides buetschilii;* plant-parasitic such as *Anguina tritici, Aphelenchoides fragariae, Ditylenchus dipsaci, D. destructor, Pratylenchus penetrans, Meloidogyne incognita, M. javanica, Cephalenchus* sp., *Criconemella xenoplax* and *Heterodera schachtii*; or animals-parasitic such as *Ostertagia ostertagi* and *O. quadrospinulatum* (Jansson et al., 1985b; Jansson, 1993). However, adhesion to the nematode is not always the result of infection which may be related to the specific properties of the cuticle for each nematode (Dijksterhuis et al., 1993). Infection by this function was not observed in third stage larvae of animals-parasitic nematodes such as *O. ostertagi, O. quadrospinulatum, O. circumcincta, Nematospiroides dubius, Trichostrongylus colubriformis* and *Haemonchus contortus* when the second stage cuticle was present (Jansson et al., 1985b; Jansson, 1993; Santos and Charles, 1995). *M. coniospora* adheres to the cuticle of *Meloidogyne* spp. but is not able to penetrate it and infect the nematode; but independent of this, the fungus was able to reduce the number of galls in tomato plants cultivated in pots under vegetable house conditions when conidia were directly inoculated in the soil under different levels of inoculum or conveyed by previously infected *P. redivivus* (Jansson et al., 1985a).

Obligate parasites are more likely to be affected by changes in population density of the host than facultative parasites and integration with control measures that reduce nematode populations can reduce the performance of this agent unless inundative applications of the agent are used, which may be impractical in large-scale agriculture (Hidalgo-Diaz and Kerry, 2007). Endoparasitic fungi have rarely been used in attempts to control nematodes (Stirling, 1988b), since their use in biological control programs is limited by the following factors (Sidiqui and Mahmood, 1996): a) locomotion and activity of zoospores depend on the presence of water in the soil, b) limited growth in culture medium in the absence of the host, which complicates mass production, c) low ability of saprophytic survival d) difficulty of establishment in new environments because of the susceptibility of its spores to mycostasis. Unlike other nematophagous fungi, endoparasitic fungi apparently are not the endophytic colonizers of the root, which may be a reflection of its slow growth and method of infecting nematodes, however although the fungi *N. pachysporus* and *H. rhossiliensis* did not penetrate barley roots, they were able to colonize their surface (Lopez-Llorca et al., 2006).

2.2. Predator Fungi

Most species of nematophagous fungi are predators and characterized by developing an extensive system of vegetative hyphae along which there is morphological differentiation in functional structures called traps that capture and retain the pre-parasitic vermiforms of nematodes, infecting them and nurturing them with their internal content (Barron, 1977; Gray, 1987). This fascinating group of nematophagous fungi possess two phases, a saprophyte phase in which they utilize a source of carbon (energy) and amino acids (nitrogen) from organic matter in soil and another parasite phase in which they feed on the bodies of captured nematodes (Stirling, 1988b; Pfister, 1997). According to Barron (2003), these fungi are saprophytic in conditions of good nutrient levels and become predatory when the level of nitrogen is deficient.

This ability to switch from the saprophytic phase to the parasitic phase by developing capture and infection structures (traps), morphologically differentiating vegetative hyphae, with the objective of capturing nematodes and using them as an additional source of nutrients, provides these fungi with a nutritional advantage in the soil (Nordbring-Hertz et al., 2006). In some species the development of these traps may be spontaneous due to genetic characteristics (Gray, 1987) and in others depend on direct contact of the nematode with the hyphae (Jansson and Nordbring-Hertz, 1980) and substances secreted by them (Burney and

Estey, 1985; Nordbring-Hertz, 1988) or under conditions of water or nutrient scarcity (Balan and Lechevalier, 1972). Hyphal differentiation may occur within 24 hours, with the production of multiple traps (Pramer, 1964). Substances such as nematode sterile filtrate, human serum, earthworm extracts, carbon dioxide, cellophane, small peptides, lectins, cattle dung, soil and soil extracts can also induce the formation of traps in predatory fungi (Bartnicki-Garcia et al., 1964; Balan and Lechevalier, 1972; Nordbring-Hertz, 1973; Kaplan et al., 1991; Dackman and Nordbring-Hertz, 1992; Dijksterhuis et al., 1994; Persmark and Nordbring-Hertz, 1997; Nordbring-Hertz et al., 2006). However, nematodes are the most effective inducers of the trap formation (Nordbring-Hertz et al., 2006), which has been directly correlated with its density (Gronvold, 1989) and motility. This is considered a substantial factor for triggering of this morphogenesis (Jansson and Nordbring-Hertz, 1980; Nansen et al., 1988; Xie et al., 2010), but differences in cuticle composition and size of the nematode may also explain the differences in how they induce the formation of traps and are captured (Mendoza-de-Gives et al., 1999, Kumar and Singh, 2006, Xie et al., 2010). The traps can also be derived directly from conidia, which are then called conidial traps (Dackman and Nordbring-hertz, 1992) and occur in almost all species forming traps when germinated on natural substrates (Persmark and Nordbring-Hertz, 1997). Although the nematode species are differently susceptible to predation by different fungal species and isolates, there is low specificity in the fungus-nematode relationship (Kerry, 2000).

Predacious fungi use complex capture devices formed at intervals along the length of their hyphae to deter vermiform nematodes (Barron, 1977; Liou and Tzean, 1992; Rubner, 1996): a) adhesive hyphae (*Arthrobotrys botryospora, A. superba, A. psychrophila, Cystopage cladospora* and *Stylopage hadra*); b) adhesive branches (*Monacrosporium gephyropagum* and *M. cionopagum*); c) adhesive hourglass knobs (*Nematoctonus robustus, N. concurrens, M. haptotylum, M. phymatopagum* and *M. ellipsosporum*); d) adhesive networks (*A. conoides, A. musiformis, A. superba, A. musiformis, A. oligospora, Duddingtonia flagrans* and *M. thaumasium*); e) constricting rings (*A. dactyloides, A. brochopaga* and *M. bembicoides*); f) non-constricting (*A. candida, M. leptosporum* and *M. haptotylum*); and g) stephanocysts formed by a cup-like basal cell and a terminal globose cell (*Hyphoderma* sp.). These remarkable structures specialized for capture, not only securely hold the nematode but also cause cuticle penetration and digestion of its contents through trophic hyphae (Veenhuis et al., 1985a). Traps such as branches, knobs and constricting rings are formed spontaneously, indicating the need for species that form traps spontaneously using the nematode as a nutrient source (Nordbring-Hertz et al., 2006). This characteristic provides greater speed and efficiency of capture for species that do not form traps spontaneously resulting in the additional benefit for survival in soil (Gray, 1987).

Phylogenetic analyses indicate that the traps are divided into two lineages: one based on constricting rings and the other based on adhesive traps (Li et al., 2005; Yang et al., 2007). While constricting rings capture the nematode mechanically by closing of the rings (Higgins and Pramer, 1967), the other traps utilize an adhesive covering containing extracellular polymers composed of proteins and carbohydrates that are important in recognition and attachment to the cuticle of the nematode (Tunlid et al., 1991). The traps possess cells with cell walls thicker than the hyphae, which contain electron-dense bodies that disappear from the invasive hypha cells at the time of penetration and may play a specific role in formation of the nematode penetration and digestion tube (Veenhuis et al., 1985b; Veenhuis et al., 1989b).

Besides the traps, predacious fungi may produce toxins that paralyze or kill the nematodes prior to penetration (Jansson and Nordbring-Hertz, 1988).

The infection process begins when the nematode is attracted by chemotaxis toward the fungal hyphae or trap in response to compounds released by these structures (Jansson and Nordbring-Hertz, 1979). Capture of the vermiform nematodes by the fungus is due to a recognition system of complementary molecules present in these microorganisms (Nordbring-Hertz, 1988), i.e. the interaction between lectins (carbohydrate-binding protein) present in cell walls of the trap cells and carbohydrate receptors in the nematode cuticle (Mattiasson et al., 1980; Zuckerman and Jansson, 1984; Tunlid et al., 1991). This recognition results in the reorganization of polymers in the trap and subsequent bonding between the nematode cuticle and fungal adhesive (Veenhuis et al. 1985a; Tunlid, 1992). Subsequently, an infective knob is formed and expands on the cuticle creating an invasive hyphae which grows inside the body of the nematode, assimilating its content and using it in the formation of vegetative mycelium (Veenhuis et al., 1985a; Jansson and Nordbring-Hertz, 1988; Murray and Wharton, 1990). A combination of hydrolytic enzymes (proteases) is probably involved in penetration of the nematode cuticle, which solubilize the cuticle macromolecules via mechanical pressure of the infective hyphae (Nordbring-Hertz et al., 2006; Lopez-Llorca et al., 2008). The proteolytic enzymes released by the fungus are considered important in penetration since the nematode cuticle contains mainly proteins, including keratin, collagen and fibers running diagonally in opposite directions from each other (Bird and Bird, 1991; Maizels et al., 1993; Bedelu et al., 1998; Tosi et al., 2002). Serine proteases were isolated and characterized from some fungi predators: PII and Aoz from *A. oligospora* (Tunlid et al., 1995; Ahman et al., 1996; Zhao et al., 2004); M1x from *Monacrosporium microscaphoides* (Wang et al., 2006); and Ds1 from *Dactylella shizishanna* (Wang et al., 2006). The action of phosphatases (Veenhuis et al. 1985a; Cruz et al., 2009) suggests that other enzymes are possibly an integral part in the process of nematode penetration.

Some predatory fungi are good saprophytic competitors but few capture nematodes, while others are effective in their capture but do not establish themselves well in soil (Gray, 1987). The density of predacious fungi that are considered effective saprophytic competitors is not related to the number of nematodes in the soil, while density-dependence was observed in fungal species with more parasitic behavior (Quinn, 1987). Rapid growth and abundant production of mycelium are two important factors for the dissemination and survival of fungi in environmental conditions, although mycelial growth is not related to the ability of an isolate to prey on nematodes (Dackman et al., 1987). The predacious fungi that produce three-dimensional networks are good saprophytes and grow well in culture media, but are less effective in capturing nematodes when compared with those that produce branches or adhesive knobs and constricting rings (Jansson and Nordbring-Hertz, 1980; Gray 1987).

According to Stirling (1991), fungi display a wide diversity in regards to their nutritional needs, where some species are certainly more efficient than others as biological control agents. Likewise, the variability is frequently observed in culture when isolates of the same taxon differ significantly in color, degree of sporulation and growth rate. The ability to capture tends to be similar among isolates that produce the same type of trap and differs between species that produce different types of traps (Hongyan et al., 2010). This variation reflects genetic differences, which may also manifest as differences in virulence, indicating that the first step in a biological control program is the selection of the most virulent strains.

For decades research has been focused on predatory fungi and remains the most studied group of agents with potential for biological control of nematodes (Kerry, 2000). Some studies have provided an optimistic basis that these organisms could be used as biocontrol agents of plant- parasitic nematode (Dias and Ferraz, 1994; Robinson and Jaffee, 1996; Ribeiro et al, 1999; Oliveira et al, 2002). Few products have been formulated based on the predacious fungi: *A. robusta* (Royal 300®) and *A. superba* (Royal 350®) in France and *A. oligospora* in Russia (Nematofagin BL®). However, the encouraging results that were obtained with the use of predacious fungi under controlled conditions were not always confirmed in field conditions (Siddiqui and Mahmood, 1996; Chen and Dickson, 2004).

Failures in achieving any measurable improvement in crop yield after inoculation of predacious fungi may be due to the fact that nematophagous activity does not coincide with the hatching period of the juvenile cyst and gall nematode (Siddiqui and Mahmood, 1996). Thus, for effective control it is necessary that the limited period of fungal activity coincides with the period of root invasion by the infective juveniles (J_2), since once inside the roots of the host plant they are protected from parasitism by fungi predators (Ferraz et al., 2010). However, according to Kerry (2001), it cannot be guaranteed that predacious fungi produce traps during the period in which J_2 are migrating to the roots. Moreover, the growth and activity of these organisms are directly influenced by environmental conditions are not always favorable (Sayre and Walter, 1991). Because these saprophytic fungi present limited saprophytic competence, are susceptible to fungistasis and dependent on the presence of organic matter in soil (Stirling, 1991). Studies have revealed that the activity of predacious fungi is strongly enhanced by association with rhizobacteria (Duponnois et al., 1993). Duponnois et al. (1995) observed that the presence of predacious fungi is linked with a high content of organic matter, and consequently with high bacteria populations. For predation of nematodes to occur, mycelial growth and formation of traps by the fungus is necessary. Both processes require energy, which can be supplied by carbohydrates present in organic matter (Stirling, 1991). Godoy et al. (1983) found that fungi are one of the main predators responsible for control of nematodes in soil amended with organic matter. Two theories have been proposed to explain the effects of organic amendments on predacious fungi (Jaffee, 2004): a) fungi are obligate parasites in nature and organic amendments that stimulate microbial activity and abundance of bacterivorous nematodes will increase populations of predacious fungi; b) fungi are facultative parasites that obtain their nitrogen from nematodes, allowing them to compete for other nutrients in the soil with nitrogen-deficient organic matter.

The trap-forming species are most abundant in soil containing organic material and this ability gives these species an advantage over species which do not form traps in soil rich in flora and fauna (Gray, 1987). Thus, the application of fungal inoculum in conjunction with any source of organic matter may increase the germination of spores, mycelial growth and predator activity, and favor its competition with other soil organisms (Kerry, 1987; Siddiqui and Mahmood, 1996; Akhtar and Malik, 2000; Ferraz et al., 2010). However, according to Kerry (2000), the relationship between the saprophytic phase, the development of traps and parasitic habit is poorly understood, with variations between the fungal species.

Predacious fungi are commonly found in soils with high organic content and present growth inversely related to the carbon:nitrogen ratio (Nguyen et al., 2007). Most species grouped in the genera *Arthrobotrys, Dactylella* and *Monacrosporium* capture nematodes mainly through three-dimensional adhesive networks (Mankau, 1980). According to Barron

(2003), the fungi species *Arthrobotrys* are probably the fungi predators most common and certainly the best known. *Arthrobotrys oligospora* was the first nematophagous fungus species described and is the predatory fungus most commonly found in soil, being the most investigated and well-known among all nematophagous fungi (Nordbring-Hertz, 2004). This species captures its prey by means of three-dimensional adhesive networks and has shown great potential to antagonize several nematodes (Nansen et al., 1988). It is an efficient competitive saprophyte capable of using various types of carbohydrates and colonizes plant roots causing changes in cell walls of the cells without affecting plant growth (Viaene et al., 2006). This response of root cells to colonization may have profound implications on the advantageous performance of nematophagous fungi as biological control agents of plant parasitic nematodes, since they generally attack the roots (Bordallo et al., 2002).

Predatory nematophagous fungi are microorganisms in which significant advances in the biological control of animal-parasitic nematodes are more likely to occur (Waller and Faedo, 1996). Since animal-parasitic nematodes spend part of their life cycle in the environment, they may be controlled by natural enemies such as predatory fungi by the capture of larvae in feces, where the development of the pre-parasitic forms of these nematodes are concentrated (Sanyal and Mukhopadhyaya, 2003). Several species of nematophagous fungi are able to rapidly colonize fresh feces deposited in the environment (Hay et al., 1997). Most biological control studies of nematodes have utilized predacious fungi from the genera *Arthrobotrys*, *Duddingtonia* and *Monacrosporium* (Larsen, 2000). The species *A. oligospora* and *D. flagrans* produce three-dimensional adhesive networks that significantly reduce the number of infective larvae both in fecal cultures as well as in feces in the pasture (Bird and Herd, 1995; Fernández et al., 1997). *Arthrobotrys oligospora* reduces the population of *Strongyloides papilosus* in cattle dung simply by spraying conidia on dung pats and this approach may be applied in confined systems of ruminant production in regions where the warm climate is suitable for development of new generations of this parasite (Chandrawathani et al., 1998). The addition of predacious fungi directly to animal feces can be in many ways superior to other approaches due to the fact that a closer contact is established between the biological control agent and the nematode in a well-defined substrate where fungistasis is not so severe.

The ability to survive passage through the gastrointestinal tract of animals and subsequently germinate in the feces to attack the developing larvae has been the main requirement in the selection of fungal biocontrol agents for animal-parasitic nematodes (Larsen, 2006). This principle has been tested in animal production systems by mixing the nematophagous agent to feed and it has shown promising results (Larsen, 2000). This fungus has already been used successfully in Australia (Larsen et al., 1994; Faedo et al., 1998), Denmark (Gronvold et al., 2000), Netherlands (Eysker et al., 2005), Lithuania (Sarkunas et al., 2000), New Zealand (Waghorn et al., 2003), Sweden (Faedo et al., 2002; Dimander et al., 2003; Waller et al., 2004), France (Paraud et al., 2007), Mexico (Mendoza-de-Gives et al., 1998), Malaysia (Chandrawathani et al., 2003, 2004), India (Sanyal and Mukhopadhyaya, 2003), United States (Fontenot et al., 2003; Terrill et al., 2004) and Brazil (Dias et al., 2007).

In recent years the studies related to biological control of ruminant-parasitic nematodes are almost exclusively focused on the species *D. flagrans* (Kerry, 2001) which is considered a promising biocontroller, principally due to the fact that it produces large quantities of spores with thick walls (chlamydospores) that may persist in the environment for long periods (Gronvold et al. 1996a; Knox et al ., 2002) and persevere in the harsh conditions of digestion (extreme pH, high temperatures and the action of digestive enzymes) during passage through the gastrointestinal tract of animals in comparison with other species of predacious fungi (Waller et al., 1994; Faedo et al., 1997; Terrill et al., 2004). Temperature in the digestive tract of grazing animals is high, between 36 °C and 39 °C (Cossins and Bowler, 1987). Two other reasons are the rapid growth in recently excreted feces and their high nematophagous capacity on the pre-parasitic larval stages before they migrate from the feces to the pasture and complete their life cycle when ingested by the grazing host animals (Larsen, 1999; Faedo et al., 2000). Gronvold et al. (1999), based on results obtained in a study to evaluate the efficacy of *D. flagrans* in different abiotic and biotic conditions, characterized it as a coprophilic fungus and concluded that it must be especially active in the surface layer of manure, an area well-ventilated that usually contains a high concentration of infective larvae of parasitic nematodes, the temperature may be high and water content low.

This fungi is effective in capturing the larvae of parasitic nematodes of different animal species including sheep (Larsen et al., 1998; Peña et al., 2002), goats (Chandrawathani et al., 2003; Chartier and Pors, 2003), cattle (Larsen et al., 1995), pigs (Nansen et al., 1996) and horses (Femandez et al., 1997). Additionally mass production of chlamydospores on a commercial scale is viable (Gillespie, 2002). Above all, no adverse effect was observed as a result of its use on free-living nematodes, earthworms, microarthropods or any other nematophagous fungus (Gronvold et al., 2000; Faedo et al., 2002; Knox et al. 2002; Yeates et al., 2002; Waller et al., 2004) and there are no reports of toxicity to animals or humans. The daily intake of chlamydospores by grazing animals for 2-3 months prevents the buildup of dangerous levels of infective larvae in pasture (Larsen, 2000). The daily dose of 1 million chlamydospores/kg body weight or greater has been used in field studies but recent data indicate that this dosage may be reduced in small ruminants (Peña et al., 2002). In a recent study on the effectiveness of *D. flagrans* in thc control of gastrointestinal nematodes of sheep fed for 30 days with 200,000 chlamydospores/kg/day, an efficacy of 92% was observed on the total parasite load, reducing parasitism by *Haemonchus contortus* and *Teladorsagia circumcincta* by 100%, *Cooperia onchopora* by 87.5 % and *Trichostrongylus axei* by 90% (Sagüés et al., 2011).

Duddingtonia flagrans presents peak growth between 25 °C and 33 °C (Morgan et al. 1997), not developing at 37 °C (Peloille, 1991). In i*n vitro* studies using 20 nematodes/cm^2 agar, Gronvold et al. (1996b) observed that the growth rate and formation of traps reaches a maximum at a constant temperature of 30 °C. At temperatures near 10 °C and 35 °C the ability to produce traps is gradually reduced. Studies showed that chlamydospores of *D. flagrans* remain stable when formulated in tablets for up to 9 months at 4°C. No effect on viability was observed when chlamydospores were exposed for a prolonged period at any ambient temperature or 40 °C in air or under atmosphere with the major ruminal gases, carbon dioxide and methane (Waller et al., 2001).

Studies with *D. flagrans* were initially undertaken in Scandinavian countries (Gronvold et al., 1993; Larsen, 2000). It was first described in Britain as *Trichothecium flagrans* by Duddington (1949), later included in the monospecific genus *Duddingtonia* by Cooke (1969) and finally in the genus *Arthrobotrys* as *A. flagrans* by Scholler et al. (1999), but to maintain continuity biological control literature, it has remained *Duddingtonia* (Skipp et al., 2002). This species widely distributed throughout the world can be isolated from cattle and horse manure, silage and decaying vegetation (Larsen et al., 1991; Mahoney and Strongman, 1994; Skipp et al., 2002). Its nematophagous activity is guaranteed by means of adhesive hyphae and three-dimensional adhesive networks (Barron, 1977; Larsen, 2000), but according to

Bogus et al. (2005), its metabolites may also significantly impair the motility of nematode larvae.

2.3. Egg-, Cyst- and Female-Parasitic Fungi

Some nematophagous fungi can directly attack eggs of animal- or plant-parasitic nematodes with the ability to penetrate and assimilate their content or indirectly cause distortions in the developing embryo (Morgan-Jones and Rodriguez-Kabana, 1988). Fungi in this group are mainly nematode antagonists of the genera *Meloidogyne* spp. *Heterodera* spp. and *Globodera* spp. (Ferraz et al., 2010). They do not parasitize mobile stages of nematodes (Nordbring-Hertz et al., 2006), but are able to parasitize sedentary stages of plant-parasitic nematodes when in the soil, on the surface of the roots or inside the root (Siddiqui and Mahmood, 1996), destroying the eggs in the uterus (Stirling, 1991). Globose females and cysts are as susceptible to fungal colonization as the egg mass of gall nematodes (Siddiqui and Mahmood, 1996). Interest in these fungi that affect sedentary nematode females increased rapidly in recent years (Akhtar and Malik, 2000).

Most of these fungi are facultative parasites and can live in soil as saprophytes, while some are obligate parasites, not developing mycelium in the soil where they remain in the form of spores with thick walls and infect the nematode females and eggs by means of zoospores (Stirling 1991; Chen and Dickson, 2004). Facultative parasitic fungi produce an expansive mycelium growth which may be increased in the rhizosphere and infect nematodes with the appressoria (Lopez-Llorca et al., 2008). Hyphae grow in the direction of the eggs and once in contact with cysts and egg masses, grow rapidly and eventually parasitize any eggs that are in the early embryonic development stage (Siddiqui and Mahmood, 1996; Nordbring-Hertz et al., 2006).

The infection process begins with the formation of an appressorium from the tip of hyphae on the surface of eggs or sedentary females (Jansson and Lopez-Llorca, 2001; Abrantes et al., 2002; Lopez-Llorca et al., 2008). An extracellular adhesive material containing carbohydrate residues is formed on the appressorium (Lopez-Llorca et al., 2008) which adheres to the surface of the host. An infection peg grows from the appressorium and penetrates the egg, gives rise to an infection bulb from which the mycelia are emitted and colonize the egg (Abrantes et al., 2002). This penetration from the appressorium is the result of joint action of physical pressure and enzymatic degradation in a small area of the host (St. Leger, 1993; Abrantes et al., 2002; Lopez-Llorca et al., 2008). The shell of the nematode egg contains a matrix of proteins embedded in chitin microfibrils (Wharton, 1980). Extracellular enzymes produced by the fungi are the main chemical constituents of the eggshell and represent a potential virulence factor involved in the penetration or digestion of the host (Huang et al., 2004; Yang et al., 2007). Proteases have been isolated and characterized from different fungal parasites of nematode eggs: P32 from *Pochonia rubescens* = *Verticillium suchlasporium* (Lopez-Llorca, 1990); VCP1 from *P. chlamydosporia* (Segers et al., 1994); PL from *Paecilomyces lilacinus* (Bonants et al., 1995); pSP-3 from *P. lilacinus* (Bonants et al., 1995); Ver112 from *Lecanicillium psalliotae* (Yang et al., 2005). The chitinolytic enzyme CHI43 was isolated from *P. rubescens* and *P. chlamydosporia* (Tikhonov et al., 2002).

Among the egg parasitic nematode, the genera *Paecilomyces, Trichoderma, Verticillium* and *Fusarium* are the most studied (Jatala, 1986; Kerry, 2000; Sharon et al., 2007, Eapen et al., 2008; Mendoza and Sikora, 2009; Kiewnick, 2009). Some of these fungi appear to be promising biocontrollers of nematodes because of their persistence in soil (Nordbring-Hertz, 1988). The species *Pochonia chlamydosporia* (= *Verticillium chlamydosporium*) and *Paecilomyces lilacinus* are considered by many researchers as the most promising fungi for biological control of plant-parasitic nematodes (Jatala, 1986; Kerry et al., 1982; Kerry, 1995; Walters and Barker, 1994; Siddiqui and Mahmood, 1996). This affirmation is based on the fact that large numbers of eggs are concentrated in the gelatinous matrix and cysts which facilitates the attack of opportunistic fungi, causing a severe impact on the reproductive capacity of the nematode since the production of large quantities of eggs is a mechanism to ensure the survival of the population of these organisms (Kerry, 1981). Once they are able to colonize adult females and affect their fertility, the multiplication of nematodes can be significantly reduced with consequent population control and subsequently lower populations in future nematode generations (Abrantes et al., 2002). An impact of similar magnitude on the population of nematodes would be impossible to be caused by predatory and endoparasitic fungi since many J_2 may escape the attack of these antagonists.

Paecilomyces lilacinus is a saprophytic fungus commonly found in soil in many parts of the world and various types of habitats (Chen and Dickson, 2004), being effective against several plant- parasitic nematodes (Kerry and Evans, 1996; Siddiqui and Mahmood, 1996; Cannayane and Sivakumar, 2001) and can infect *Meloidogyne*, *Tylenchulus*, *Nacobbus*, *Radopholus*, *Heterodera*., *Globodera* and *Pratylenchus* (Chen and Dickson, 2004). It principally attacks the sedentary stages of nematode cysts and galls, having little effect on juveniles (Roumpos, 2005). Its efficacy has been proven in greenhouse and field conditions (Holland et al., 2003; Kiewnick and Sikora, 2004), being one of the biological control agents widely tested in the control of phytonematodes (kins et al., 2005; Kiewnick and Sikora, 2006). Jatala et al. (1979) was the first to report the effective parasitism of this fungus on *M. incognita* and *G. pallida*. Its control has been shown effective against *M. incognita* (Kiewnick and Sikora, 2006; Oclarit and Cumagun, 2009), *M. javanica* (Nasr Esfahani and Ansari Pour, 2006; Siddiqui et al., 2000) and *G. rostochiensis* (Cannayane and Sivakumar, 2001).

Commercial formulations containing this fungus have been used to control plant-parasitic nematodes in the form of bionematicidas due to its ability to spread rapidly in agricultural soils, rapidly becoming the dominant species where it is applied (Ferraz et al., 2010). However, *P. lilacinus* is an opportunistic fungus and some isolated occasions have been registered in which it cause infections in both immunocompetent and in immunocompromised humans, commonly affecting the eyes, nails, skin and soft tissue (Rhonda et al., 2008). In immunocompromised patients, this fungus can infect any organ or system (Sigler and Kennedy, 1999; Martin et al., 2002; Van Schooneveld et al., 2008). Although reports of human infections are rare in immunocompetent patients without an identifiable risk factors (Carey et al., 2003) and most cases occur in those who have compromised immune systems (Itin et al., 1998; Blackwell et al., 2000), there are concerns related to human health risks from use of a biopesticide based on some strains of this fungus, which has limited the registration of such products (Stirling, 1991). However, despite these facts there are several products based on this fungus in various countries: Philippines (Biocon®), Germany (BioAct WG®, BioAct WP® and NemaChek®), China (Soybean Root Bioprotectant®), Australia (Paecil®), United States (MeloCon WG® and NemOut®), Netherlands (Bio-Sol Nematode®), India (Paecilo®, Shakti Paecil® and Phytopali®, Bionemtex® and Yorker®), Columbia (Biostat WP®) and South Africa (Pl Plus). The active strain of most of these products is *P. lilacinus*

251 which was isolated from infected nematode eggs in the Philippines and taxonomically described in 1974. It meets the toxicity requirements for mammals of the Food Quality Protection Act of 1996, sufficient for registration of this microorganism as a nematicide (EPA, 2005).

Recently, representatives of the genus *Verticillium* capable of infecting nematode eggs were transferred to a new genre denominated *Pochonia* whose teleomorph is located within the genus *Cordyceps* (Ascomycota), based on morphological and molecular characteristics (Gams and Zare, 2001; Zare et al., 2001). *Pochonia chlamydosporia* has been studied as a potential biological control agent which parasites the eggs of *Meloidogyne* spp., *Nacobbus* spp., *Heterodera* spp. and *Globodera* spp. (Mauchline et al., 2004). The association between this fungus and nematodes was first documented by Willcox and Tribe (1974), having been isolated from eggs of *H. schachtii*. Soon after it was isolated from eggs of *H. avenae* (Kerry, 1975) and years later was considered as the main reason for the declining population of *H. avenae* in susceptible cereal cultivars in a monoculture system (Kerry et al., 1982). This fungus has also been reported as a parasite of eggs and females of *M. arenaria* (Morgan-Jones et al., 1981; Godoy et al., 1983) and several other species of this nematode genus (Freire and Bridge, 1985; Hidalgo-Diaz et al., 2000; Sun et al., 2006). According to Monfort et al. (2005), this nematode antagonist also promotes the growth of plants since it can reduce colonization of the root and the damage caused by pathogenic fungi which makes it a dual purpose biological control microorganism with great potential in agriculture. *Pochonia chlamydosporia* is not phytotoxic to corn and bean plants because it does not affect germination, growth and survival when applied and does not accumulate in tissues of the aerial portion of those presenting endophytic growth on the roots (Garcia et al., 2008b).

This fungus is associated with suppression of nematodes in the soil and acts as a natural enemy of cyst and gall nematodes (Kerry, 1993; Abrantes et al., 2002), and has considerable potential as a biological control agent for use in soils infested with nematodes (Kerry and Jaffee, 1997; Bourne and Kerry, 2000) due to its ability to colonize the rhizosphere of many plant species even in the absence of nematodes, and to use organic matter as a source of nutrients so it can be reisolated from the soil several weeks after application (Stirling, 1991; Bourne et al., 1996; Viaene and Abawi , 2000). The fungus does not reduce initial invasion of roots by J_2 or the damage they cause to plant growth. However, egg parasitism can dramatically reduce multiplication of the nematode and control its population. Therefore subsequent generations of gall nematodes in the soil will be lower, which is more significant after the second life cycle, possibly requiring more than one fungal application and harvest to reduce populations to non-harmful levels (Kerry, 2001).

Pochonia chlamydosporia is more abundant on the surface of the galls during the egg laying period of *Meloidogyne* spp. than on healthy roots (Bourne et al., 1996). It colonizes plant roots without causing damage or affecting growth (Lopez-Llorca et al., 2002), however there are differences in the ability to support growth by plant species, with no increase in density in those which are not good hosts (Bourne et al., 1999). Only isolates that grow in the rhizosphere and rapidly colonize the nematode eggs are capable of controlling populations of gall nematodes (Kerry, 2001). Reduction in the nematode populations is greater in plants that are not good hosts for nematodes, but that support an abundant growth of the fungus in its rhizosphere (Bourne, 2001; Abrantes et al., 2002). There are also differences in the saprophytic competence between different isolates of *P. chlamydosporia* in the soil or rhizosphere of plants or even a single isolate in the different plant species (Bourne et al.,

1996; Mauchline et al., 2004). Thus, the host plant influences the effectiveness of *P. chlamydosporia* as a biocontrol agent (Bourne et al., 1996). Based on this a practical management method to maintain effectiveness of this antagonist on the gall nematodes in organic production systems for vegetables is the rotation of crops that include poor host plants for nematodes, but that support colonization of the fungus in their roots (Atkins et al., 2003; Puertas and Hidalgo-Diaz, 2007). Its impact on the nematode is maximized when applied around these plants and at temperatures below 30 °C (Kerry, 2001).

The chlamydospore is the preferred type of inoculum for application of *P. chlamydosporia*, since this is its survival structure in the soil (Abrantes et al., 2002). This structure allows that the fungus is easily formulated and colonizes the rhizosphere of good hosts plants when added to soil in aqueous suspension without nutrients (Kerry, 2001). According to Peteira et al. (2005), the incorporation of *P. chlamydosporia* in the soil, either in the form of chlamydospores or a colonized substrate, results in a similar population control of *Meloidogyne* species. Generally, the inoculum of this fungus is applied at a rate of 5,000 chlamydospores per gram of soil (Kerry, 2001), but other quantities may allow for control of *Meloidogyne* spp. (De Leij et al., 1992b; Lopes et al., 2007). The use of *P. chlamydosporia* is considered safe since no clinical pathogens were observed in studies of acute toxicity/pathogenicity in rats, rabbits and quail (García et al., 2004ab; García et al., 2009). In an ecotoxicological evaluation this fungus was considered harmless to the earthworms *Eisenia andrei* and insect species *Apis mellifera* and *Chrysopa exterior* (García et al., 2008a). Because of these various attributes, promising bioproducts based on isolates of *P. chlamydosporia* have been formulated and marketed in Brazil (Rizotec®), Cuba (Klamic®) and Portugal (PcMR-1®) through technologies that enable the large scale production of chlamydospores.

Studies related to biological control of animal-parasitic nematodes with egg parasitic fungi have been confined to laboratory interaction tests. Because hatching of the eggs of most nematodes that parasitize domestic animals occurs within hours after being released in the feces, use of the fungi would be directed at controlling the species that need to stay within the egg environment for long periods to complete their embryonic development. Egg parasitic fungi may exert ovicidal or ovistatic effects on eggs of animal-parasitic nematodes with different levels of efficiency (Basualdo et al., 2000). Fungi from the genera *Acremonium*, *Aspergillus*, *Chrysosporium*, *Fusarium*, *Humicola*, *Mortierella*, *Mucor*, *Paecilomyces* and *Penicillium* may parasitize eggs of *Toxocara* (Gortari et al., 2007). Ciarmela et al. (2010) reported that a large number of eggs of *T. canis* were parasitized by the fungi *C. merdarium, F. oxysporum, F. moniliforme* and *F. sulphureum. Fusarium pallidoroseum* displays a high ovicidal action on eggs of *T. canis* (Ciarmela et al., 2002). The fungus *P. chlamydosporia* and *P. lilacinus* are potential biological control agents of *T. canis* (Carvalho et al. 2010). According to Basualdo et al. (2000), *Paecilomyces lilacinus* and *P. marquandii* which attack *T. canis* eggs showed a high and intermediate parasitism, respectively. Eggs of *Ascaris suum* may also be parasitized by *P. chlamydosporia* (Araújo et al., 2008).

2.4. Fungi That Produce Toxic Metabolites

Toxic metabolites produced by fungi can have an effect on hatching, mobility and penetration capacity of nematodes on the host, or even alter the physiology of the plant, making it less attractive to nematodes (Khan et al., 1984). Metabolites produced by nematophagous fungi have attracted considerable interest, since these can be one of the strategies used by nematophagous fungi to infect and kill nematodes. However, knowledge on the effect of these possible nematotoxinas to control nematodes is still limited, requiring further investigation (Ferraz et al., 2010).

Filtrates from the fungi genera *Aspergillus*, *Pleurotus*, *Penicillium*, *Trichoderma*, *Myrothecium, Fusarium* and *Paecilomyces* presented nematoxic effects in some studies (Khan et al., 1984; Ameen, 1991; Hallmann and Sikora, 1996; Nitao et al., 1999; Chen et al., 2000; Costa et al., 2001). Purified extract of *Penicillium* sp., *P. oxalicum*, *P. anatolicum* and *Aspergillus niger*, at 100 and 200 ppm (Molina and Davide, 1986) of *Fusarium* spp. (Ciancio et al., 1988) showed high nematicidal activity against *M. incognita*. The fungus *T. virens*, biocontrol agent of phytopathogenic fungi, produces substances capable of inhibiting the emergence and mobility of *M. incognita* juveniles (Meyer et al., 2000). The nematicidal activity of compounds produced by *P. chlamydosporia* has been rarely reported and discussed. Khambay et al. (2000) isolated and identified a toxin produced by this fungus that showed nematicidal activity against *M. incognita* mainly due to phomalactone which may be able to suppress the invasion of roots of the host plants by nematodes. Niu et al. (2010) suggested that the Aurovertins metabolites produced by *P. chlamydosporia* may be one of the pathogenic factors involved in suppression of nematodes. Extracts of the fungus *P. lilacinus*, *Cylindrocarpon destructans* and *Chaetomium murorum* reduce the motility of J_2 from *H. schachtii* after 24 and 48 hours of incubation (Ayatollahy and Fatemy, 2010). Leucinostatins in *P. lilacinus* are indicators of nematicidal activity (Park et al., 2004).

The prospecting for nematicidal substances produced by fungi is an approach that should be encouraged and which may result in efficient and less toxic products (Kerry, 1990). There is a product on the U.S. market based on fermentation product of *Myrothecium verrucaria* (Ditera DF®) which has shown effective in controlling *Meloidogyne, Heterodera, Globodera, Pratylenchus, Trichodorus, Belonolaimus* and *Radopholus similis* (Grau et al., 1996; Hafez et al., 1996; Warrior et al., 1999).

3. Bacteria

Some bacteria species have been identified as biological control agents of plant-parasitic nematodes and have received increased attention by the scientific community in recent years.

3.1. Pasteuria Penetrans

Bacteria of the genus *Pasteuria,* obligate parasites of plant-parasitic nematodes, are gram-positive producers of endospores and pseudo-mycelium septate (Starr and Sayre, 1988). The first report of nematode parasitism by *P. penetrans* was made by Cobb in 1906 when observing its spores in *Dorylaimus bulbiferus*, however he mistakenly classified it as a sporozoan (Chen and Dickson, 1998). Thorne (1940) also considered a protozoan when observing the parasitized *Pratylenchus pratensis*, though he described it as *Duboscquia penetrans* due to the similarity it had with a nematode parasitic organism described by

Micoletzky in 1929. Results of ultra-structural studies in electron microscopy led Mankau (1975) to its reclassification as a prokaryotic bacteria due to its nature, renaming it *Bacillus penetrans*. Sayre et al. (1977) pointed out that this bacterium was closer to the genus *Pasteuria*. Later, when analyzing *P. ramosa* described by Metchnikoff in 1888, they observed that both strains had morphological similarities and suggested that there was a common relationship between them (Sayre et al., 1979). After several studies, Sayre and Starr (1985) stated that this bacterium should be called *P. penetrans*, describing it as a parasite of *M. incognita* and probably other gall nematodes. This genetic proximity was confirmed by studies of cloning and sequencing of the 16 S region of ribosomal DNA from *P. ramosa* and *P. penetrans* (Anderson et al., 1999). Charles et al. (2005), by means of phylogenetic analyses based on 27 housekeeping genes, suggested that *P. penetrans* is ancestral of *Bacillus* spp., both saprophytic (*B. subtilis* and *B. halodurans*) and pathogenic (*B. cereus* and *B. anthracis*).

Pasteuria penetrans is a bacterial endoparasite able to form resistant spores (endospores). It presents several features that place it at the level of bacterial agent with the greatest potential for biological control of nematodes: survival for long periods in soil, resistance to heat and desiccation, high reproductive potential, harmless to humans and other animals, compatible with many pesticides, fertilizers and other biocontrol organisms, is not affected by cultural practices commonly adopted by farmers, endospore storage for over 11 years without loss of viability and does not have any natural enemies described so far (Dutky and Sayre, 1978; Stirling and Wachtel, 1980; Jatala, 1986; Stirling et al., 1986; Sayre and Starr, 1988; Stirling, 1988a; Oostendorp et al., 1990; Giannakou et al., 1997; Campos et al., 1998; Cho et al., 2000).

Pasteuria penetrans is the most widespread species and the principal parasite of *Meloidogyne* spp. However, there arc other species within the genus which parasitize nematodes, such as *P. nishizawae*, which parasites *Heterodera* and *Globodera*; *P. thornei* and *P. usgae*, which parasitize nematodes of the genera *Pratylenchus* and *Belonolaimus*, respectively (Sayre et al., 1991; Sayre et al., 1988; Giblin-Davis et al., 2003). Bishop et al. (2007), based on morphological and analysis of the 16S rRNA region, described a new species of *Pasteuria*, the *P. hartismeri* infecting *M. ardenensis*.

Although *Pasteuria* spp. possesses specific hosts, its life cycle in different hosts is similar. The endospores adhere to the cuticle during movement of the nematode in soil. Juveniles, even with large numbers of spores are still capable of penetrating the root, however, their movement is affected. Inside the root, the spores produce germ tubes that penetrate the body of the nematode, then forming vegetative micro-colonies where the sporangia and endospores are formed (Chen and Dickson, 2004). The parasite nematode develops normally, but its reproductive capacity is compromised. Adult females of the parasitized gall nematodes contain about two million endospores, which are released into the soil when plants and nematodes die. The endospores can survive in soil for several years, but success in infecting other nematodes depends on the distribution of endospores in the soil, which in turn is influenced by soil type, moisture, temperature and cultural practices (Viaene et al. 2006).

The first step of infection occurs when the endospores adhere to the cuticle of the nematode, however, the adhesion of *Pasteuria* to the nematode host has shown a high degree of variation (Davies et al., 2000). Monoclonal antibodies on the surface of endospores reveal the high degree of heterogeneity of the endospore populations (Davies and Redden, 1997) and some biochemical studies indicate that recognition and adhesion between bacteria-nematode involves lectin/carbohydrate interaction (Davies and Danks, 1993; Spiegel et al., 1996). It has been suggested that recognition of the nematode cuticle by bacteria is mediated by parasporal fibers, which interact hydrophobically with a receiver similar to fibronectin present in the nematode cuticle (Davies et al., 1996; Mohan et al., 2001).

In a study comparing the adherence of endospores of six populations of *P. penetrans* to the cuticle of J_2 of several gall nematode species, adherence was not related to the phylogeny of the nematode (Davies et al., 2001). In another study, Davies et al. (2008) observed a variation between the litters from each female descendent of the line of a single J_2 of both nematodes *M. incognita* and *M. haplo*, suggesting that the offspring of a single individual can differ in the adhesion of spores in nematode species with sexual and asexual reproduction. Special mechanisms that produce these functional differences on the surface of the cuticle may be involved in nematodes that reproduce sexually and asexually as a strategy to overcome the infection by these specialized hyper-parasites.

Carneiro et al. (2004) found a high variability in adherence among isolates of *P. penetrans* from different geographical regions to different species and populations of *Meloidogyne*. The isolates showed pathogenicity to *M. incognita*, *M. arenaria* and *M. paranaensis*, but were incompatible with *M. graminicola* and *M. mayaguensis*. Tzortzakakis (2008) evaluated the adhesion of three isolates of *P. penetrans* to the cuticle of *M. incognita* and *M. javanica* populations which differ in their ability to parasitize tomatoes containing the Mi gene, which confers resistance to gall nematodes. The number of adhered spores did not differ between virulent and avirulent populations within each species, so it is possible that the adherence rate probably reflects the specificity of *P. penetrans* to different populations of the same species within a country or shared area.

Besides the high specificity of the nematode cuticle to spores of *P. penetrans* by biochemical recognition, soil moisture and temperature have been the most important factors that influence the adhesion ability of *P. penetrans* spores to juveniles of *Meloidogyne* spp. (Mateille et al., 2009). Temperature affects adhesion, germination of endospores, pathogenicity and development of the life cycle of this bacterium. *Pasteuria penetrans* is a mesophilic bacterium, which shows optimal growth between 28 °C and 35 °C (Hatz and Dickson, 1992; Serracin et al., 1997). The minimum growth temperature is between 16.7 °C and 18.5 °C (Chen and Dickson, 1997; Giannakou and Gowen, 2004), temperatures up to 30 °C increased the adhesion of endospores (Freitas et al., 1997; Giannakou et al., 1999) and above 30 °C adhesion is reduced (Chen and Dickson, 1998). However, different temperature requirements may exist for different isolates of the bacteria due to its cosmopolitan distribution.

Samaliev (2009) conducted an experiment to estimate the influence of different temperatures (16 °C, 20 °C, 24 °C, 28 °C and 32 °C) on the adhesion and development of a mixture of populations of *P. penetrans* in *M. incognita* breed 1. The highest rate of endospore adhesion in J_2 occurred between 28 °C and 32 °C. Extreme temperatures affect this parasite-host relationship, since there was no adhesion of endospores to the cuticle of *M. arenaria* juveniles after being incubated for four days at 10 °C and 50 °C in soil infested by *P. penetrans*. This may be due the temperatures to which the nematodes were submitted that may have hampered their movement in the soil and complicated contact with bacterium endospores (Freitas, 1997).

Studies that assessed the effect of humidity on the life cycle of *Pasteuria* are few. Soil moisture, especially the alternations of high and low moisture contents, has a significant effect on this bacterium, although endospores of *P. penetrans* can survive for long periods in dry soils (Brown and Smart, 1984; Oostendorp et al., 1990; Stirling et al., 1990), however, the reproduction of nematodes and *P. penetrans* is limited in waterlogged soils. According to Davies et al. (1988), not only the mobility of the nematode host is impaired in the soil at field capacity, reducing the penetration of roots, but the development of bacteria is also compromised due to low oxygen levels in the soil. Therefore, a proper irrigation system and soils that allow better drainage are favorable to the bacteria.

The distribution of *P. penetrans* in the soil may play an important role in its effectiveness and success as a biocontrol agent. Endospores of *P. penetrans* measure about 4 mm in length and have no active movement, whereas juvenile nematodes are mobile and measure about 10 mm in diameter. Therefore, porosity and water content in the soil will determine the active and passive transport of both soil organisms (Mateille et al., 2010). The percolation of endospores in the soil by action of water from rain or irrigation may affect their distribution and even remove them from the 30 cm surface layer of soil where the bacteria are most efficient (Cetintas and Dickson, 2005).

The epidemiology of *P. penetrans* was studied in a field planted with *Solanum Ethiopicum* infested with *M. javanica* in Senegal. Biotic and abiotic parameters were analyzed from 100 soil samples. In the finest fraction (< 20 μM), spores of *P. penetrans* were most abundant in moderately irrigated areas, where the proportion of juveniles correlated to the presence of finer texture and high water retention capacity. Based on these field results, the authors conducted laboratory tests to evaluate the influence of irrigation and soil texture on the dispersion of *P. penetrans*. The results showed that 53% of inoculated spores were leached by water flow in sandy soil, but only 14% in sandy-clayey soils and 0.1% in the clayey soil. In comparison of different soil combinations containing clay, the best balance between percolation/retention of endospores occurred in soils with 10-30% clay (Dabiré et al., 2007).

There are two strategies for implementation of *P. penetrans* in field conditions for control of *Meloidogyne* spp. (Stirling, 1991): inundative and inoculative. The inundative strategy consists of frequent applications of large numbers of endospores to reduce nematode populations directly and immediately. However, because there is no satisfactory production of this bacterium *in vitro*, multiplication occurs *in vivo* in inoculated tomatoes with juveniles containing adhered endospores in a greenhouse. This mass production is slow and laborious, resulting in a reduced amount of inoculum of this bacterium which justifies the need to develop *in vitro* production techniques for production on a commercial scale. Therefore, the strategy has been adopted for application of this biocontrol agent is inoculated in field experiments, i.e., few endospores of the bacteria in suspension are applied to soil near the root systems so that multiplication of the bacteria on the nematode populations occurs and infest the soil. Even when the endospores are applied at low densities, there may be the multiplication of these bacteria in the soil when in ideal conditions to the point where they become suppressive to nematodes (Kasumimoto et al., 1993). The powder resulting from grinding dried roots of tomato plants infected by nematodes parasitized by *P. penetrans* can be incorporated directly to soil of the fields or used as substrate for seedlings. Mixed with water it can also be applied on the soil surface by irrigation water. It is necessary to use an isolate of *P. penetrans* with broad host range or else a mixture of strains to counteract the diversity of nematode populations encountered in field conditions.

Until recently, the obligate nature of the life cycle of *Pasteuria* as a nematode parasite has hampered its axenic cultivation (Davies, 2009). However, an *in vitro* method for cultivating *Pasteuria* spp. was developed that would make formulation and marketing of this bacteria viable as a biopesticide (Luc et al., 2010). However, until now, no product has been introduced to the market.

3.2. Rhizobacteria

Roughly 7 to 15% of the total root surface is occupied by bacteria (Gray and Smith, 2005). Bacteria are not distributed randomly in the rhizosphere, but aggregated, especially in the intercellular epidermis regions, since these are active exudation areas (Bowen and Rovira, 1976; Bowen and Foster, 1978). Colonization of newly formed roots is minimal, but after a few days micro-colonies appear in association with organic matter that the root tips encounter as they grow (Stirling, 1991). The bacteria commonly found in the rhizosphere and/or rhizoplane of plants colonizing the root environment, also known as rhizobacteria (Kloepper et al., 1990), may promote plant growth and/or protect them against phytopathogens (Ferraz et al. 2010). They use various mechanisms to survive in the rhizosphere and suppress the pathogen attack on plants, where more than one mechanism may be used by a rhizobacteria. By promoting the growth of plants, these bacteria are also known as PGPR (plant growth-promoting rhizobacteria) (Kloepper and Schroth, 1981). Acting directly, the PGPR may produce growth phytohormones (cytokinins and gibberellins), increase permeability of the roots, fixate nitrogen, mineralize organic matter, solubilize and provide nutrients such as phosphorus, iron and nitrogen, among others (Glick and Bashan, 1997; El-Khawas and Adachi, 1999; Lucy et al., 2004; Zahir et al., 2004).

However, the major effect of these rhizobacteria is to suppress pathogens and bacteria harmful to plant growth through the production of hydrocyanic acid and various antibiotics, induction of systemic resistance or other biocontrol mechanisms (Luz, 1996; Gray and Smith, 2005; Pieterse et al., 2005; Ferraz et al., 2010). The PGPR can prevent proliferation of fungi and other pathogens by production of siderophores (compounds with high affinity for iron) that bond to a large portion of the FeIII in the rhizosphere of the plant, making it unavailable to the pathogen. The plants are not affected by the localized shortage of iron in the soil, since most can grow at much lower iron concentrations (about 1000 times less) than the microorganisms (Siddiqui et al., 2007). These bacteria can act directly on nematodes through the production of metabolites at the location of establishment and indirectly by triggering reactions in the plant (induced systemic resistance) that prevent the formation of giant cells or result in changes in root exudates, interfering in the process of root recognition by the nematode. In both cases there is a deterrent or inhibition to hatching and mobility of J_2, with consequent reduction in penetration of the plant roots (Oostendorp and Sikora, 1990; Kerry, 2000; Ramamorthy et al., 2000).

The PGPR must possess three intrinsic characteristics (Barea, 2005): a) capable of colonizing the root, b) survive and multiply in microhabitats associated with the root surface, in competition with other microbiota, at least for the time necessary to express its promotion/protection activity of plants, c) promote plant growth. They have been divided into two groups: those involved in nutrient cycling and phytostimulation, and those involved in the biocontrol of plant pathogens (Bashan and Hoguin, 1998). However, recently Gray and

Smith (2005) proposed new concepts to categorize or divide PGPRs: a) iPGPR, bacteria that reside within the cells of plant roots, producing nodules, specialized structures for nitrogen fixation in legumes (*Rhizobium, Bradyrhizobium, Sinorhizobium, Mesorhizobium* and *Allorhizobium*) b) ePGPR, bacteria that develop extracellularly in the tissues of various plant roots, not producing nodules, but with ability to promote plant growth via production signals or specific substances (*Bacillus, Pseudomonas, Serratia* and *Burkholderia*).

Microbiolization, seed treatment with bacterial suspension, is one of the main methods for applying rhizobacteria in biological control programs. The seed germination process releases carbohydrates and amino acids in abundance in the form of seed exudates (Lynch, 1978; Subrahmanyan et al., 1983). Thus, these organisms introduced to the soil with the seeds utilize the exudates as a source of nutrition and colonize the roots as they emerge (Ferraz et al., 2010). Rhizobacteria isolates that have the greatest ability to use seed exudates have a selective advantage in colonization of roots (Kloepper et al., 1985). The density of bacteria colonizing the rhizosphere is independent of the density of initial inoculum applied to seed, but depends on the source of substrate from the roots (Bennett AND Lynch, 1981; Scher et al., 1984; Kloepper et al., 1985; Juhnke et al., 1989).

Several researchers reported the effect of rhizobacteria on nematode control, especially the genera *Bacillus* and *Pseudomonas* (Kloepper et al. 1990; Stirling, 1991). Recent studies have shown that 2,4-diacetylphloroglucinol and extracellular protease of *P. fluorescens* CHA0 (Siddiqui and Shaukat, 2003b; Siddiqui et al., 2005) and hydrogen cyanide from *P. aeruginosa* PAO1 (Gallagher and Manoil, 2001) are the main factors of these bacteria for biocontrol of nematodes. In studies using *P. fluorescens* CHA0 and *Pseudomonas* sp. BG33R, the components GacS/GacA were identified as key regulators in the production of microbial determinants involved in antagonism to phytonematodes (Siddiqui et al., 2005; Wechter et al., 2001). Lee et al. (2011) studied the factors involved in the production and regulation of nematicidal activity of *P. chlororaphis* O6. Pirrolnitrina and phenazine were reported as the key factors and GacS acts as a regulator of the expression of nematicidal components and enzymes in *P. chlororaphis* O6.

To assess whether *B. thuringiensis* could be used to control nematodes, Li et al. (2008) conducted a study to express the activity component of *B. thuringiensis* (Cry5B) within tomato roots. *Cry5B* expression in the roots was examined at the level of mRNA and quantified at the protein level. Characterization and quantification of gall nematode infection in the control versus the expression of *cry5B* indicated that this acts on the nematode, inhibiting the nematode in order to establish or maintain its feeding sites and to reduce the reproduction by roughly three times.

Bacillus firmus was assessed to control *M. incognita* in tomato plants under laboratory, greenhouse and field conditions. In laboratory tests, an aqueous suspension of the formulation at concentrations of 0.5%, 1%, 1.5% and 2% reduced egg hatching by 98 to 100% 24 hours after treatment. In tests in the greenhouse, the application of 8g/pot (1200 cm^3 of soil) containing a tomato seedling reduced the formation of galls, population level and number of eggs by 91%, 76% and 45% respectively, and resulted in an increase in biomass 50 days after application. In tests conducted under field conditions, applications of 200 and 400 kg ha^{-1} were effective in reducing the number of galls (75-84%) in addition to promoting the increase in height (29-31%) and weight (20-40 %) of the plants when compared to the control (Terefe et al., 2009).

Rhizobacteria can control more than one group of pathogens. Ten isolates with proven induction of resistance to *P. syringae* pv. *tomato* were tested against *M. javanica* and *M. incognita* on tomato. Plants treated with rhizobacteria showed reductions of up to 77.4% in the number of eggs and 66.4% in the number of galls compared to untreated plants (Neves et al., 2000). The isolates that showed greater control of *M. javanica* were not the most efficient for *M. incognita* and vice versa, indicating specificity to the nematode species.

Plants have developed a series of defense mechanisms against attack by pathogens, including resistance (Odjakova and Hadjiivanova, 2001). Induction of resistance means all types of elicited responses in the plant leading to their protection against diseases, including local and systemic responses (Hammerschmidt, 2007). There are two types of induced resistance: systemic acquired resistance (SAR) and induced systemic resistance (ISR). The SAR and ISR are distinct phenomena (Sticher et al., 1997), initiated and coordinated by different biochemical mechanisms. In SAR there occurs the accumulation of pathogenesis-related proteins (PR-proteins) as a defense mechanism of the plant, where its induction is performed via salicylate, usually triggered by pathogens or chemical activators, resulting in visual alterations. In the case of ISR, there is no accumulation of PR-proteins, the plant suffering induction does not display changes, the elicitor agent is usually a nonpathogenic microorganism and the signaling pathway is associated with jasmonate and ethylene. However, SAR and ISR are phenotypically similar phenomena in the sense that after exposure to an inducing agent the plants have their defense mechanisms activated not only at the site of induction, but also generally activated in more distant regions (Romero and Garcia, 2009). In the late twentieth century, two groups independently discovered that the ISR is a mode of PGPR action for suppressing diseases (Van Peer et al., 1991; Wei et al., 1991).

Several bacterial components induce ISR, such as lipopolysaccharides, flagella, salicylic acid, siderophores, cyclic lipopeptides, Phl antifungal factor, signaling molecule AHL and volatile blends produced by *B subtilis* GB03 (Iavicoli et al., 2003; Ryu et al., 2003; Shuhegge et al., 2006; Ongena et al., 2007; Van Loon, 2007). Only when specific components of inducers are active in stimulating resistance in the plant, but inactive in antagonism of the *in vitro* pathogen in different types of medium, can data be inferred regarding induction of resistance. Induction of resistance by living organisms always requires proof that the bodies are not in contact with one another, a condition that can be found when a rhizobacteria remains confined to the roots and the pathogen colonizes the leaves only. Under such circumstances, the inductive bacteria cause the roots to produce a local signal that moves to the leaves to active the increase in systematic defense capabilities (Van Loon, 2007). Plants exposed to an inducing agent respond more rapidly to infection by a pathogen, due to conditioning by the elicitor agent (Sticher et al., 1997). It is assumed that contact between the elicitor and plant tissue triggers the synthesis of substances that act as biochemical signals, spreading throughout the plant and activating the previously inactive genes (Romero and Garcia, 2009).

Fabry et al. (2007) evaluated the effect of *R. etli* isolate G12, obtained from the rhizosphere of potato plants, on induction of systemic resistance in tomato plants against *M. javanica*. The isolate did not promote plant growth, but the number of galls and eggs per root system was reduced by 35.3% and 38.8%, respectively, compared to treatment without bacterial inoculation. The split root technique was used by the authors to demonstrate the induction of systemic resistance by bacteria, and from the application of this methodology it was possible to determine the reduction in number of galls and eggs on plants treated with bacteria and with spatial separation between the antagonist and nematode, confirming the hypothesis of induced systemic resistance.

Although few specific studies have been conducted to demonstrate the mode(s) of action of rhizobacteria on the nematodes, it is suggested that some mechanisms act on particular life stages of the nematode (Ferraz et al., 2010):

a) Antibiotics and toxins produced by bacteria in the rhizosphere are diffused in the soil and can be absorbed by the eggs of the nematodes, maintaining their cells and preventing embryonic development and juvenile formation. Some microorganism genera that are known to produce antibiotics are *Streptomyces, Bacillus, Pseudomonas, Penicillium, Trichoderma* and *Aspergillus*.
b) The hypothesis to explain the inhibition of hatching is that rhizobacteria degrade the root exudates that act as a hatching factor for many nematode species, as suggested by Oostendorp and Sikora (1990). It is still possible that the compounds absorbed by the egg inactivate the nematode or cause deformities in its development that would prevent them from exiting the egg;
c) The nematode presents a random movement in the soil until it detects the root exudates of its host via its sensory organs. This directs and stimulates its movement towards the root. The transformation of root exudate in sub-products by the metabolism of rhizobacteria can make the nematode simply not recognize the chemotropic stimulus, and therefore it would continue to move randomly until depleting its energy reserves and die without entering the root. If the nematode recognizes the root exudates and is directed towards the roots, some bacterial products can present nematostatic characteristics and reduce the mobility of the nematode, preventing it from reaching the roots.
d) Some phytonematode genera have a restricted number of hosts, so the correct recognition of the host is a key factor for their survival. When finding the root, the nematode punctures the root epidermis to test the content of root cells. Substances produced by rhizobacteria are absorbed by roots and can alter their chemical composition so that the nematodes do not recognize the host. According to Zuckerman (1983), the recognition process is controlled by interactions between lectins on the surface of the root and carbohydrates on the cuticle of the nematode. It is also believed that the rhizobacteria bind to lectins on the surface of roots and thus interfere with the recognition process of the host plant by plant-parasitic nematodes (Stirling, 1991).
e) Toxic or repellents substances, in high concentration in the region of the rhizoplane or in the cellular content of the root epidermis, may discourage the penetration of endoparasitic nematodes in the host plant, or even the feeding of nematode ectoparasites.
f) Endoparasitic nematodes such as those of the genera *Meloidogyne, Heterodera* and *Globodera* feed on special cells known as giant and syncytial cells. Without these cells the nematode does not reach adulthood. The rhizobacteria, or their metabolic products, can be absorbed by the plant and trigger a hypersensitivity reaction in the giant cells, which is the main mechanism of host resistance to nematodes of the genus *Meloidogyne*.

g) When giant and syncytial cells are abnormal in shape, size and content, which occurs in poor host plants, the nematode females cannot obtain sufficient energy reserves to produce eggs. Because some rhizobacteria have a greater effect in reduction of eggs than in reducing the number of galls, this may be one of the mechanisms at work. Although not an ideal mechanism of resistance, since it allows damage to the parasitized crop, interruption of nematode reproduction involves minor damage during the life of the host plant, even in annual crops because the gall or cyst nematodes have more than one life cycle.

When the action mechanisms of rhizobacteria against nematodes are properly identified, *in vitro* methods for isolation and selection of effective agents can be developed. Some of the selection parameters can be quite specific, such as production of nematode repellent substances or inactivation of eggs. However, it seems unlikely that certain rhizobacteria display only one mode of action, but a series of events that result in the effectiveness of an isolate. Therefore, the best method of selection is still the cumbersome and slow mass method in greenhouse and subsequent confirmation of efficiency of the most promising bacteria in greater soil volume and greater number of repetitions (Ferraz et al., 2010). It may also be that selected isolates in the greenhouse are not effective in field conditions when sterilized soil is used in the mass selections. For a rhizobacteria to be efficient in real cultivation conditions, it must be competent in the rhizosphere, which is the ability to colonize the root system of the host plant and be able to compete with native bacteria of the most diverse soils. This attribute is influenced by a series of factors such as soil texture, percolation of bacterial cells with rain water or irrigation, and the influence of fungi or bacteria that may present antibiosis (Ferraz et al., 2010). Some rhizobacteria may have a consistent behavior in various soil types, as observed by Hackenberg et al. (2000) in controlling *P. penetrans* by *P. chlororaphis*. Because these factors alter the antagonism of bacteria on nematodes, ideally the rhizobacteria selected in a greenhouse be tested in field conditions and at various locations before developing a commercial product.

Isolation of rhizobacteria from the rhizosphere of plants antagonistic to nematodes may increase by up to six times the odds of obtaining isolates efficient for control (Kloepper et al., 1991, 1992). Heating the soil may also be useful to increase the chances of isolating rhizobacteria effective for controlling nematodes. Solarization of the soil in the field significantly reduced populations of *P. fluorescens*, however, after the sowing of tomato plants in solarized soils, the bacteria rapidly colonized the rhizosphere and plant roots. Populations of *P. fluorescens* increased up to 130 times more in solarized soils in comparison with non-solarized soil (Gamliel and Katan, 1991). In another study, the population level of *Bacillus* spp. was similar in lettuce roots grown in soil previously solarized and non-solarized, while the number of *P. fluorescens* was 6 to 10 times greater in lettuce roots in solarized soil than in non-solarized soil (Gamliel and Stapleton, 1993). Freitas et al. (2005) performed soil heating to obtain bacteria for control of gall nematodes. From the soil mixture taken from the rhizosphere of tomato plants with dried and chopped cabbage leaves, heated in microwave for 4 minutes and incubated for 24 h at 26°C, the authors obtained a bacterial isolate which, when applied to tomato seeds, reduced the number of galls of *M. javanica* by 59%. A 55% reduction in the number of galls was achieved with another bacterial isolate obtained from the same mixture, but autoclaved for 60 minutes at 120 °C, and incubated for 24 h at 26 °C.

Some bacteria based products are available on the world market: Israel (*B. firmus* = Bionem WP® and Biosafe®), United States (*P. usgae* = Econem®, *B. chitinosporus, B. latorosporus, B. licheniformis* = Biostart Rhizoboost®) and Brazil (*Bacillus* spp. = Nemix®).

4. Combination of Fungal and Bacterial Antagonists of Nematodes

Most approaches applied to biological control of plant diseases have been based on using a single antagonist against a target pathogen (Jatala, 1986; Siddiqui and Shaukat, 2002). However, it is likely that in most cases of natural occurring biological control, this event is the result of the mixture of antagonists, much more than a high population of only one antagonist. In soils suppressive to plant diseases, the mixture of antagonist populations of is considered a major factor contributing to this phenomenon (Lemanceau and Alabouvette, 1991, Stirling, 1991). Thus, the introduction of a mixture of antagonists would probably increase the effectiveness and reliability of control, since the expansion of the spectrum of activity may possess multiple mechanisms of action against the target pathogen (Siddiqui and Shaukat, 2003a). Some advantages of the combination of antagonists in biological control may be listed (Lemanceau and Alabouvette, 1991; Meyer and Roberts, 2002): a) multiple modes of action against the target pathogen, b) ability to affect more than one stage in the life cycle of the target pathogen, c) activity at different times during the growing season, d) increased consistency of performance over a wide range of soil conditions, e) potential to select organisms that affect more than one pathogen or pest, increasing the usage spectrum of the product.

The declining population of *H. avenae*, under susceptible monoculture conditions in England was studied in detail by Kerry (1975). The author concluded that the high population of the nematophagous fungi *Nematophthora gynophila* and *P. chlamydosporia* found in the planted areas was principal responsible for suppresion. The combination of biological control agents has been investigated by several researchers and the beneficial effect of such combinations on plant vigor and suppression in the number of females, eggs, juveniles and galls has not been frequently verified, mainly when involving species of *Meloidogyne* (Stirling, 1991; De Leij et al., 1992a; Duponnois et al., 2002). Parasitic activity of *P. chlamydosporia* on *M. incognita* was enhanced in the joint application with *Rhizobium* sp., *T. harzianum* and *Glomus clarum*, suggesting the possibility of combined use of these microorganisms in the management of gall nematodes (Puertas et al., 2006). The joint application of *P. chlamydosporia* (parasite eggs and females) and *P. penetrans* (parasite juveniles and prevent egg production) was more efficient in controlling *M. incognita* in tomato plants than the isolated application of the antagonists (De Leij et al., 1992a). The same additive effect occurred with the interaction with *P. penetrans - P. lilacinus* for control of *M. incognita* and *M. javanica* (Dube and Smart Jr., 1987; Maheswari and Mani, 1988; Zaki and Maqbool, 1991).

The integrated use of *P. penetrans*, *B. subtilis*, *Talaromyces flavus* and *P. lilacinus* controlled *M. incognita* in okra, with particular emphasis on the combination *P. penetrans - P. lilacinus* (Zaki and Maqbool, 1991). However, especially in combinations involving *P. penetrans*, the authors took into account the reduction in number of galls and eggs after only one or two plant cycles, thus it is not known if over time the isolated application of *P. penetrans* would no longer be effective, since it is dependent on high population density of gall nematodes to become soil suppressive. Additionally, *P. penetrans* can remain in the soil for longer periods than the fungi, since endospores are dormant and not dependent on organic matter or compete for substrate for survival (Freitas and Carneiro, 2000).

Rhizobacteria can improve the adhesion of endospores to the cuticle and reproduction of *P. penetrans* inside gall nematodes (Ferraz et al., 2010). Duponnois et al. (1999) isolated bacteria from the rhizosphere of tomato plants from soil in the field highly infested with *M. javanica* and *P. penetrans*. Some of these isolates increased the percentage of J_2 infested with *P. penetrans* and the number of endospores adhered per J_2, compared with the control without rhizobacteria. This effect is probably a consequence of the action of rhizobacteria in degradation of *P. penetrans* sporangia, exposing the parasporal fibers (Duponnois et al., 1999). Two of these species of rhizobacteria were identified as *Enterobacter cloacae* and *Pseudomonas mendoncina*, and considered PGRP by inducing plant growth. These bacteria, when applied without *P. penetrans*, reduced the number of egg masses produced by nematodes per root system, and when applied with *P. penetrans*, presented additional control effects. The author reported that *Enterobacter cloacae* favored the increase in number of *P. penetrans* endospores produced by tomato roots.

The joint application of rhizobacteria with nematophagous fungi may also be beneficial. For example, the combination of the predatory fungus *Arthrobotrys* sp. with fluorescent rhizobacteria from the genus *Pseudomonas* reduced the populations of *Meloidogyne* spp. and *R. reniformis* in tobacco plants, and increased growth of the shoots (Duponnois et al., 2002). Siddiqui et al. (2007) studied the biocontrol of *M. javanica* in lentil using *P. putida*, *P. alcaligenes*, *Paenibacillus polymyxa*, *B. pumilus* and *Rhizobium* sp. Among the PGPR, *P. putida* alone and in combination with *Rhizobium* were the treatments which obtained the best results.

The integration between nematophagous fungi, rhizobacteria and organic matter may promote the establishment of these organisms in the soil, especially fungi, as well as enhance control of nematodes, releasing nematicidal compounds and improving the nutrition of the host plant due to the release of nutrients (Cannayane and Rajendran, 2001). The integrated use of oil cakes such as neem, castor, peanut and mustard with *P. lilacinus* has been effective in the control of different nematode species (Reddy et al., 1991; Rao and Reddy, 1994; Khan and Saxena, 1997). The integration of *P. chlamydosporia* with neem leaves increased the size of tomato seedlings, the percentage of egg colonization and density of the bioagent on roots of seedlings, and reduced the number of galls of *M. incognita* and nematode populations in soil (Reddy and Rao, 1999). The biomass of *Eucalyptus camaldulensis*, *Azadirachta indica* and *Acacia holosericea* applied together with *A. oligospora* controlled *Meloidogyne mayaguensis* in tomato (Duponnois et al., 2001). Tomato plants treated with *R. etli* and earthworm castings presented increased height and mass of shoots, and a reduced number of galls of *M. incognita* by up to 64% when compared with the treatment without the application of humus (Fabry et al., 2008). In the absence of *R. etli*, the incorporation of humus did not reduce the number of galls caused by nematodes.

Despite the benefits and promising perspectives for the combination of microorganisms and/or organic matter as a new mentality in the control of nematodes, the effect of such combinations must be known, since it may cause incompatibilities between antagonists and/or organic waste with the release of substances that inhibit the growth of control agents (Ferraz et al., 2010). For example, the combined application of the fungus *H. rhosiiliensis* and *P. chlamydosporia* with the aim of controlling *M. haplo* in lettuce was not efficient, unlike the individual application of both fungi (Viane and Abawi, 2000). Other problems may arise from technical difficulties in the formulation of more than one microorganism that may lead to increased processing costs, more rigorous quality control to avoid contamination and other factors. Therefore, according to Kerry (2001) there is a need for more experimentation to demonstrate if the combined applications of agents compete or act synergistically to improve the control achieved by the addition of a single agent.

5. Final Consideration

Biological control of animal- or plant-parasitic nematodes, although complex, is a reality that must be part of a series of measures directed at reducing the damage caused by these parasites on the productivity of livestock and agriculture, since the largest proportion of biomass of these pathogens resides in livestock and agriculture. Fungi and bacteria with nematophagous ability have been the major microorganisms used in biological control of nematodes, acting both directly (predation and parasitism) or indirect (antibiosis and competition).

The modern strategy for biological control of plant-parasitic nematodes is focused on using fungi and/or bacteria that establish themselves in the rhizosphere of plants and act on the eggs and female nematodes; and due to the impact they have on their population, there is a resultant smaller generation of offspring for each life cycle of the nematode. The approach with the best results in the control of animal-parasitic nematodes utilized predacious fungi that are resistant to digestive environment of animals and capable of germinating in the feces, and reduces the number of infective larvae of these nematodes before they migrate to pastures and consequently infect new hosts. These approaches were based on better understanding of these agents which has been obtained from studies focused on biology, ecology, evolution, behavior and mode of action since success depends on a detailed knowledge of the agent, the host and their interactions with the environment. Molecular and biochemical analysis are generating relevant data, especially regarding the production of enzymes/toxic metabolites and their role as important pathogenic factors for nematodes. However the potential of nematoxins for biological control of animal-parasitic nematodes remains unexplored.

Commercial formulations containing fungi and bacteria as the active ingredient are being developed and made available on the market, but unfortunately only for the control of plant-parasitic nematodes. Research groups and companies have invested in developing these products, since chemical control is often inefficient, costly and uneconomical. Its large-scale use in the environment still requires additional scientific and technological progress in production, storage, preparation and application. Additionally, more detailed evaluations should be performed under field conditions, since once formulated, the agent must be stable, viable and efficient for a longer period. Few studies have been conducted on management methods of nematophagous fungi for animals, despite the increasing problems of nematode resistance to anthelmintics.

Integrated management is a current requirement of import markets that have stringent requirements for quality and sustainability, emphasizing primarily the protection of biodiversity and conservation of soils in agricultural ecosystems, food security, working conditions, human and animal health and economic viability. In this context, although they do not replace the chemicals, the products based on nematophagous agents should be considered an integral part in the management of nematodes since they may very well offer a cheaper and sustainable approach to reducing the impact of plant- and animal-parasitism, as well as the use of nematicides and anthelmintics.

Despite the large amount of knowledge acquired until now, there is still much to be discovered and elucidated by researchers that should be viewed as a necessity in the face of climate changes that may have an influence on the multiplication of pathogens in the environment and the relationship with the host and its antagonists. The costs of investigation efforts and development of technologies for production scale of nematode antagonists for biological control is insignificant in the face of its promising future prospects.

Acknowledgments

The authors thank the National Council for Scientific and Technological Development (CNPq), an agency dedicated to the promotion of scientific and technological research and to the formation of human resources for research in the Brazil, for its financial support and the Federal University of Viçosa (UFV) for its professionalism and competence in developing education, research and extension, without whom this work would not have been accomplished.

References

Abrantes, I., Santos, S.A., Bourne, J., Ciancio, A., Lopez-Llorca, L., Kerry, B., Mota, M., Tzortzakakis, E., and Verdejo-Lucas, S. (2002). The use of *Verticillium chlamydosporium* as a biological control agent. In: Kerry, B.R. and Bourne, J.M. (Eds.). *A manual for research on Verticillium chlamydosporium, a potential biological control agent for root-knot nematodes.* (pp. 1-12). Gent: IOBC/WPRS.

Ahman, J., Ek, B., Rask, L., and, Tunlid, A. (1996). Sequence analysis and regulation of a gene encoding a cuticle-degrading serine protease from the nematophagous fungus *Arthrobotrys oligospora. Microbiology, 142,* 1605-1616.

Ahman, J., Johansson, T., Olsson, M., Punt, P.J., Van den Hondel, C.A.M.J.J. and Tunlid, A. (2002). Improving the pathogenicity of a nematode-trapping fungus by genetic engineering of a subtilisin with nematotoxic activity. *Applied and Environmental Microbiology, 68,* 3408-3415.

Ahrén, D., Ursing, B.M., and Tunlid, A. (1998). Phylogeny of nematode-trapping fungi based on 18S rDNA sequences. *FEMS Microbiology Letters,* 158, 179-184.

Akhtar, M., and Malik, A. (2000). Roles of organic soil amendments and soil organisms in the biological control of plant-parasitic nematodes: a review. *Bioresource Technology, 74*, 35-47.

Alexopoulos, C.J., Mims, C.W., and Blackwell, M. (1996). *Introductory mycology.* New York: John Wiley and Sons.

Ameen, H.H. (1991). Nematicidal effect of *Aspergillus ochraceous* filtrate. *Bulletin Faculty Agriculture, Cairo University, 42,* 963-970.

Anderson, J.M., Preston, J.F., Dickson, D.W., Hewlett, T.E., Williams, N.H., and Maruniak, J.E. (1999). Phylogenetic analysis of *Pasteuria penetrans* by 16 S rRNA gene cloning and sequencing. *Journal of Nematology, 31,* 319-325.

Araújo, J.V., Braga, F.R., Milani, J.A., Silva, A.S., and Tavela, A.O. (2008). *In vitro* evaluation of the effect of the nematophagous fungi *Duddingtonia flagrans, Monacrosporium sinense* and *Pochonia chlamydosporia* on *Ascaris suum* eggs. *Parasitology Research, 102,* 787-790.

Atkins, S.D., Clark, I.M., Pande, S., Hirsch, P.R., and Kerry, B.R. (2005). The use of real-time PCR and species-specific primers for the identification and monitoring of *Paecilomyces lilacinus. FEMS Microbiology Ecology, 51,* 257-264.

Atkins, S.D., Hidaldo-Diaz, L., Kalisz, H., Mauchline, T.H., Hirsch, P.R., and Kerry, B.R. (2003). Development of a new management strategy for the control of root-knot nematodes (*Meloidogyne* spp.) in organic vegetable production. *Pest Management Science, 59,* 183-189.

Ayatollahy, E., and Fatemy, S. (2010). *In vitro* assessment of pathogenicity and culture filtrates of fungi against *Heterodera schachtii. Applied Entomology and Phytopathology, 77,* 15-26.

Balan, J., and Lechevalier, H.A. (1972). The predaceous fungus *Arthrobotrys dactyloides*: induction of trap formation. *Mycologia, 64,* 919-922.

Barea, J.M., Pozo, M.J., Azcón, R., and Azcón-Aguilar, C. Microbial co-operation in the rhizosphere. *Journal of Experimental Botany, 56,* 1761-1778.

Barnett, H.L., and Hunter, B.B. (1998). *Illustrated genera of imperfect fungi.* St. Paul: APS Press.

Barron, G.L. (1977). *The nematode-destroying fungi. Topics in mycobiology.* Guelph: Canadian Biological Publications.

Barron, G.L. (2003). Predatory fungi, wood decay, and the carbon cycle. *Biodiversity, 4,* 3-9.

Bartnicki-Garcia, S., Eren, J., and Pramer, D. (1964). Carbon dioxide dependent morphogenesis in *Arthrobotrys conoides* (nematode trapping fungus). *Nature, 204,* 804.

Bashan, Y., and Holguin, G. (1998). Proposal for the division of plant growth-promoting rhizobacteria into two classifications: biocontrol-PGPB (plant growth-promoting bacteria) and PGPB. *Soil Biology and Biochemistry, 30,* 1225-1228.

Basualdo, J.A., Ciarmela, M.L., Sarmiento, P.L., and Minvielle, M.C. (2000). Biological activity of *Paecilomyces* genus against *Toxocara canis* eggs. *Parasitology Research, 86,* 854-859.

Bedelu, T., Gessesse A., and Abate, D. (1998). Relation of protease production to nematode degrading ability of two *Arthrobotrys* spp. *World Journal of Microbiology and Biotechnology, 14,* 731-734.

Bekal, S., and Becker, S. (2000). Population dynamics of the sting nematode in California turf grass. *Plant Disease, 84,* 1081-1084.

Bennett, R.A. and Lynch, J.M. (1981). Colonization potential of bacteria in the rhizosphere. *Current Microbiology, 6,* 137-138.

Bilgram, A.L. (1997). Evaluation of the predation abilities of the mite *Hypoaspis calcuttaensis*, predaceous on plant and soil nematodes. *Fundamental Applied of Nematoly, 20,* 96-98.

Bilgrami, A.L., and Gaugler, R. (2004). Feeding behaviours. In: Gaugler, R., and Bilgrami, L. (Eds.). *Nematode behaviour*. (pp. 91-126). Wallingford: CAB Publishing.

Bird, A.F., and Bird, A.F. (1991). *The structure of nematodes.* San Diego: Academic Press.

Bird, D.M., and Kaloshian, I. (2003). Are roots special? Nematodes have their say. *Physiological and Molecular Plant Pathology, 62*, 115-123.

Bird, J., and Herd, R.P. (1995). *In vitro* assessment of two species of nematophagous fungi (*Arthrobotrys oligospora* and *Arthrobotrys flagrans*) to control the development of infective cyathostome larvae from naturally infected horses. *Veterinary Parasitology, 56,* 181-187.

Bishop, A.H., Gowen, S.R., Pembroke, B., and Trotter, J.R. (2007). Morphological and molecular characteristics of a new species of *Pasteuria* parasitic on *Meloidogyne ardenensis*. *Journal of Invertebrate Pathology, 96,* 28-33.

Blackwell, V., Ahmed, K., O'Docherty, C., and Hay, R.J. (2000). Cutaneous hyalohyphomycosis caused by *Paecilomyces lilacinus* in a renal transplant patient. *British Journal of Dermatology, 143,* 873-875.

Bogus, M.I., Czygier, M., Kedra, E., and Samborski, J. (2005). *In vitro* assessment of the influence of nutrition and temperature on growing rates of five *Duddingtonia flagrans* isolates, their insecticidal properties and ability to impair *Heligmosomoides polygyrus* motility. *Experimental Parasitology, 109,* 115-123.

Bonants, P.J.M., Fitters, P.F.L., Thijs, H., Den Belder, E., Waallwijk, C., Willem, J., and Henfling, D.M. (1995). A basic serine protease from *Paecilomyces lilacinus* with biological activity against *Meloidogyne hapla* eggs. *Microbiology, 141,* 775-784.

Boogert, P.H.J.F., Dijksterhuis, J., Velvis, H., and Veenhuis, M. (1992). Adhesive knob formation by conidia of the nematophagous fungus *Drechmeria coniospora*. *Antonie van Leeuwenhoek, 61,* 221-229.

Bordallo, J.J., Lopez-Llorca, L.V., Jansson, H.B., Salinas, J., Persmark, L., and Asensio, L. (2002). Colonization of plant roots by egg-parasitic and nematode-trapping fungi. *New Phytologist, 154,* 491-499.

Bourne, J.M. and Kerry, B.R. (2000). Observation on the survival and competitive ability of the.nematophagous fungus *Verticillum chlamydosporium* in soil. *International Journal of Nematology, 10,* 9-18.

Bourne, J.M. (2001). Making a soil suppressive to root-knot nematodes by applications of *Verticillium chlamydosporium*. Tri-trophic interactions in the rhizosphere. *IOBC/WPRS Bulletin, 24,* 25-30.

Bourne, J.M., Kerry, B.R., and De Leij, F.A.A.M. (1996). The importance of the host plant on the interaction between root-knot nematodes (*Meloidogyne* spp.) and the nematophagous fungus, *Verticillium chlamydosporium* Goddard. *Biocontrol Science and Technology, 6,* 539-548.

Bourne, Joanna y Kerry, B.R. (1999): Effect of the host plant on the efficacy of *Verticillium chlamydosporium* as a biological control agent of root-knot nematodes at different nematode densities and fungal application rates. *Soil Biology and Biochemistry, 31,* 75-84.

Bowen, G.D., and Foster, R.C. (1978). Dynamics of microbial colonization of plant roots. In Broughton, W.J. and John, C.J. (Eds.). *Symposium of Soil Microbiology and Plant Nutrition.* (pp 14-31). Kuala Lumpur: University of Malaya Press.

Bowen, G.D., and Rovira, A.D. (1976). Microbial colonization of plant roots. *Annual Review of Phytopathology, 14,* 121-144.

Brown, S.M., and Smart Jr., G.C. (1984). Attachment of *Bacillus penetrans* to *Meloidogyne incognita. Nematropica, 14,* 171-172.

Bundy, D.A.P. (1994). The global burden of intestinal nematode disease. *Transactions of the Royal Society of Tropical Medicine and Hygiene, 88,* 259-261.

Burney, K., and Estey, H. (1985). A note on the non-difusibility of a substance produced by nematodes that induced *Arthrobotrys oligospora* to form nematode traps. *Phytoprotection, 66,* 163-164.

Campos, A.K. (2006). *Fungos nematófagos no controle de nematóides gastrintestinais de ruminantes.* Tese de Doutorado. Belo Horizonte: Universidade Federal de Minas Gerais, Brasil.

Campos, V.P., Souza, J.T., Souza, R.M. (1998). Controle de fitonematóides por meio de bactérias. *Revisão Anual de Patologia de Plantas, 6,* 285-327.

Cannayane, I. and Rajendran, G. (2001). Application of biocontrol agents and oil cakes for the managment for *Meloidogyne incognita* in brinjal (*Solanum melongena* L.). *Current Nematology, 12,* 51-55.

Cannayane, I., and Sivakumar, C.V. (2001). Nematode egg - parasitic fungus 1: *Paecilomyces lilacinus* - a review. *Agricultural Review, 22,* 79-86.

Canning, E.U. (1973). Protozoal parasites as agents for biological control of plant-parasitic nematodes. *Nematologica, 19,* 342-348.

Carey, J., D'Amico, R., Sutton, D.A., and Rinaldi, M.G. (2003). *Paecilomyces lilacinus* vaginitis in an immunocompetent patient. *Emerging Infectious Diseases, 9,* 1155-1158.

Carneiro, R.M.D.G., Tigano, M.S., Lopes Jorge, C., Oliveira Teixeira, A.C., and Cordeiro, M.C. (2004). Selection and polymorphism of *Pasteuria penetrans* isolates in relation to *Meloidogyne* spp. from coffee. *Nematology, 6,* 37-47.

Carvalho, R.O., Araújo, J.V., Braga, F.R., Araújo, J.M., and Alves, C.D.F. (2010). Ovicidal activity of *Pochonia chlamydosporia* and *Paecilomyces lilacinus* on *Toxocara canis* eggs. *Veterinary Parasitology*, *169,* 123-127.

Cetintas, R., and Dickson, D.W. (2005). Distribution and downward movement of *Pasteuria penetrans* in field soil. *Journal of Nematology, 37,* 155-160.

Chandrawathani, P., Jamnah, O., Waller, P.J., Larsen, M. and Gillespie, A.T. (2004). Field studies on the biological control of nematode parasites of sheep in the tropics, using the microfungus *Duddingtonia flagrans*. *Veterinary Parasitology, 120,* 177-187.

Chandrawathani, P., Jamnah, O., Waller, P.J., Larsen, M., Gillespie, A., and Zahari, W.M. (2003). Biological control of nematode parasites of small ruminants in Malaysia using the nematophagous fungus *Duddingtonia flagrans*. *Veterinary Parasitology, 117,* 173-183.

Chandrawathani, P., Omar, J., and Waller, P.J. (1998). The control of free living stages of *Strongyloides papillosus* by nematophagous fungus *Arthrobotrys* spp. *Veterinary Parasitology, 76,* 321-325.

Chandrawathani, P., Yusoff, N., Wan, L.C., Ham, A., and Waller, P.J. (2004). Total anthelmintic failure to control nematode parasites of small ruminants on government breeding farms in Sabah, East Malaysia. *Veterinary Research Communications, 28,* 479-489.

Charles, L., Carbone, I., Davies, K.G., Bird, D., Burke, M., Kerry, B.R., and Opperman, C.H. (2005) Phylogenetic analysis of *Pasteuria penetrans* by use of multiple genetic loci. *Journal of Bacteriology, 187,* 5700-5708.

Chartier, C., and Pors, I. (2003). Effect of the nematophagous fungus, *Duddingtonia flagrans*, on the larval development of goat parasitic nematodes: a plot study. *Veterinary Research, 34,* 221-230.

Chauhan, J.B., Sanyal, P.K., and Subramanian, R.B. (2005). The nematode-trapping efficacy of two chlamydospore-forming fungi against *Haemonchus contortus* in sheep. *Journal of Helmintology, 79,* 315-319.

Chen, S., and Dickinson, D.W. (2004). Biological control of nematodes by fungal antagonists. In: Chen, Z., Chen, S., and Dickinson, D.W (Eds). *Nematology – Advances and perspectives, v.2: Nematode management and utilization.* (pp. 979-1039). Cambridge: CABI Publishing.

Chen, S.Y., and Liu, X.Z., 2005. Control of the soybean cyst nematode by the fungi *Hirsutella rhossiliensis* and *Hirsutella minnesotensis* in greenhouse studies. *Biological Control 32,* 208-219.

Chen, S.Y., and Reese, C.D. (1999). Parasitism of the nematode Helerodera glycines by the fungus *Hirsutella rhossiliensis* as influenced by crop sequence. *Journal of Nematology, 31,* 437-444.

Chen, S.Y., Dickson, D.W., and Mitchell, D.J. (2000). Viability of *Heterodera glycines* exposed to fungal filtrates. *Journal of Nematology, 32,* 190-197.

Chen, Z.X., and Dickson, D.W. (1997). Minimal growth temperature of *Pasteuria penetrans. Journal of Nematology, 25,* 635-639.

Chen, Z.X., and Dickson, D.W. (1998). Review of *Pasteuria penetrans*: Biology, ecology, and biological control potential. *Journal of Nematology, 30,* 313-340.

Cho, M.R., Na, S.Y., Yiem, M.S., Cho, M.R., Na, S.Y.; and Yiem, M.S. (2000). Biological control of *Meloidogyne arenaria* by *Pasteuria penetrans. Journal of Asia Pacific Entomology, 3,* 71-76.

Ciancio, A., Logricco, A., Lamberti, L., and Bottalico, A. (1988). Nematicidal effects of some *Fusarium* toxins. *Nematologia Mediterranea, 16,* 137-138.

Ciarmela, M.L., Arambarri, A.M., Basualdo, J.A., and Minvielle, M.C. (2010). Effect of saprotrophic soil fungi on *Toxocara canis* eggs. *Malaysian Journal of Microbiology, 6* , 75-80.

Ciarmela, M.L., Minvielle, M.C., Lori, G., and Basualdo, J.A. (2002). Biological interaction between soil fungi and *Toxocara canis* eggs. *Veterinary Parasitology*, *103*, 251-257.

Cooke, R.C. (1969). Two nematode-trapping Hyphomycetes, *Duddingtonia flagrans* gen. et comb. nov. and *Monacrosporium mutabilis* sp. nov. *Transactions of the British Mycological Society, 53,* 315-319.

Cooke, R.C., and Godfrey, B.E.S. (1964). A key of nematode-destroying fungi. *Transactions of the British Mycological Society, 47,* 61-74.

Cossins, A. R., and Bowler, K. (1987). Temperature biology of animals. London: Chapman and Hall.

Costa, M.J.N., Campos, V.P., Pfenning, L.H., and Oliveira, D.F. (2001). Toxicidade de filtrados fúngicos a *Meloidogyne incognita*. *Fitopatologia Brasileira, 26,* 749-755.

Crompton, D.W. (2000). The public health importance of hookworm disease. *Parasitology, 121,* 39-50.

Cruz, D.G., Silva, C.P., Carneiro, C.N.B., Retamal, C.A., Thiébaut, J.T.L., DaMatta, R.A., and Santos, C.P. (2009). Acid phosphatase activity during the interaction of the nematophagous fungus *Duddingtonia flagrans* with the nematode *Panagrellus* sp. *Journal of Invertebrate Pathology, 102,* 238-244.

Dabiré, R.K., Ndiaye, S., Mounport, D., and Mateille, T. (2007). Relationships between abiotic soil factors and epidemiology of the biocontrol bacterium *Pasteuria penetrans* in a root-knot nematode *Meloidogyne javanica*-infested field. *Biological Control, 40,* 22-29.

Dackman, C., and Nordbring-Hertz, B. (1992). Conidial traps - a new survival structure of the nematode-trapping fungus *Arthrobotrys oligospora*. *Mycology Research, 96,* 194-198.

Dackman, C., Olsson, S., Jansson; H.B., Lundgren B., and Nordbring-Hertz, B. (1987). Quantification of predatory and endoparasitic nematophagous fungi and their activies in soil. *Microbial Ecology, 13,* 89-93.

Davies KG, Fargette M, Balla G, Daudi, A., Duponnois, R., Gowen, S.R., Mateille, T., Phillips, M.S., Sawadogo, A., Trivino, C., Vouyoukalou, E., and Trudgill, D.L. (2000). Cuticle heterogeneity as exhibited by *Pasteuria* spore attachment is not linked to the phylogeny of parthenogenetic root-knot nematode (*Meloidogyne* spp.). *Parasitology,122,* 111-120.

Davies, K.G. (2009). Understanding the interaction between an obligate hyperparasitic bacterium, *Pasteuria penetrans*, and its obligate plant parasitic host, *Meloidogyne* spp. *Advances in Parasitology, 68,* 211-245.

Davies, K.G., and Danks, C. (1993). Carbohydrate/protein interactions between the cuticle of infective juveniles of *Meloidogyne incognita* and spores of the obligate hyperparasite *Pasteuria penetrans. Nematologica, 39,* 54-64.

Davies, K.G., and Redden, M. (1997) Diversity and partial characterisation of putative virulence determinants in *Pasteuria penetrans*, the hyperparasite of root-knot nematodes. *Journal of Applied Microbiology, 83,* 227-235.

Davies, K.G., Afolabi, P., and O'Shea, P.S. (1996). Adhesion of *Pasteuria penetrans* to the cuticle of root-knot nematodes (*Meloidogyne* spp) inhibited by fibronectin: a study of electrostatic and hydrophobic interactions. *Parasitology, 112,* 553-559.

Davies, K.G., Fargette, M., Balla, G., Daudi, A., Duponnois, R., Gowen, S.R., Mateille, T., Phillips, M.S., Sawadogo, A., Trivino, C., Vouyoukalou, E., and Trudgill, D.L. (2001). Cuticle heterogeneity as exhibited by *Pasteuria* spore attachment is not linked to the phylogeny of parthenogenetic root-knot nematodes (*Meloidogyne* spp.). *Parasitology, 122,* 111-120.

Davies, K.G., Flynn, C.A., and Kerry, B.R. (1988). The life-cycle and pathology of the root-knot nematode parasite *Pasteuria penetrans*. In: *Proceedings of the Brighton Crop Protection Conference - Pests and Diseases.* (pp. 1221-1226). Brighton: BCPC.

Davies, K.G., Rowe, J.A., and Williamson, V.M. (2008). Inter and intra specific cuticle variation between amphimictic and parthenogenetic species of root-knot nematode (*Meloidogyne* spp.) as revealed by a bacterial parasite (*Pasteuria penetrans*). *International Journal for Parasitology, 38,* 851-859.

De Leij, F.A.A.M., Davies, K.G., and Kerry, B.R. (1992a). The use of *Verticillium chlamydosporium* Goddard and *Pasteuria penetrans* (Thorne) Sayre and Starr alone and in combination to control *Meloidogyne incognita* on tomato plants. *Fundamental and Applied Nematology, 15,* 235-242.

De Leij, F.A.A.M., Kerry, B.R., and Dennehy, J.A. (1992b). The effect of fungal application rate and nematode density on the effectiveness of *Verticillium chlamydosporium* as a biological control a gent for *Meloidogyne incognita. Nematologica, 38,* 112-122.

Decker, H. (1989). *Plant nematodes and their control: Phytonematology.* New York: E.J. Brill.

Descazeaux, J. (1939a). Action de champignons Hypomicétes prédateurs sur les larves de certains nématodes parasites des ruminants. *Bulletin de la Societe de Pathologie Exotique, 32,* 457-459.

Descazeaux, J. (1939b). Stérilisation biologique du crottin parasité par des larves de nematódes. *Bulletin de l'Academie Veterinaire de France, 12,* 136-139.

Descazeaux, J., and Capelle, R. (1939). Contribuition à l'etude des champignons prédateurs des larves de nématodes parasites des animaux domestiques. *Bulletin de l'Academie Veterinaire de France, 12,* 284-288.

Deschiens, R. (1939a). Capture et destruction de larves de *Strongyloides* du singe et du boeuf par des Hyphomycetes. *Bulletin de la Societe de Pathologie Exotique, 32,* 394-398.

Deschiens, R. (1939b). Conditions de capture des larves de *Dictyocaulus* par des Hyphomycétes prédateurs. *Bulletin de la Societe de Pathologie Exotique, 32,* 698-700.

Deyn, G.B., Raaijmakers, C.E., and van der Putten, W.H. (2004). Plant community development is affected by nutrients and soil biota. *Journal of Ecology, 92,* 824-834.

Dias, A.S., Araújo, J.V., Campos, A.K., Braga, F.R., and Fonseca, T.A. (2007). Application of a formulation of the nematophagous fungus *Duddingtonia flagrans* in the control of cattle gastrointestinal nematodiosis. *World Journal of Microbiology and Biotechnology, 23,* 1245-1252.

Dias, W.P., and Ferraz, S. (1994). Avaliação de espécies de *Arthrobotrys* para o controle de *Meloidogyne incognita. Fitopatologia Brasileira, 19,* 189-192.

Dijksterhuis, J., Veenhuis, M., and Harder, W. (1993). Conidia of the nematophagous fungus *Drechmeria coniospora* adhere to but barely infect *Acrobeloides buetschilii. FEMS Microbiology Letters, 113,* 183-188.

Dijksterhuis, J., Veenhuis, M., Harder, W., and Nordbring-Hertz, B. (1994). Nematophagous fungi: physiological aspects and structure-function relationships. *Advances in Microbial Physiology, 36,* 111-143.

Dimander, S.O., Höglund, J., Uggla, A., Spörndly, E., and Waller, P.J. (2003). Evaluation of gastro-intestinal nematode parasite control strategies for first-season grazing cattle in Sweden. *Veterinary Parasitology, 111,* 193-209.

Doncaster, C.C., and Hooper, D.J. (1961). Nematodes attacked by protozoa and tardigrades. *Nematologica, 6,* 333-335.

Drechsler, C. (1937). Some hyphomycetes that prey on free-living terricolous nematodes. *Mycologia, 36,* 382-399.

Drechsler, C. (1941). Predaceous fungi. *Biological Reviews, 16,* 265-290.

Dube, B., and Smart Jr, G.C. (1987). Biological control of *Meloidogyne incognita* with *Paecilomyces lilacinus* and *Pasteuria penetrans. Journal of Nematology, 19,* 222-227.

Duddington, C.L. (1949). A new predaceous species of *Trichothecium. Transactions of the British Mycological Society, 32,* 284-287.

Duddington, C.L. (1956). The predaceous fungi: Zoopagales and Moniliales. *Biological Reviews of the Cambridge Philosophical Society, 31,* 152-193.

Duponnois, R, Garbaye, J., Bouchard, D. and Churin, J.L. (1993). The fungus-specificity of mycorrhization helper bacteria (MHBs) used as an alternative to soil fumigation for ectomycorrhizal inoculation of bare-root Douglas-fir planting stocks with *Laccaria laccata. Plant and Soil, 157,* 257-262.

Duponnois, R., Bâ, A.M., and Mateille, T. (1999). Beneficial effects of *Enterobacter cloacae* and *Pseudomonas mendocina* for biocontrol of *Meloidogyne incognita* with the endospore-forming bacterium *Pasteuria penetrans. Nematology, 1,* 95-101.

Duponnois, R., Chotte, J.L., and Cadet, S.S.P. (2001). The effects of organic amendments and the interactions between nematophagous fungus *Arthrobotrys oligospora* and the root-knot nematode *Meloidogyne mayaguensis* parasitizing tomato plants. *Biology and Fertility of Soils, 34,* 1-6.

Duponnois, R., Founoune, H., and Thioulouse, J. (2002). Interactions between some bacterial strains belonging to the fluorescent *Pseudomonas* group and the nematophagous fungus *Arthrobotrys* sp. isolate T41: effect on fungal development and activity against plant-parasitic nematodes. *Nematologia Brasileira, 26,* 141-151.

Duponnois, R., Mateille, T. and Gueye, M. (1995). Biological characteristics and effects of two strains of Arthrobotrys oligospora from Senegal on Meloidogyne species parasitizing tomato plants. *Biocontrol Science and Technology, 5,* 517-525.

Dutky, E.M., and Sayre, R.M. (1978). Some factors affecting infection of nematodes by the bacterial spore parasite *Bacillus penetrans. Journal of Nematology, 10,* 285.

Eapen, S.J., Beena, B., and Ramana, K.V. (2008). Evaluation of fungal bioagents for management of root-knot nematodes in ginger and turmeric fields. *Journal of Spices and Aromatic Crops, 17,* 122-127

Eisenback, J.D. (1993). Interactions between nematodes in cohabitance. In: Khan, M.W. (Ed.). *Nematode interactions.* (pp. 134-174). London: Chapman and Hall.

El-Khawas, H., and Adachi, K. (1999). Identification and quantification of auxins in culture media of *Azospirillum* and *Klebsiela* and their effect on rice roots. *Biology and Fertility of Soils, 28,* 377-381.

EPA, 2005. Biopesticides registration action document. *Paecilomyces lilacinus* strain 251 (PC Code 028826). 06/07/2005. Biopesticides and Pollution Prevention Division. Office of Pesticide Programs. U.S. Environmental Protection Agency.

Eysker, M., Bakker, N., Kooyman, F.N.J., Van Der Linden, D., Schrama, C., and Ploeger, H.W. (2005). Consequences of the unusually warm and dry summer of 2003 in the Netherlands: poor development of free living stages, normal survival of infective larvae and long survival of adult gastrointestinal nematodes of sheep. *Veterinary Parasitology, 133,* 313-321.

Fabry, C.F.S., Freitas, L.G., Godinho, M.T., Neves, W.S., and Ferraz, S. (2007). Resistência sistêmica a *Meloidogyne javanica* induzida por *Rhizobium etli. Nematologia Brasileira, 31,* 86-91.

Fabry, C.F.S., Freitas, L.G., Lopes, E.A., Neves, W.S., Dallemole-Giaretta, R., and Ferraz, S. (2008). Efeito da aplicação de húmus e *Rhizobium etli* sobre *Meloidogyne incognita* e *Meloidogyne javanica. Revista Trópica-Ciencias Agrárias e Biológicas, 2,* 3-7.

Faedo, M., Barnes, E.H., Dobson, R.J., and Waller, P.J. (1998). The potential of nematophagous fungi to control the free-living stages of nematode parasites of sheep: pasture plot study with *Duddingtonia flagrans. Veterinary Parasitology, 76,* 129-135.

Faedo, M., Larsen, M. and Thamsborg, S.M. (2000). Effect of different times of administration of the nematophagous fungus *Duddingtonia flagrans* on the transmission of ovine parasitic nematodes on pasture - a plot study. *Veterinary Parasitology, 94,* 55-65.

Faedo, M., Larsen, M., Dimander, S.O., Yeates, G.W., Höglund, J.H., and Waller, P.J. (2002). Growth of the fungus *Duddingtonia flagrans* in soil surrounding feces deposited by cattle or sheep fed the fungus to control nematode parasites. *Biological Control, 23,* 64-70.

Faedo, M.,Waller, P.J., and Larsen, M. (1997). The potential of nematophagous fungi to control the free-living stages of nematode parasites of sheep: comparison between Australian isolates of *Arthrobotrys* spp. and *Duddingtonia flagrans*. *Veterinary Parasitology, 72,* 149-155.

Femandez, A.S., Larsen, M., Nansen, P., Grenvold, J., Henriksen, S.A., and Wolstrup, J. (1997). Effect of the nematode-trapping fungus *Duddingtonia flagrans* on the free-living stages of horse parasitic nematodes: a plot study. *Veterinary Parasitology, 73,* 257-266.

Ferraz, S., and Freitas, L.G. (2004). Use of antagonistic plants and natural products. In: Chen, Z., Chen, S., and Dickinson, D.W (Eds). *Nematology – Advances and perspectives, v.2: Nematode management and utilization.* (pp. 931-977). Cambridge: CABI Publishing.

Ferraz, S., Freitas, L.G., Lopes, E.A., Dias-Arieira, C.R. (2010). Manejo sustentável de fitonematóides. Viçosa: Editora UFV.

Fontenot, M.E., Miller, J.E., Larsen, M. and Gillespie, A. (2003). Efficiency of feeding *Duddingtonia flagrans* chlamydospores to grazing ewes on reducing availability of parasitic nematode larvae on pasture. *Veterinary Parasitology, 118,* 203-213.

Freire, F.C.O., and Bridge, J. (1985). Parasitism of eggs, females and juveniles of *Meloidogyne incognita* by *Paecilomyces lilacinus* and *Verticillium chlamydosporium*. *Fitopatologia Brasileira, 10,* 577-596.

Freitas, L.G., and Carneiro, R.M.D.G. (2000). Controle biológico de nematóides por *Pasteuria* spp.. In: Melo, I.S. (Ed.). *Controle Biológico.* (197-216). Jaguariuna: EMBRAPA.

Freitas, L.G., Mitchell, D.J., and Dickson, D.W. (1997). Temperature effects on the attachment of *Pasteuria penetrans* endospores to *Meloidogyne arenaria* race 2. *Journal of Nematology, 29,* 547-555.

Freitas, L.G., Neves, W.D., Fabry, C.S., Marra, B.M., Coutinho, M.M., Romeiro, R.S., and Ferraz, S. (2005). Isolamento e seleção de rizobactérias para o controle do nematóide das galhas (*Meloidogyne* spp.) na cultura do tomateiro. *Nematologia Brasileira, 29,* 215-220.

Gallagher, L.A., and Manoil, C. (2001). Pseudomonas aeruginosa PAO1 kills *Caenorhabditis elegans* by cyanide poisoning. *Journal of Bacteriology, 183,* 6207-6214.

Gamliel, A., and Katan, J. (1991). Involvement of fluorescent pseudomonads and other microorganisms in increased growth response of plants in solarizes soils. *Phytopathology, 81,* 494-502.

Gamliel, A., and Stapleton, J.J. (1993). Effect of chicken compost or ammonium phosphate and solarization on pathogen control, rhizosphere microorganisms, and lettuce growth. *Plant Disease, 77,* 886-891.

Gams, W., and Zare, R. (2001). A revision of *Verticillium* sect. Prostrata. III. Generic classification. *Nova Hedwigia, 72,* 47-55.

García, L., Bulnes, C., Melchor, G., Vega, E., Ileana, M., Oca, N.M., Hidalgo, L., and Marrero, E. (2004a). Safety of *Pochonia chlamydosporia* var catenulata in acute oral and dermal toxicity/pathogenicity evaluations in rats and rabbits. *Veterinary and Human Toxicology, 46,* 248-250.

García, L., Bulnes, C., Melchor, L., Hidalgo, L., Figueredo, J., Vega, E., and Marrero, E. (2009). Safety of *Pochonia chlamydosporia* var. *catenulata* in quails. *Revista de Salud Animal, 31,* 40-45.

García, L., Melchor, G., Arévalos, G., and Hidalgo, L. (2008a). Ecotoxicological evaluation of *Pochonia chlamydosporia* var. catenulata in terrestrial invertebrates. *Revista de Toxocología en Línea, 14,* 39-50.

García, L., Melchor, G., Arévalos, G., and Hidalgo, L. (2008b). Evaluación de la fitotoxicidad de la cepa IMI SD 187 *Pochonia chlamydosporia* var. *catenulata* sobre *Zea mays* L. y *Phaseolus vulgaris* L. *Revista de Protección Vegetal, 23,* 38-42.

García, L., Melchor, G., Montes de Oca, N., and Hidalgo, L. (2004b). Estudio de la irritación ocular y dérmica de *Pochonia chlamydosporia* var. *catenulata. Revista de Toxología, 21,* 103-107.

Gaspard, J.T., and Mankau, R. (1986). Nematophagous fungi associated with *Tylenchulus semipenetrans* and the citrus rhizosphere. *Nematologica, 32,* 359-363.

Giannakou, I.O., and Gowen, S.R. (2004). Factors affecting biological control effectiveness of *Pasteuria penetrans* in *Meloidogyne javanica* and the bacterial development in the nematode. *Nematropica, 34,* 153-163.

Giannakou, I.O., Pembroke, B., Gowen, S.R., and Davies, K.G. (1997). Effects of long-term storage and above-normal temperatures on spore adhesion of *Pasteuria penetrans* and infection of the root-knot nematode *Meloidogyne javanica. Nematologica, 43,*185-192.

Giannakou, I.O., Pembroke, B., Gowen, S.R., and Douloumpaka, S. (1999). Effects of fluctuating temperatures and different host plants on development of *Pasteuria penetrans* in *Meloidogyne javanica. Journal of Nematology, 31,* 312-318.

Giblin-Davis, R.M., Willams, D.S., Bekal, S., Dickson, D.W., Brito, J.A., Becker, J.O., and Preston, J.F. (2003). 'Candidatus *Pasteuria usgae*' sp. nov., an obligate endoparasite of the phytoparasitic nematode *Belonolaimus longicaudatus. International Journal of Systematic and Evolutionary Microbiology, 53,* 197-200.

Gillespie, A. (2002). *Duddingtonia flagrans* for control of parasites in farm animals: a commercial perspective. In: *Biological control of nematode parasites of small ruminants in Asia. Final proceedings of FAO technical co-peration project in Malysia TCP/MAL/0065 (T).* (pp. 41-43). Rome: FAO animal production and health paper.

Giuma, A.Y., and Cooke, R.C. (1974). Potential of Nematoctonus conidia for biological control of soil-borne phytonematodes. *Soil Biology and Biochemistry, 6,* 217-220.

Giuma, A.Y., Hackett, A.M., and Cooke, R.C. (1973). Thermostable nematoxins produced by germinating conidia of some endozoic fungi. *Transactions of the British Mycological Society, 60,* 49-56.

Glick, B.R, and Bashan, Y. (1997). Genetic manipulation of plant growth-promoting bacteria to enhance biocontrol of phytopathogens. *Biotechnology Advances, 15,* 353-378.

Glickman, L., and Magnaval, J. (1993). Zoonotic roundworm infections. In: Maguire, J. and Keystone, J. (Eds.). *Infectious disease clinics of North America* (pp. 717-732). Philadelphia: WB Saunders.

Godoy, G., Rodiguez-Kabana, R., and Morgan-Jones, G. (1983). Fungal parasites of *Meloidogyne arenaria* in Alabama soil. A mycological survey and green house studies. *Nematropica, 13,* 201-213.

Grau, P.A., Hopkins, R., Radewald, J.D., and Warrior, P. (1996). Efficacy of DiTera® biological nematicide for root-knot nematode suppression on carrot. *Nematropica, 26,* 268.

Gray, E.J., and Smith, D.L. (2005). Intracellular and extracelular PGPR: commonalities and distinctions in the plant-bacterium signaling processes. *Soil Biology and Biochemistry, 37,* 395-412.

Gray, N.F. (1987). Nematophagous fungi with particular reference to their ecology. *Biological Reviews, 62,* 245-304.

Griffin D.H. (1994). *Fungal physiology.* New York: Wiley-Liss.

Gronvold, J. (1989). Induction of nematode-trapping organs in the predacious fungus *Arthroborys oligospora* (hyphomycetales) by infective larvae of *Ostertagia ostertagi* (trichostrongylidae). *Acta Veterinaria Scandinavica,* 30, 77-87.

Gronvold, J., Henriksen, S.A., Larsen, M., Nansen, P., and Wolstrup, J. (1996a). Biological control. Aspects of biological control with special reference to arthropods, protozoans and helminths of domesticated animals. *Veterinary Parasitology, 64,* 47-64.

Gronvold, J., Nansen, P., Henriksen, S.A., Larsen, M., Wolstrup, J., Bresciani, J., Rawat, H., and Fribert, L. (1996). Induction of traps by *Ostertagia ostertagi* larvae, chlamydospore production and growth rate inb the nematode trapping fungus *Duddingtonia flagrans. Journal of Helminthology, 70,* 291-297.

Gronvold, J., Wolstrup, J., Nansen, P., and Henriksen, S.A. (1993). Nematode-trapping fungi against parasitic cattle nematodes. *Parasitology Today, 9,* 137-140.

Gronvold, J., Wolstrup, J., Nansen, P., Larsen, M., and Bjorn, H. (2000). Absence of obvious short term impact of the nematode-trapping fungus *Duddingtonia flagrans* on survival and growth of the earthworm *Aporrectodea longa*. *Acta Veterinaria Scandinavica, 41,* 147-151.

Gronvold, J., Wolstrup, J., Nansen, P., Larsen, M., Henriksen, S.A., Bjorn, H., Kirchheiner, K., Lassen, K., Rawat, H., and Kristiansen, H.L. (1999). Biotic and abiotic factors influencing growth rate and production of traps by the nematode trapping fungus *Duddingtonia flagrans* when induced by *Cooperia oncophora* larvae. *Journal of Helminthology, 73,* 129-136.

Hackenberg, C., Muehlchen, A., Forge, T., and Vrain, T. (2000). *Pseudomonas chlororaphis* strain Sm3, bacterial antagonist of *Pratylenchus penetrans*. *Journal of Nematology, 32,* 183-189.

Hafez, S.L., Weiser, G.C., Grau, P.A., and Warrior, P. (1996). Efficacy of two formulations of ABG-9008, a biological nematicide, on sugar-beets grown in soil infested with *Heterodera schachtii. Nematropica, 26,* 270.

Hallmann, J. and Sikora, R.A. (1996). Toxicity of fungal endophyte secondary metabolites to plant-parasitic nematodes and soil-borne plant pathogenic fungi. *European Journal of Plant Pathology, 102,* 155-162.

Hammerschmidt, R. (2007). Introduction: definitions and some history. In: Walters, D., Newton, A., and Lyon, G. (Eds.). *Induced resistance for plant defence: a sustainable approach to crop protection.* (pp. 1-8). Oxford: Blackwell Publishers.

Hatz, B., and Dickson, D.W. (1992). Effect of temperature on attachment, development, and interaction of *Pasteuria penetrans* on *Meloidogyne arenaria*. Journal of Nematology, 24, 512-521.

Hay, F.S., Niezen, J.H., Ridley, S., Bateson, L., Milller, C., and Robertson, H. (1997). The influence of pasture species and time of deposition of sheep dung on infestation by nematophagous fungi. *Applied Soil Ecology, 6,* 181-186.

Hidalgo-Diaz, L., Bourne, J.M., Kerry, B.R., and Rodríguez, M.G. (2000). Nematophagous *Verticillium* spp. in soils infested with *Meloidogyne* spp. in Cuba: isolation and screening. *International Journal of Pest Management, 46,* 277-284.

Higgins, M.L., and Pramer, D. (1967). Fungal morphogenesis: ring formation and closure by *Arthrobotrys dactyloides*. *Science, 155,* 345-346.

Hildalgo-Diaz, L., and Kerry, B.R. (2007). Integration of biological control with other methods of nematode management. In: Ciancio, A., and Mukerji, K.G. (Eds.). *Integrated management and biocontrol of vegetable and grain crops nematodes.* (pp. 29-49). Dordrecht: Springer.

Holland, R.J., Williams, K.L., and Nevalainen, H.K.M. (2003). *Paecilomyces lilacinus* strain Bioact251 is not a plant endophyte. *Australasian Plant Pathology, 32,* 473-478.

Hongyan, X., Aminuzzaman, F.M., Lingling, X., Yiling, L., Feng, L., and Xingzhong, L. (2010). Trap induction and trapping in eight nematode-trapping fungi (Orbiliaceae) as affected by juvenile stage of *Caenorhabditis elegans*. *Mycopathologia, 169,* 467-473.

Huang, X., Zhao, N., and Zhang, K. (2004). Extracellular enzymes serving as virulence factors in nematophagous fungi involved in infection of the host. *Research in Microbiology, 155,* 811-816.

Iavicoli, A., Boutet, E., Buchala, A., and Métraux, J.P. (2003). Induced systemic resistance in Arabidopsis thaliana in response to root inoculation with *Pseudomonas fluorescens* CHA0. *Molecular Plant Microbe Interactions, 16,* 851-858.

Imbriani, I., and Mankau, R. (1983). Studies on *Lasioseius scapulatus*, a mesostigmatid mite predaceous on nematodes. *Journal of Nematology, 15,* 523-528.

Inserra, R.N., and Davis, D.W. (1983). Hypoaspis nr. aculeifer: a mite predacious on root-knot and cyst nematodes. *Journal of Nematology, 15,* 324-325.

Itin, P.H., Frei, R., Lautenschlager, S., Buechner, S.A., Surber, C., Gratwohl, A., and Widmer, A.F. (1998). Cutaneous manifestations of *Paecilomyces lilacinus* infection induced by a contaminated skin lotion in patients who are severely immunosuppressed. *Journal of the American Academy of Dermatology, 39,* 401-409.

Jackson, F. (2004). Anthelmintics - Whats the alternative? *Brazilian Journal of Veterinary Parasitology, 13,* 62-67.

Jaffee, B.A., Phillips, R., Muldoon, A., and Mangel, M. (1992). Density-dependent host-pathogen dynamics in soil microcosms. *Ecology, 73,* 495-506.

Jaffee, B.A. (2004). Do organic amendments enhance the nematode-trapping fungi *Dactylellina haptotyla* and *Arthrobotrys oligospora*? *Journal of Nematology, 36,* 267-275.

Jaffee, B.A., and Muldoon, A.E. (1989). Suppression of cyst nematode by natural infestation of a nematophagous fungus. *Journal of Nematology, 21,* 505-510.

Jaffee, B.A., Gaspard, J.T., Ferris, H., and Muldoon, A.E. (1988). Quantification of parasitism of the soil-borne nematode *Criconemella xenoplax* by the nematophagous fungus *Hirsutella rhossiliensis*. *Soil Biology and Biochemistry, 20,* 631-636.

Jaffee, B.A., Muldoon, A.E., and Westerdahl, B.B. (1996). Failure of a mycelial formulation of the nematophagous fungus Hirsutella rhossiliensis to suppress the nematode *Heterodera schachtii. Biological Control, 6,* 340-346.

Jaffee, B.A., Tedford, E.C., and Muldoon, A.E. (1993). Tests for density-dependent parasitism of nematodes by nematode-trapping and endoparasitic fungi. *Biological Control, 3,* 329-336.

Jansson, H.B. (1982a). Attraction of nematodes to endoparasitic nematophagous fungi. *Transactions of the British Mycological Society, 79,* 25-29.

Jansson, H.B. (1982b). Predacity by nematophagous fungi and its relation to the attraction of nematodes. *Microbiol Ecology, 8,* 237-240.

Jansson, H.B. (1993). Adhesion to nematodes of conidia from the nematophagous fungus *Drechmeria coniospora. Journal of General Microbiology, 139,* 1899-1906.

Jansson, H.B., and Lopez-Llorca, L.V. (2001). Biology of nematophagous fungi. In: Misra, J.K., and Horn, B.W. (Eds). *Mycology: Trichomycetes, other fungal groups and mushrooms.* (145-173). Enfield: Science Publishers.

Jansson, H.B., and Lopez-Llorca, L.V. (2004). Control of nematodes by fungi. In: Arora, D.K. (Ed.). *Fungal biotechnology in agriculture, food, and environmental applications.* (pp. 205-215). New York: Dedder.

Jansson, H.B., and Nordbring-Hertz, B. (1979). Attraction of nematodes to living mycelium of nematophagous fungi. *Journal of General Microbiology, 112,* 89-93.

Jansson, H.B., and Nordbring-Hertz, B. (1980). Interactions between nematophagous fungi and plant parasitic nematodes: attraction, induction of trap formation and capture. *Nematologica, 26,* 383-389.

Jansson, H.B., and Nordbring-Hertz, B. (1983). The endoparasitic fungus *Meria coniospora* infects nematodes specifically at the chemosensory organs. *Journal of General Microbiology, 129,* 1121-1126.

Jansson, H.B., and Nordbring-Hertz, B. (1984). Involvement of sialic acid in nematode chemotaxis and infection by an endoparasitic nematophagous fungi. *Journal of General Microbiology, 130,* 39-43.

Jansson, H.B., and Nordbring-Hertz, B. (1988). Infection events in the fungus-nematode system. In: Poinar, G.O., and Jansson, H.B. (Eds.). *Diseases of nematodes.* (pp. 59-72). Boca Raton: CRC Press.

Jansson, H.B., and Poinar, O.G. (1986). Some possible fossil nematophagous fungi. *Transactions of the British Mycological Society,* 87, 471-474.

Jansson, H.B., Jeyaprakash, A., and Zuckerman, B.M. (1985a). Control of root-knot nematodes on tomato by the endoparasitic fungus *Meria coniospora. Journal of Nematology, 17,* 327-329.

Jansson, H.B., Jeyaprakash, A., and Zuckerman, B.M. (1985b). Differential adhesion and infection of nematodes by the endoparasitic fungus *Meria coniospora* (Deuteromycetes). *Applied and Environmental Microbiology, 49,* 552-555.

Jatala, P. (1986). Biological control of plant-parasitic nematodes. *Annual Review of Phytopathology, 24,* 453-489.

Jatala, P., Kaltenbach, R., and Bocangel, M. (1979). Biological control of *Meloidogyne incognita acrita* and *Globodera pallida* on potato. *Journal of Nematology, 11,* 303

Juhnke, M.E., Mathre, D.E. and Sands, D.C. (1989). Relationship between bacterial seed inoculum density and rhizosphere colonization of spring wheat. *Soil Biology and Biochemistry, 21,* 591-595.

Kaplan, D.T., Davis, E.L., and Walter, D.E. (1991). Influence of lectins on constricting ring formation by *Arthrobotrys dactyloides*. *Journal of Nematology, 23,* 264-266.

Kasumimoto, T., Ikeda, R., and Kawada, H. (1993). Dose response of *Meloidogyne incognita* infected cherry tomatoes to application of *Pasteuria penetrans*. *Japanese Journal of Nematology 23,* 10-17.

Kerry, B. (1980). Biocontrol: Fungal Parasites of Female Cyst Nematodes. *Journal of Nematology, 12,* 253-259.

Kerry, B.R. (1975). Fungi and the decrease of cereal cyst-nematode populations in cereal monoculture. *EPPO Bulletin, 5,* 353-361.

Kerry, B.R. (1981). Progress in the use of biological agents for control of nematodes. In: Papavizas, G.C. (Ed.). *Biological Control in Crop Production.* (79-90). Allanheld: Osmum.

Kerry, B.R. (1987). Biological control. In: Brown, R.H., and Kerry, B.R. (Eds.). *Principles and practice of nematode control in crops.* (pp. 233-263). London: Academic Press.

Kerry, B.R. (1989). Fungi as biological control agents for plant parasitic nematodes. In: Whipps, J.M., and Lumsden, R.D. (Eds). *Biotechnology of fungi for improving plant growt.* (pp. 153-170). Cambridge: Cambridge University Press.

Kerry, B.R. (1993). The use of microbial agents for the biological control of plant-parasitic nematodes. In: Jones, D.G. (Ed.). *Exploitation of microorganisms.* (pp. 81-104). London: Chapman and Hall.

Kerry, B.R. (1995). Ecological considerations for the use of the nematophagous fungus, *Verticillium chlamydosporium*, to control plant parasitic nematodes. *Canadian Journal of Botany, 73*, 65-70.

Kerry, B.R. (2000). Rhizosphere interactions and the exploitation of microbial agents for the biological control of plant parasitic nematodes. *Annual Review of Phytopathology, 38*, 423-441.

Kerry, B.R. (2001). Exploitation of nematophagous fungal *Verticillium chlamydosporium* Goddard for the biological control of root-knot nematodes (*Meloidogyne* spp.). In: Butt, T.M., Jackson, C., and Magan, N. (Eds.). *Fungi as biocontrol agents: Progress, problems and potential.* (pp. 155-167). Wallingford: CAB International.

Kerry, B.R. 1990. An assessment of progress toward microbial control of plant-parasitic nematodes. *Journal of Nematology, 22,* 621-631.

Kerry, B.R., and Evans, K. (1996). New strategies for the management of plant parasitic nematodes. In: Hall, R. (Ed.). *Principles and practice of managing soilborne plant pathogens.* (pp. 134-152). St. Paul: APS Press.

Kerry, B.R., and Jaffee, B.A. (1997). Fungi as biological control agents for plant parasitic nematodes. In: Wicklow, D.T., and Söderström, B. (Eds.). *The Mycota.* (203-218). Berlin: Springer.

Kerry, B.R., Crump, D.H., and Mullen, L.A. (1982). Studies of the cereal cystnematode, *Heterodera avenae* under continuous cereals, 1975-1978. II. Fungal parasitism of nematode females and eggs. *Annals of Applied Biology, 100,* 489-499.

Khambay, B.P.S., Bourne, J.M., Cameron, S. and Kerry, B.R. (2000). Communication to the editor a nematicidal metabolite from Verticillium chlamydosporium. *Pest Management Science, 56,* 1098-1099.

Khan, T.A., and Saxena, S.K. (1997). Integrated management of root-knot nematode *Meloidogyne javanica* infecting tomato using organic materials and *Paecilomyces lilacinus*. *Bioresource Technology, 61,* 247-250.

Khan, T.A., Azam, M.F., and Husain, S.I. (1984). Effect of fungal filtrates of *Aspergillus niger* and *Rhizoctonia solani* on penetration and development of root-knot nematodes and the plant growth to tomato var. marglobe. *Indian Journal of Nematology, 14,* 106-109.

Kiewnick, S. (2009). Importance of multitrophic interactions for successful biocontrol of plant parasitic nematodes with *Paecilomyces lilacinus* strain 251. In: Gisi, U., Chet, I., and Gullino, M.L. (Eds). *Recent developments in management of plant diseases.* (81-92). Dordrecht: Springer.

Kiewnick, S., and Sikora, R.A. (2004). Optimizing the efficacy of *Paecilomyces lilacinus* (strain 251) for the control of root-knot nematodes. *Communications in Agricultural and Applied Biological Sciences, 69,* 373-380.

Kiewnick, S., Sikora, R.A. (2006). Biological control of the root-knot nematode *Meloidogyne incognita* by *Paecilomyces lilacinus* strain 251. *Biological Control, 38,* 179-187.

Kim, K.S., and Riggs, R.D. (1992). Cytopathological reactions of resistant soybean plants to nematode invasion. In: Wrather, J.A., and Riggs, R.D. (Eds.). *Biology and management of the soybean cyst nematode.* (pp. 157-168). St. Paul: APS Press.

Kimpinski, J., and Sturz, A.V. (2003). Managing crop root zone ecosystems for prevention of harmful and encouragement of beneficial nematodes. *Soil and Tillage Research, 72,* 213-221.

Kloepper, J. W., Zablotowixz, R.M., Tipping, E.M., and Lifshitz, R. (1990). Plant growth-promotion mediated by bacterial rhizosphere colonizers. In: Keister, D.L., and Cregan, P.B. (Eds.). The rhizosphere and plant growth. (pp. 315-326). Dordrecht: Kluwer Academic Publishers.

Kloepper, J.W., and Schroth, M.N. (1981). Relationship of *in vitro* antibiosis of plant growth-promoting rhizobacteria to plant growth and the displacement of root microflora. *Phytopathology, 71,* 1020-1024.

Kloepper, J.W., Rodríguez-Kábana, R., Mcinroy, J.A., and Collins, D.J. (1991). Analysis of populations and physiological characterization of microorganisms in the rhizosphere of plant with antagonistic properties to phytopathogenic nematodes. *Plant and Soil, 136,* 95-102.

Kloepper, J.W., Rodríguez-Kábana, R., Mcinroy, J.A., and Young, R.W. (1992). Rhizosphere bacteria antagonistic to soybean cyst (*Heterodera glycines*) and root-knot (*Meloidogyne incognita*) nematodes: Identification by fatty acid analysis and frequency of biological control activity. *Plant and Soil, 139,* 75-84.

Kloepper, J.W., Scher, F.M., Laliberte, M., and Zaleska, I. (1985). Measuring the spermosfere colonizing capacity (spermosphere competence) of bacterial inoculants. *Canadian Journal of Microbiology, 31,* 926-929.

Knox, M.R., Josh, P.F., and Anderson, L.J. (2002). Deployment of *Duddingtonia flagrans* in an improved pasture system: dispersal, persistence, and effects on free-living soil nematodes and microarthropods. *Biological Control, 24,* 176-182.

Koziak, A.T.E.; Cheng, K.C., and Thorn, R.G. (2007). Phylogenetic analyses of *Nematoctonus* and Hohenbuehelia (Pleurotaceae). *Canadia Journal of Botany, 85,* 762-773.

Kumar, D., and Singh, K.P. (2006). Assessment of predacity and efficacy of *Arthrobotrys dactyloides* for biological control of root knot disease of tomato. *Journal of Phytopathology, 154,* 1-5.

Lackey, B.A., Jaffee, B.A., and Muldoon, A.E. (1992). Sporulation of the nematophagous fungus Hirsutella rhossiliensis from hyphae produced in vitro and added to soil. *Phytopathology, 82,* 415-420.

Lackey, B.A., Jaffee, B.A., and Muldoon, A.E. (1994). Effect of nematode inoculum on suppression of root-knot and cyst nematodes by the nematophagous fungus *Hirsutella rhossiliensis. Phytopathology, 84,* 415-420.

Larsen, M. (1999). Biological control of helminths. *International Journal for Parasitology, 29,* 139-146.

Larsen, M. (2000). Prospects for controlling animal parasitic nematodes by predacious micro fungi. *Parasitology, 120,* 121-131.

Larsen, M. (2006). Biological control of nematode parasites in sheep. *Journal of Animal Science, 84,* 133-139.

Larsen, M., Faedo, M., and Waller, P.J. (1994). The potential of nematophagous fungi to control the free-living stages of nematode parasites of sheep: survey for the presence of fungi in fresh faeces of grazing livestock in Australia. *Veterinary Parasitology, 53,* 275-281.

Larsen, M., Faedo, M., Waller, P.J., and Hennessy, D.R. (1998). The potential of nematophagous fungi to control the free-living stages of nematode parasites of sheep: studies with *Duddingtonia flagrans. Veterinary Parasitology, 76,* 121-128.

Larsen, M., Nansen, P., Wolstrup, J., Gronvold, J., Henriksen, S.A., and Zorn, A. (1995). Biological control of trichostrongyles in calves by the fungus *Duddingtonia flagrans* fed to animals under natural grazing conditions. *Veterinary Parasitology, 60,* 321-330.

Larsen, M., Wolstrup, J., Henriksen, S.A., Dackman, C., Gronvold, J., and Nansen, P. (1991). *In vitro* stress selection of nematophagous fungi for biocontrol of parasitic nematodes in ruminants. *Journal of Helminthology, 65,* 193-200.

Lee, J.H., Ma, K.C., Ko, S.J., Kang, B.R, Kim, I.S., Kim, Y.C. (2011). Nematicidal activity of a nonpathogenic biocontrol bacterium, *Pseudomonas chlororaphis* O6. *Current Microbiology, 62,* 746-751.

Lee, Q., and Widden, P. (1996). *Folsomia candida*, a "fungivorous" collembolan, feeds preferentially on nematodes rather than soil fungi. *Soil Biology and Biochemistry, 28,* 689-690.

Lehman, P.S., and Reid, J.W. (1993). *Phyllognathopus viguieri* (*Crustacea: Harpacticoida*), a predaceous copepod of phytoparasitic, entomophatogenic and free-living nematodes. *Soil and Crop Science Society of Florida Proceedings, 52,* 78-82.

Lemanceau, P., and Alabouvette, C. (1991). Biological control of *Fusarium* diseases by fluorescent *Pseudomonas* and non-pathogenic *Fusarium*. *Crop Protection, 10,* 279-286.

Li, T.F., Zhang, K.Q., and Liu, X.Z. (2000). *Taxonomy of nematophagous fungi.* Beijing: Chinese Scientific and Technological Publications.

Li, X.Q., Tan, A., Voegtline, M., Bekele, S., Chen, C.S., Aroian, R.V. (2008). Expression of Cry5B protein from *Bacillus thuringiensis* in plant roots confers resistance to root-knot nematode. *Biological Control, 47,* 97-102.

Li, Y., Hyde, K.D., Jeewon, R., Cai, L., Vijaykrishna, D., and Zhang, K. (2005). Phylogenetics and evolution of nematode-trapping fungi (Orbiliales) estimated from nuclear and protein coding genes. *Mycologia*, *97,* 1034-1046.

Linford, M.B. (1937). Stimulated activity of natural enemies of nematodes. *Science, 85,* 123-124.

Linford, M.B., and Yap, F. (1939). Root-knot nematode injury restricted by a fungus. *Phytopathology 29,* 596-609.

Linford, M.B., Yap, F., and Oliveira, J.M. (1938). Reduction of soil populations of the root-knot nematode during decomposition of organic matter. *Soil Science, 45,* 127-141.

Liou, J.Y., and Tzean, S.S. (1992). Stephanocysts as nematode-trapping and infecting propagules. *Mycologia, 84,* 786-790.

Liu, X.Z., Xiang, M.C., and Che, Y.S. (2009). The living strategy of nematophagous fungi. *Mycoscience, 50,* 20-25.

Lopes, E.A., Ferraz, S., Ferreira, P.A., Freitas, L.G., Dhingra, O.D., Gardiano, C.G., and Carvalho, S.L. (2007). Potencial de isolados de fungos nematófagos no controle de *Meloidogyne javanica*. *Nematologia Brasileira, 31,* 78-84.

Lopez-Llorca, L.V. (1990). Purification and properties of extracellular proteses produced by the nematophagous fungus. *Canadian Journal of Microbiology, 36,* 530-537.

Lopez-Llorca, L.V., Bordallo, J.J., Salinas, J., Monfort, M.L., and Lopez-Serna, M.L. (2002). Use of light and scanning electron microscopy to examine colonization of barley rhizosphere by the nematophagous fungus *Verticillium chlamydosporium*. *Micron, 33,* 61-67.

Lopez-Llorca, L.V., Jansson, H.B., Maciá-Vicente, J.G., and Salinas, J. (2006). Nematophagous fungi as root endophytes. In: Schulz B.J.E., Boyle, C.J.C., and Sieber, T.N. (Eds.). *Soil biology: Microbial root endophytes.* (pp. 191-206). Berlin: Springer.

Lopez-Llorca, L.V., Maciá-Vicente, J.G., and Jansson, H.B. (2008). Mode of action and interactions of nematophagous fungi. In: Ciancio, A., and Mukerji, K.G. (Eds.). *Integrated management and biocontrol of vegetable and grain crops nematodes.* (pp. 51-76). Dordrecht: Springer.

Luc, J.E., Crow, W.T., McSorley, R., and Giblin-Davis, R.M. (2010). Suppression of *Belonolaimus longicaudatus* with *in vitro*-produced *Pasteuria* sp. endospores. *Nematropica, 40,* 217-225.

Lucy, M., Reed, E., and Glick, B.R. (2004). Applications of free living plant growth-promoting rhizobacteria. *Antonie Van Leeuwenhoek, 86,* 1-25.

Luo, H., Li, X., Li, G., Pan, Y., and Zhang, K. (2006) Acanthocytes of *Stropharia rugosoannulata* function as a nematode-attacking device. *Applied and Environmental Microbiology, 72,* 2982-2987.

Luo, H., Liu, Y., Fang, L., Li, X., Tang, N., and Zang, K. (2007). *Coprinus comatus* damages nematode cuticles mechanically with spiny balls and produces potent toxins to immobilize nematodes. *Applied and Environmental Microbiology, 73,* 3916-3923.

Luz, W.C. (1996). Rizobactérias promotoras de crescimento de plantas e bioproteção. *Revisão anual de Patologia de Plantas, 4,* 1-47.

Lynch, J.M. (1978). Microbial interaction around imbibed seeds. *Annals of Applied Biology, 89,* 165-167.

Lysek, H. (1963). Effect of certain soil organisms on the eggs of parasitic roundworms. *Nature, 199,* 925.

Maciel, A.S., Araújo, J.V., Campos, A.K., Lopes, E.A., and Freitas, L.G. (2009). Predation of *Ancylostoma* spp. dog infective larvae by nematophagous fungi in different conidial concentrations. *Veterinary Parasitology, 161,* 239-247.

Maheswari, T.U., and Mani, A. (1988). Combined efficacy of *Pasteuria penetrans* and *Paecilomyces lilacinus* on the biocontrol of *Meloidogyne javanica* on tomato. *International Nematology Network Newsletter, 5,* 10-11.

Mahoney, C.J., and Strongman, D.B. (1994). Nematophagous fungi from cattle manure in four state of decomposition at three sites in Nova Scotia, Canada. *Mycologia, 86,* 371-375.

Maizels, R.M., Blaxter, M.L., and Selkirk, M.E. (1993). Forms and functions of nematode surfaces. *Experimental Parasitology, 77,* 380-384.

Mankau, R. (1975). Bacillus penetrans n. comb. causing virulent disease of plant-parasitic nematodes. *Journal of Invertebrate Pathology, 26,* 333-339.

Mankau, R. (1980). Biocontrol: Fungi as nematode control agents. *Journal of Nematology, 12,* 244-252.

Mankau, R. (1981). Microbial control of nematodes. In: Zuckerman, B. M., and Rohde R. (Eds.). *Plant parasitic nematodes* (pp. 475-494). New York: Academic Press.

Mankau, R., and Imbriani, I.L. (1978). Studies on the mite *Lasioseius scapulatus* - a predator on soil nematodes. *Nematropica, 8,* 15.

Martin, C.A., Roberts, S. and Greenberg, R.N. (2002). Voriconazole treatment of disseminated Paecilomyces infection in a patient with acquired immunodeficiency syndrome. *Clinical Infectious Diseases, 35,* 78-81.

Mateille, T., Dabiré, K.R., Fould, S., and Diop M.T. (2010). Host-parasite soil communities and environmental constraints: Modelling of soil functions involved in interactions between plant-parasitic nematodes and *Pasteuria penetrans. Soil Biology and Biochemistry, 42,* 1193-1199.

Mateille, T., Fould, S., Dabiré, K.R., Diop, M.T., and Ndiaye, S. (2009). Spatial distribution of the nematode biocontrol agent *Pateuria penetrans* as influenced by its soil habitat. *Soil Biology and Biochemistry, 41,* 303-308.

Mattiasson, B., Johansson, P.A., and Nordbring-Hertz, B. (1980). Host-microorganism interaction: studies of the molecular mechanism behind the capture of nematodes by nematophagous fungi. *Acta Chemica Scandinavica, 34,* 539-540.

Mauchline, T.H., Kerry, B.R., and Hirsch, P.R. (2004). The biocontrol fungus *Pochonia chlamydosporia* shows nematode host preference at the infraspecific level. *Mycological Research, 108,* 161-169.

Mendoza, A.R., and Sikora, R.A. (2009). Biological control of *Radopholus similis* in banana by combined application of the mutualistic endophyte *Fusarium oxysporum* strain 162, the egg pathogen *Paecilomyces lilacinus* strain 251 and the antagonistic bacteria *Bacillus firmus*. Biocontrol, 54, 263-272.

Mendoza-de-Gives, P., Crespo, J.F., Rodriguez, D.H., Prats, V.V., Hernandez, E.L., and Fernandez, G.E.O. (1998). Biological control of *Haemonchus contortus* infective larvae in ovine faeces by administering an oral suspension of *Duddingtonia flagrans* chlamydospores to sheep. *Journal of Helminthology, 72,* 343-347.

Mendoza-de-Gives, P., Davies, K.G., Clark, S.J., and Behnke, J.M. (1999). Predatory behaviour of trapping fungi against srf mutants of *Caenorhabditis elegans* and different plant and animal parasitic nematodes. *Parasitology, 119,* 95-104.

Mendoza-de-Gives, P., Zavaleta-Mejia, E., Herrera-Rodriguez, D., and Quiróz-Romero, H. (1994). *In vitro* trapping capability of *Arthrobotrys* spp. on infective larvae of *Haemonchus contortus* and Nacobbus aberrans. *Journal of Helminthology, 68,* 223-229.

Meyer, S.L.F., and Roberts, D.P. (2002). Combinations of biocontrol agents for management of plant-parasitic nematodes and soilborne plant pathogenic fungi. *Journal of Nematology, 34,* 1-8.

Meyer, S.L.F., Massoud, S.L., Chitwood, D.J., and Roberts, D.P. (2000). Evaluation of *Trichoderma virens* and *Burkholderia cepacia* for antagonistic activity against root-knot nematode, *Meloidogyne incognita. Nematology, 2,* 871-879.

Mohan, S., Fould, S., and Davies, K.G. (2001). The interaction between the gelatine binding domain of fibronectin and the attachment of *Pasteuria penetrans* endospores to nematode cuticle. *Parasitology, 123,* 271-276.

Molina, G.C., and Davide, R.G. (1986). Evaluation of microbial extracts for nematicidal acitivity against parasitic nematodes *Meloidogyne incognita* and *Radopholus similis. Philippine Agriculturist, 69,* 173-186.

Monfort, E., Lopez-Llorca, L.V., Jansson, H.B., Salinas, J., Park, J.O., and Sivasithamparam, K. (2005). Colonisation of seminal roots of wheat and barley by egg-parasitic nematophagous fungi and their effects on *Gaeumannomyces graminis* var. *tritici* and development of root-rot. *Soil Biology and Biochemistry, 37,* 1-7.

Morgan, M., Behnke, J.M., Lucas, J.A., and Peberdy, J.F. (1997). *In vitro* assessment of the influence of nutrition temperature and larval density on trapping of the infective larvae of *Heligmosomoides polygyrys* by *Arthrobotrys oligospora, Duddingtonia flagrans* and *Monacrosporium megalosporum*. *Parasitology, 115,* 303-310.

Morgan-Jones, G., and Rodrigues-Kábana, R. (1988). Fungi colonizing cysts and eggs. In: Poinar, G.O., and Jansson, H.B. (Eds.). *Diseases of nematodes.* (pp. 39-58). Boca Raton: CRC Press.

Morgan-Jones, G., Godoy, G., Rodríguez-Kábana, R. (1981). *Verticillium chlamydosporium* fungal parasite of *Meloidogyne arenaria* females. *Nematropica, 11,* 115-120.

Müller, J. (1982). The influence of fungal parasites on the population dynamics of *Heterodera schachtii* on oil radish. *Nematologica, 28,* 161.

Murray, D.S., and Wharton, D.A. (1990). Capture and penetration processes of the freeliving juveniles of Trichostrongylus colubriformis (Nematoda) by the nematophagous fungus, *Arthrobotrys oligospora. Parasitology, 101,* 93-100.

Nansen, P., Gronvold, J., Henriksen, S.A., and Wolstrup, J. (1988). Interactions between the predacious fungus *Arthrobotrys oligospora* and third-stage larvae of a series of animal-parasitic nematodes. *Veterinary Parasitology, 26,* 329-337.

Nansen, P., Larsen, M., Roepstorff, A., Gronvold, J., Wolstrup, J., and Henriksen, S.A. (1996). Control of *Oesophagostomum dentatum* and *Hyostrongylus rubidus* in outdoor reared pigs through daily feeding with the microfungus *Duddingtonia flagrans. Parasitology Research, 82,* 580-584.

Nasr Esfahani, M., and Ansari Pour, B. (2006). The effects of *Paecilomyces lilacinus* on the pathogenesis of *Meloidogyne javanica* and tomato plant growth parameters. *Iran Agricultural Research, 24-25,* 67-76.

Neves, W.S., Freitas, L.G., Romeiro, R.S., and Silva Alves, H.S. (2000). Controle de *Meloidogyne javanica* e *M. incognita* em tomateiro por bactérias endofíticas. *Fitopatologia Brasileira, 25,* 102-103.

Nguyen, V.L., Bastow, J.L., Jaffee, B.A., and Strong, D.R. (2007). Response of nematode-trapping fungi to organic substrates in a coastal grassland soil. *Mycological Research, 111,* 856-862.

Nitao, J.K., Meyer, S.L.F., and Chitwood, D. (1999). *In vitro* assays of *Meloidogyne incognita* and *Heterodera glycines* for detection of nematode antagonistic fungal compounds. *Journal of Nematology, 31,* 172-183.

Niu, X.M., Wang, Y.L., Chu, Y.S., Xue, H.X., Li, N., Wei, L.X., Mo, M.H., and Zhang, K.Q. (2010). Nematodetoxic aurovertin-type metabolites from a root-knot nematode parasitic fungus *Pochonia chlamydosporia*. *Journal of Agricultural and Food Chemistry, 58,* 828-834.

Nordbring-Hertz, B. (1973). Peptide-induced morphogenesis in the nematode-trapping fungus *Arthrobotrys oligospora*. *Physiologia Plantarum, 29,* 223-233.

Nordbring-Hertz, B. (1988). Nemathophagus fungi: strategies for nematode exploitation and for survival. *Microbiology Science, 5,* 108-168.

Nordbring-Hertz, B. (2004). Morphogenesis in the nematode-trapping fungus *Arthrobotrys oligospora* - an extensive plasticity of infection structures. *Mycologist, 18,* 125-133.

Nordbring-Hertz, B., Jansson, H.B., and Tunlid, A. (2006). *Nematophagous fungi. Encyclopedia of Life Sciences.* Chichester: John Wiley and Sons.

Oclarit, E.L., and Cumagun, C.J.R. (2009). Evaluation of efficacy of *Paecilomyces lilacinus* as biological control agent of *Meloidogyne incognita* attacking tomato. *Journal of Plant Protection Research, 49,* 337-340.

Odjakova, M., and Hadjiivanova, C (2001). The complexity of pathogen defense in plant. *Bulgarian Journal of Plant Physiology, 27,* 101-109.

Oliveira, R.D.L., Ferraz, S., and Dias-Arieira, C.R. (2002). Eficácia de isolados de *Arthrobotrys* spp. no controle de *Meloidogyne incognita*, *M. javanica* e *Heterodera glycines*. *Nematologia Brasileira, 26,* 49-57.

Ongena, M., Jourdan, E., Adam, A., Paquot, M., Brans, A., Joris, B., Arpigny, J.L., Thonart, P. (2007). Surfactin and fengycin lipopeptides of Bacillus subtilis as elicitors of induced systemic resistance in plants. *Environmental Microbiology, 9,* 1084-1090.

Oostendorp, M., and Sikora, R.A. (1990). *In vitro* interrelationships between rhizosphere bacteria and *Heterodera schachtii*. *Revue de Nematologie, 13,* 269-274.

Oostendorp, M., Dickson, D.W. and Mitchell, D.J. (1990). Host range and ecology of isolates of *Pasteuria* spp. from the southeastern United States. *Journal of Nematology, 22,* 525-531.

Pandey, R.; Sikora, R.A., Kalra, A., Singh, H.B., and Pandey, S. (2003). Plants and their products act as major nematode inhibitory agents. In: Trivedi, P.C. (Ed.). *Nematode management in plants.* (pp. 103-131). Jodhpur: Scientific Publishers.

Paraud, C., Pors, I., and Chartier, C. (2007). Efficiency of feeding *Duddingtonia flagrans* chlamydospores to control nematode parasites of first-season grazing goats in France. *Veterinary Research Communications, 31,* 305-315.

Park, J.O., Hargreaves, J.R., Mcconville, E.J., Stirling, G.R., Ghisalberti, E.L., and Sivasithamparam, K. (2004). Production of leucinostatins and nematicidal activity of Australian isolates of *Paecilomyces lilacinus* (Thom) Samson. *Letters in Applied Microbiology, 38,* 271-276.

Peloille, M. (1981). Les hyphomycètes prédateurs de nématodes: phénomène de prédation; écologie; utilisation en lutte biologique. *Agronomie, 1,* 331-337.

Peña, M.T., Miller, J.E., Fontenot, M.E., Gillespie, A., and Larsen, M. (2002). Evaluation of *Duddingtonia flagrans* in reducing infective larvae of *Haemonchus contortus* in faeces of sheep. *Veterinary Parasitology, 103,* 259-265.

Persmark, L., and Nordbring-Hertz, B. (1997). Conidial trap formation of nematode-trapping fungi in soil and soil extracts. *FEMS Microbiology Ecology, 22,* 313-324.

Peteira, B., Puertas, A., Hidalgo-Díaz, L., Hirsch, P.R., Kerry, B.R. and Atkins, S.D. (2005). Real-time PCR to monitor and assess the efficacy of the nematophagous fungus *Pochonia chlamydosporia* var. *catenulata* against root-knot nematode populations in the field. *Biotecnología Aplicada, 22,* 261-266.

Peterson, E.A., and Katznelson, H. (1965). Studies on the relationships between nematodes and other soil microorganisms. IV. Incidence of nematode-trapping fungi in the vicinity of plant roots. *Canadian Journal of Microbiology, 11,* 491-495.

Petit, T.H., Olson, R.J., Foos, F.Y., and Martin, W.J. (1980). Fungal endophthalmitis following intraocular lens implantation. *Archives of Ophthalmology, 98,* 1025-1039.

Pfister D.H. (1997). Castor, Pollux and life histories of fungi. *Mycologia, 89,* 1-23.

Pieterse, C.M.J., van Pelt, J.A., van Wess, S.C.M., Ton, J., Verhagen, B.W.M., Leon-Kloosterziel, K., Hase, S., De Vos, M., Oosten, V.V., Pozo, M., Spoel, S., Koornneef, S.V.E.A., Chalfun-Junior, A., Resende, M.L.V., Van Loon, L.L. (2005). Indução de resistência sistêmica por rizobactérias e comunicação na rota de sinalização para uma defesa refinada. *Revisão Anual de Patologia de Plantas, 13,* 277-295.

Pramer, D. (1964). Nematode-trapping fungi. *Science, 144,* 382-388.

Puertas, A., and Hidalgo-Diaz, L. (2007). Influência de la planta hospedante y su interacción con *Meloidogyne incognita* sobre la efectividad de *Pochonia chlamydosporia* var. *catenulata*. *Revista de Protección Vegetal, 22, 104-109.*

Puertas, A., Noval, B.M., Martínez, B., Miranda, I., Fernández, F., and Hidalgo-Díaz, L. (2006). Interacción de *Pochonia chlamydosporia* var. *catenulata* con *Rhizobium* sp., *Trichoderma harzianum* y *Glomus clarum* em el control de *Meloidogyne incognita*. *Revista de Protección Vegetal, 21,* 80-89.

Quinn, M.A. (1987). The influence of saprophytic competition on nematode predation by nematode-trapping fungi. *Journal of Invertebrate Pathology, 49,* 170-174.

Rao, M.S., and Reddy, P.P. (1994). A method for conveying *Paecilomyces lilacinus* to soil for the management of root-knot nematodes on egg-plant. *Nematologia Mediterranea, 22,* 265-267.

Reddy, P.P., and Rao, M.S. (1999). Eco-friendly management of *Meloidogyne incognita* on tomato by integration of *Verticillium chlamydosporium* with neem and calotropis leaves. *Zeitschrift fur Pflanzenkrankheiten und Pflanzenschutz, 106,* 503-533.

Reddy, P.P., Khan, R.M., and Rao, M.S. (1991). Integrated management of the citrus nematode *Tylenchulus semipenetrans* using oil-cakes and *Paecilomyces lilacinus*. *Afro-Asian Journal of Nematology, 1,* 221-222.

Rhonda, V., Fleming, M.D., Anaissie, E.J. (2008). Hyalohyphomycosis - Infection due to hyaline moulds. In: Hospenthal, D.R., and Rinaldi, M.G. (Eds.). *Diagnosis and treatment of human mycoses.* (201-214). Totowa: Humana Press.

Ribeiro, R.C.F., Ferraz, S., and Mizobutsi, E.H. (1999). Avaliação da eficácia de isolados de *Monacrosporium* spp. no controle de *Meloidogyne javanica* e *Heterodera glycines. Nematologia Brasileira, 23,* 48-61.

Robinson, A.F., and Jaffee, B.A. (1996). Repulsion of *Meloidogyne incognita* by alginate pellets containing hyphae of *Monacrosporium cionapagum, M. ellipsosporum* or *Hirsutella rhossiliensis. Journal of Nematology, 28,* 133-147.

Romeiro, R.S., and Garcia, F.A.O. (2009). Indução de resistência em plantas a patógeno por eliciadores de natureza bacteriana. In: Bettiol, W., and Morandi, M.A.B. *Biocontrole de doenças de plantas: uso e perspectivas.* (pp. 85-99). Jaguariúna: Embrapa Meio Ambiente.

Roubaud, E., and Descazeaux, J. (1939). Action de certains champignons prédateurs sur les larves de Strongylidés du cheval. *Bulletin de La Societe de Pathologie Exotique, 32,* 290-294.

Roubaud, E., and Descazeaux, J. (1941a). Action des Hyphomycetes prédateurs sur les larves de Synthétocaules et de Bunostomes. *Bulletin de La Societe de Pathologie Exotique, 34,* 127-130.

Roubaud, E., and Descazeaux, J. (1941b). Essais relatifs à la prophylaxie de l'aguillulose du mouton par l'usage des Hyphomycètes prédateurs du sol. *Comptes Rendus des Seances de la Societe de Biologie, 135,* 687-690.

Roubaud, E., and Deschiens, R. (1939). Destruction des larves infectieuses d'Anguillules intestinales par *Dactylella ellipsospora. Bulletin de La Societe de Pathologie Exotique, 32,* 160-165.

Roumpos, C. (2005). *Ecological studies on paecilomyces lilacinus strain 251 and their importance for biocontrol of plant-parasitic nematodes and environmental risk assessment.*Göttingen: Cuvillier Verlag.

Rubner, A. (1996). Revision of predacious hyphomycetes in the *Dactylella-Monacrosporium* complex. *Studies in Mycology, 39,* 1-134.

Ryu, C.M., Farag, M.A., Hu, C.H., Reddy, M.S., Wie, H.X., Paré, P.W., and Kloepper, J.W. (2003). Bacterial volatiles promote growth of *Arabidopsis. Proceedings of the National Academy of Sciences of the United States of America, 100,* 4927-4932.

Saberhagen, C., Klotz, S.A., Bartholomew, W., Drews, D., and Dixon, A. (1997). Infection due to *Paecilomyces lilacinus*: a challenging clinical identification. *Clinical Infectious Diseases, 25,* 1411-1413.

Sagüés, M.F., Fusé, L.A., Fernández, A.S., Iglesias, L.E., Moreno, F.C., and Saumell, C.A. (2011). Efficacy of an energy block containing *Duddingtonia flagrans* in the control of gastrointestinal nematodes of sheep. *Parasitology Research.* (article in press, DOI:10.1007/s00436-011-2302-y).

Samaliev, H. (2009). Influence of the temperature on the attachment and development of *Pasteuria penetrans* on *Meloidogyne incognita. Agrarni Nauki, 1,* 37-43.

Santos, C.P. and Charles, T.P. (1995). Effect of an endoparasitic fungus, *Drechmeria coniospora*, in faecal cultures containing eggs of *Haemonchus contortus. Brazilian Journal of Veterinary and Animal Sciences, 47,* 123-128.

Sanyal, P.K., and Mukhopadhyaya, P.N. (2003). Top dressing of feed with desiccated chlamydospores of *Duddingtonia flagrans* for biological control of the pre-parasitic stages of ovine *Haemonchus contortus*. *Veterinary Research Communications, 27,* 381-390.

Sarkunas, M., Larsen, M., Nansen, P., and Hansenet, J.W. (2000). Biological control of trichostrongyle infections in calves on pasture in Lithuania using *Duddingtonia flagrans,* a nematode-trapping fungus. *Journal of Helminthology, 74,* 355-359.

Sayre, R. M., Adams, J. R., and Wergin, W. P. (1979). Bacterial parasite of a cladoceran: morphology, development *in vivo*, and taxonomic relationships with *Pasteuria ramosa* Metchnikoff 1888. *Internation Journal of Systematic Bacteriology, 29,* 252-262.

Sayre, R.M. (1971). Biotic influences in soil environment. In: Zuckerman, B. M., and Rhode, R. A. (Eds.). *Plant parasitic nematodes*. (pp. 235-256). New York: Academic Press.

Sayre, R.M. (1986). Pathogens for biological control of nematodes. *Crop Protection, 5,* 268-276.

Sayre, R.M., and Starr, M.P. (1985). *Pasteuria penetrans* (ex Thorne, 1940) nom. rev., comb. n., sp. n., a mycelial and endospore-forming bacterium parasitic in plant-parasitic nematodes. *Journal of the Helminthological Society of Washington*, *52,*149-165.

Sayre, R.M., and Starr, M.P. (1988). Bacterial disease and antagonism of nematodes. In: Poinar, G.O., and Jansson, H.B. (Eds.). *Diseases of nematodes.* (pp. 69-101). Boca Raton: CRC Press.

Sayre, R.M., and Walter, D.E. (1991). Factors affecting the efficacy of natural enemies of nematodes. *Annual Review of Phytopathology, 29,* 149-166.

Sayre, R.M., and Wergin, W.P. (1979). The use of SEM to classify and evaluate the parasites and predators of pest nematodes. *Scanning Electron Microscopy, 3,* 89-96.

Sayre, R.M., and. Powers, E.M. (1966). A predacious soil turbellarian that feeds on free-living and plant-parasitic nematodes. *Nematologica, 12,* 619-629.

Sayre, R.M., Starr, M.P., Golden, A.M., Wergin, W.P., and Endo, B.Y. (1988). Comparison of *Pasteuria penetrans* from *Meloidogyne incognita* with a related mycelial and endospore-forming bacterial parasite from *Pratylenchus brachyurus*. *Journal of the Helminthological Society of Washington, 55,* 28-34.

Sayre, R.M., W.P. Wergin, J.M. Schmidt, and Starr, M.P. (1991). *Pasteuria nishizawae* sp. nov. a mycelial and endospore forming bacterium parasitic on cyst nematodes of genera *Heterodera* and *Globodera*. *Research in Microbiology, 142,* 551-564.

Sayre, R.M., Wergin, W.P., and Davis, R.E. (1977). Occurrence in *Moina rectirostris* (Cladocera: Daphnidae) of a parasite morphologically similar to *Pasteuria ramosa* (Metchnikoff 1888). *Canadian Journal of Microbiology, 23,* 1573-1579.

Scher, F.M., Ziegle, J.S., and Kloepper, J.W. (1984). A method for assessing the root colonizing capacity of bacteria on maize. *Canadian Journal of Microbiology, 30,* 151-157.

Schmidt, A.R., Dorfelt, H., and Perrichot, V. (2007). Carnivorous fungi from Cretaceous amber. *Science, 318,* 1743.

Scholler, M., Hagedorn, G., and Rubner, A.A. (1999). A reevaluation of predatory orbiliaceous fungi. II. A new generic concept. *Sydowia, 51,* 89-113.

Segers, R., Butt, T.M., Kerry, B.R., and Peberdy, F. (1994). The nematophagous fungus *Verticillium chlamydosporium* Goddard produces a chymoelastase-like protease which hydrolyses host nematode proteins in situ. *Microbiology, 140,* 2715-2723.

Serracin, M., Schuerger, A.C., Dickson, D.W., and Weingartner, D.P. (1997). Temperature-dependente development of *Pasteuria penetrans* in *Meloidogyne arenaria*. *Journal of Nematology, 29,* 228-238.

Sharon, E., Chet, I., Viterbo, A., Bar-Eyal, M., Nagan, H., Samuels, G.J., and Spiegel, Y. (2007). Parasitism of *Trichoderma* on *Meloidogyne javanica* and role of the gelatinous matrix. *European Journal of Plant Pathology, 118,* 247-258.

Shuhegge, R., Ihring, A., Gantner, S., Bahnweg, G., Knappe, C., Vogg, G., Hutzler, P., Schmid, M., van Breusegem, F., Eberl, L., Hartmann, A., and Langebartels, C. (2006). Induction of systemic resistance in tomato by N-acyl-L-homoserine lactone-producing rhizosphere bacteria. *Plant, Cell and Environment, 29,* 909-918.

Siddiqui, I.A. and Shaukat, S.S. (2002). Mixtures of plant disease suppressive bacteria enhance biological control of multiple tomato pathogens. *Biology and Fertility of Soils, 36,* 260-268.

Siddiqui, I.A. and Shaukat, S.S. (2003a). Combination of *Pseudomonas aeruginosa* and *Pochonia chlamydosporia* for control of root-knot infecting fungi in tomato. *Journal of Phytopathology, 151*, 215-222.

Siddiqui, I.A., and Shaukat, S.S. (2003b). Suppression of root-knot disease by *Pseudomonas fluorescens* CHA0 in tomato: importance of bacterial secondary metabolite, 2,4-diacetylpholoroglucinol. *Soil Biol Biochem, 35,* 1615-1623.

Siddiqui, I.A., Haas, D., Heeb, S. (2005). Extracellular protease of *Pseudomonas fluorescens* CHA0, a biocontrol factor with activity against the root-knot nematode *Meloidogyne incognita. Applied and Environmental Microbiology, 71,* 5646-5649.

Siddiqui, I.A., Qureshi, S.A., Sultana, V., Ehteshamul-Haque, S., and Ghaffar, A. (2000). Biological control of root-rot disease complex of tomato. *Plant Soil, 227,* 163-169.

Siddiqui, Z.A., and Mahmood, I. (1996). Biological control of plant parasitic nematodes by fungi: a review. *Bioresource Technology, 58,* 229-239.

Siddiqui, Z.A., and Mahmood, I. (1999). Role of bacteria in the management of plant parasitic nematodes: a review. *Bioresource Technology, 69,* 167-179.

Siddiqui, Z.A., Baghel, G., and Akhtar, M.S. (2007). Biocontrol of Meloidogyne javanica by Rhizobium and plant growth-promoting rhizobacteira on lentil. *World Journal of Microbiology and Biotechnology, 23,* 435-441.

Sigler, L., and Kennedy, M.J. (1999). *Aspergillus*, *Fusarium*, and other opportunistic moniliaecious fungi. In: Murray, P.R., Baron, E.J., Pfaller, M.A., Tenover, F.C., and Yolken, R.H. (Eds). *Manual of clinical microbiology.* (1231-1235). Washington: ASM Press.

Skipp, R.A., Yeates, G.W., Chen, L.Y., and Glare, T.R. (2002). Occurrence, morphological characteristics and ribotyping of New Zealand isolates of *Duddingtonia flagrans*, a candidate for biocontrol of animal parasitic nematodes. *New Zealand Journal of Agricultural Research, 45,* 187-196.

Small, R.W., and Grootaert, P. (1983). Observations on the predation abilities of some soil dwelling predatory nematodes. *Nematologica, 29,* 109-118.

Spiegel, Y., Mor, M., and Sharon, E. (1996) Attachment of *Pasteuria penetrans* endospores to the surface of *Meloidogyne javanica* second-stage juveniles. *Journal of Nematology, 28,* 328-334.

St. Leger, R.J. (1993). Biology and mechanism of insect-cuticle invasion by deuteromycete fungal pathogens. In: Beckage, N.E., Thompson, S.N., and Federici, B.A. (Eds.). *Parasites and pathogens of insects. Pathogens.* (211-229). San Diego: Academic Press.

Starr, M.P., and Sayre, R.M. (1988). *Pasteuria thornei* sp. nov. and *Pasteuria penetrans* sensu stricto emend., mycelial and endospore-forming bacteria parasitic, respectively, on plant-parasitic nematodes of the genera *Pratylenchus* and *Meloidogyne. Annals of the Intitute Pasteur of Microbiology,* 139, 11-31.

Stear, M.J., Doligalska, M., Donskow-Schmelter, K. (2007). Alternatives to anthelmintics for the control of nematodes in livestock. *Parasitology 134,* 139-151.

Sticher, L., Mauch Mani, B., and Metraux, J.P. (1997). Systemic acquired resistanve. *Annual Review of Phytopathology, 35,* 235-270.

Stirling, G.R. (1988a). Biological control of plant parasitic nematodes. In: Poinar, G.O., and Jansson H.B. (Eds.). *Diseases of nematodes.* (pp. 93-139). Boca Raton: CRC Press.

Stirling, G.R. (1988b). Prospects for the use of fungi in nematode control. In: Burge, M.M. *Fungi in biological control systems.* (pp. 188-210). Mancherter: Manchester University Press.

Stirling, G.R. (1991). *Biological control of plant parasitic nematodes: progress problems, and prospects.* Wallingford: CAB Iternational.

Stirling, G.R., and Wachtel, M.F. (1980). Mass production of *Bacillus penetrans* for the biological control of root-knot nematodes. *Nematologica, 26,* 308-312.

Stirling, G.R., Bird, A.F., and Cakurs, A.B. (1986). Attachment of *Pasteuria penetrans* spores to the cuticles of root-knot nematodes. *Revue de Nématologie, 9,* 251-260.

Stirling, G.R., Sharma, R.D., and Perry, J. (1990). Attachment of *Pasteuria penetrans* spores to the root knot nematode *Meloidogyne javanica* in soil and its effects on infectivity. *Nematologica, 36,* 246-252.

Subrahmanyan, P., Reddy, M.N., and Rao, A.S. (1983). Exudation of certain organic compounds from seeds of groundnut. *Seed Science and Technology, 11,* 267-272.

Sun, M.H., Gao, L., Shi, Y.X., Li, B.J., and Liu, X.Z. (2006). Fungi and actinomycetes associated with *Meloidogyne* spp. eggs and females in China and their biocontrol potential. *Journal of Invertebrate Pathology, 93,* 22-28.

Tedford, E.C., Jaffee, B.A., Muldoon, A.E., Anderson, C.E., and Westerdahl, B.B. (1993). Parasitism of Heterodera schachtii and Meloidogyne javanica by Hirsutella rhossiliensis in microplots over two growing seasons. *Journal of Nematology, 25,* 427-433.

Terefe, M., Tefera, T., and Sakhuja, P.K. (2009). Effect of a formulation of *Bacillus firmus* on root-knot nematode *Meloidogyne incognita* infestation and the growth of tomato plants in the greenhouse and nursery. *Journal of Invertebrate Pathology, 100,* 94-99.

Terrill, T.H., Larsen, M., Samples, O., Husted, S., Miller, J.E., Kaplan, R.M., and Gelaye, S. (2004). Capability of the nematode-trapping fungus *Duddingtonia flagrans* to reduce infective larvae of gastrointestinal nematodes in goat feces in the southeastern United States: dose titration and dose time interval studies. *Veterinary Parasitology, 120,* 285-296.

Thomason, I.J. (1987). Challenges facing nematology: Environmental risks with nematicides and the need for new approaches. In: Veech, J.A., and Dickson, D.W. (Eds.). *Vistas on nematology.* (pp. 469-476). Hyattsville: Society of nematologists.

Thorn, R.G., and Barron, G.L. (1986). *Nematoctonus* and the Tribe Resupinateae in Ontario Canada. *Mycotaxon, 25,* 321-454.

Thorne, G. (1940). *Duboscquia penetrans*, n. sp. (Spermatozoa, Microspodia nematodiae) a parasite of the nematode *Pratylenchus pratensis* (de Man) Filip-jev. *Proceedings of the Helminthological Society of Washington*, 7, 51-53.

Tian, B., Yang, J., and Zhang, K.Q. (2007). Bacteria used in the biological control of plant-parasitic nematodes: populations, mechanisms of action, and future prospects. *FEMS Microbiology Ecology, 61,* 197-213.

Tikhonov, V.E., Lopez-Llorca, L.V., Salinas, J., and Jansson, H.B. (2002). Purification and characterization of chitinases from the nematophagous fungi *Verticillium chlamydosporium* and *V. suchlasporium. Fungal Genetics and Biology, 35,* 67-78.

Timper, P., and Brodie, B.B. (1993). Infection of *Pratylenchus penetrans* by nematode-pathogenic fungi. Journal of Nematology, 25, 297-302.

Tosi, S., Annovazzi, L., Tosi, I., Iadarola, P., and Caretta, G. (2002). Collagenase production in an Antarctic strain of *Arthrobotrys tortor* Jarowaja. *Mycopathologia, 153,* 157-162.

Tunlid, A., Jansson, H.B., and Nordbring-Hertz, B. (1992). Fungal attachments to nematodes. *Mycological Research, 96,* 401-412.

Tunlid, A., Johansson, T., and Nordbring-Hertz, B. (1991). Surface polymers of the nematode-trapping fungus *Arthrobotrys oligospora. Journal of General Microbiology, 137,* 1231-1240.

Tunlid, A., Rosén, S., Ek, B., and Rask, L. (1995). Purification and characterization of an extracellular serine protease from the nematode-trapping fungus *Arthrobotrys oligospora. Microbiology, 140,* 1687-1695.

Tzortzakakis, E.A. (2008). Attachment of *Pasteuria penetrans* spores on populations of *Meloidogyne javanica* and *M. incognita* virulent and avirulent on the Mi gene of resistant tomato. *Helminthologia, 45,* 54-56.

Van Loon, L.C. (2007). Plant responses to plant growth-promoting rhizobacteria. *European Journal of Plant Pathology, 119,* 243-254.

Van Peer, R., Niemann, G.J., and Schippers, B. (1991). Induced resistance and phytoalexin accumulation in biological control of *Fusarium wilt* of carnation by *Pseudomonas* sp. strainWCS417r. *Phytopathology, 81,* 728-734.

Van Schooneveld, T., Freifeld, A., Lesiak, B., Kalil, A., Sutton, D.A., and Iwen, P.C. (2008). *Paecilomyces lilacinus* infection in a liver transplant patient: case report and review of the literature. *Transplant Infectious Disease, 10,* 117-122.

Veenhuis, M., Nordbring-Hertz, B. and Harder, W. (1985a). An electron-microscopical analysis of capture and initial stages of penetration of nematodes by *Arthrobotrys oligospora. Antonie van Leeuwenhoek, 51,* 385-398.

Veenhuis, M., Nordbring-Hertz, B. and Harder, W. (1985b). Development and fate of electron dense microbodies in trap cells of the nematophagous fungus *Arthrobotrys oligospora. Antonie van Leeuwenhoek, 51,* 399-407.

Veenhuis, M., Sjollema, K., Nordbring-Hertz, B., and Harder, W. (1989a). An improved method for light and electron microscopial studies of nematode/fungal interactions. *Antonie Van Leeuwenhoek, 55,* 361-368.

Veenhuis, M., Van Wijk, C., Wyss, U., Nordbring-Hertz, B. and Harder, W. (1989b). Significance of electron dense microbodies in trap cells of the nematophagous fungus *Arthrobotrys oligospora. Antonie van Leeuwenhoek, 56,* 251-261.

Velvis, H., and Kamp, P. (1996). Suppression of potato cyst nematode root penetration by the endoparasitic nematophagous fungi *Hirsutella rhossiliensis. European Journal of Plant Pathology, 102,* 115-122.

Viaene, N., Coyne, D.L., and Kerry, B.R. (2006). Biological and cultural management. In: Perry, R.N., and Moens, M. (Eds.). *Plant nematology.* (pp. 346-369). Wallingford: CAB International.

Viaene, N.M., and Abawi, G.S. (2000). *Hirsutella rhossiliensis* and *Verticillium chlamydosporium* as biocontrol agents of the root-knot nematode *Meloidogyne hapla* on lettuce. *Journal of Nematology, 32,* 85-100.

Vlassoff, A., Leathwick, D.M., and Heath, A.C.G. (2001). The epidemiology of nematode infections of sheep. *New Zealand Veterinary Journal, 49,* 213-221.

Waghorn, T.S., Leathwick, D.M., Chen, L.Y, Gray, R.A.J., and Skipp, R.A. (2002). Influence of nematophagous fungi, earthworms and dung burial on development of the free-living stages of *Ostertagia (Teladorsagia) circumcinta* in New Zealand. *Veterinary Parasitology, 104,* 119-129.

Waghorn, T.S., Leathwick, D.M., Chen, L.Y., and Skipp, R.A. (2003). Efficacy of the nematode-trapping fungus *Duddingtonia flagrans* against three species of gastro-intestinal nematodes in laboratory faecal cultures from sheep and goats. *Veterinary Parasitology, 118,* 227-234.

Waller, P.J., and Faedo, M. (1996). The prospects for biological control of the free-living stages of nematode parasites of livestock. *International Journal for Parasitology, 26,* 915-925.

Waller, P.J., and Larsen, M. (1993). The role of nematophagous fungi in the biological control of nematode parasites of livestock. *International Journal for Parasitology, 23,* 539-546.

Waller, P.J., Faedo, M., and Ellis, K. (2001). The potential of nematophagous fungi to control the free-living stages of nematode parasites of sheep: towards the development of a fungal controlled release device. *Veterinary Parasitology, 102,* 299-308.

Waller, P.J., Larsen, M., Faedo, M., and Hennessy, D.R. (1994). The potential of nematophagous fungi to control the free-living stages of nematode parasites of sheep: *in vitro* and *in vivo* studies. *Veterinary Parasitology, 51,* 289-299.

Waller, P.J., Schwan, O., Ljungström, B.L., Rydzik, A., and Yeates, G.W. (2004). Evaluation of biological control of sheep parasites using *Duddingtonia flagrans* under commercial farming conditions on the island of Gotland, Sweden. *Veterinary Parasitology, 126,* 299-315.

Walters, S.A., and Barker, K.R. (1994). Current distribution of five major *Meloidogyne* species in the United States. *Plant Disease 78,* 772-774.

Wang, R.B., Yang, J.K., Lin, C., Zhang, Y., and Zhang, K.Q. (2006). Purification and characterization of an extracellular serine protease from the nematode-trapping fungus *Dactylella shizishanna. Letters in Applied Microbiology, 42,* 589-594.

Ward, S., Thomson, N., White, J.G., and Brenner, S., (1975). Electron microscopical reconstruction of the anterior sensory anatomy of the nematode *Caenorhabditis elegans. Journal of Comparative Neurology, 160,* 313-337.

Warrior, P., Rehberger, L.A., Beach, M., Grau, P.A., Kirfman, G.W., and Conley, J.M. (1999). Commercial development and introduction of DiTera®, a new nematicide. *Pesticide Science, 55,* 376-379.

Wechter, W.P., Glandorf, D.C.M., Derrick, W.C., Leverentz, B., Kluepfel, D.A. (2001) Identification of genetic loci in a rhizosphere-inhabiting, species of *Pseudomonas* involved in expression of a phytoparasitic nematode ovicidal factor. *Soil Biology and Biochemistry, 33,* 1749-1758.

Wei, G., Kloepper, J.W., and Tuzun, S. (1991). Induction of systemic resistance of cucumber to *Colletotrichum orbiculare* by select strains of plant growth-promoting rhizobacteria. *Phytopathology, 81,* 1508-1512.

Wharton, D.A. (1980), Nematode egg-shells. *Parasitology, 81,* 447-463.

Willcox, J., and Tribe, H.T. (1974). Fungal parasitism in cysts of *Heterodera*. I Preliminary investigations. *Transactions of the Britsh Mycological Society, 62,* 585-594.

Xie, H., Aminuzzaman, F.M., Xu, L., Lai, Y., Li, F., and Liu, X. (2010). Trap induction and trapping in eight nematode-trapping fungi (Orbiliaceae) as affected by juvenile stage of *Caenorhabditis elegans*. *Mycopathologia, 169,* 467-473.

Yang, J., Baoyu, T., Liang, L., and Zhang, K.Q. (2007). Extracellular enzymes and the pathogenesis of nematophagous fungi. *Applied Microbiology and Biotechnology, 75,* 21-31.

Yang, J., Huang, X., Tian, B., Wang, M., Niu, Q., and Zhang, k. (2005). Isolation and characterization of a serine protease from the nematophagous fungus *Lecanicillium psalliotae*, displaying nematicidal activity. *Biotechnology Letters, 27,* 1123-1128.

Yang, Y., Yang, E., An, Z., and Liu, X. (2007). Evolution of nematode-trapping cells of predatory fungi of the Orbiliaceae based on evidence from rRNA-encoding DNA and multiprotein sequences. *Proceedings of the National Academy of Sciences*, *104,* 8379-8384.

Yeates, G.W., and Wardle, D.A. (1996). Nematodes as predators and prey: Relationships to biological control and soil processes. *Pedobiologia, 40*, 43-50.

Yeates, G.W., Bongers, T., Goede, R.G.M., Freckman, D.W., and Georgieva, S.S. (1993). Feeding habits in soil nematode families and genera - an outline for soil ecologists. *Journal of Nematology, 25*, 315-331.

Yeates, G.W., Dimander, S.O., Waller, P.J., and Höglund, J. (2002). Environmental impact on the soil nematodes following the use of either ivermectin sustained release boluses or the nematophagous fungus *Duddingtonia flagrans* to control nematode parasites of cattle in Sweden. *Acta Agriculturae Scandinavica Section A: Animal Science, 52,* 233-242.

Zahir, Z.A., Arshad, M., Frankenberger, W.T. (2004). Plant growth promoting rhizobacteria: applications and perspectives in agriculture. *Advances in Agronomy, 81,* 97-168.

Zaki, M.J., and Maqbool, M.A. (1991). Combined effect of *Pasteuria penetrans* and other biological control agents on the control of root-knot nematode on okra. *Pakistan Journal of Nematology, 9,* 49-52.

Zare, R., Gams W., and Evans, H.C. (2001). A revision of *Verticillium* section *Prostrata*. V. The genus *Pochonia*, with notes on *Rotiferophthora*. *Nova Hedwigia 73,* 51-86.

Zehr, E.I. (1985). Evaluation of parasites and predators of plant parasitic nematodes. *Journal of Agricultural Entomology, 2,* 130-134.

Zhang, L., Yang, E., Xiang, M., Liu, X., and Chen, S. (2008). Population dynamics and biocontrol efficacy of the nematophagous fungus *Hirsutella rhossiliensis* as affected by stage of the soybean cyst nematode. *Biological Control, 47,* 244-249.

Zhao, M., Mo, M., and Zhang, Z. (2004). Characterization of a serine protease and its full-length cDNA from the nematode-trapping fungus *Arthrobotrys oligospora. Mycology, 96,* 16-22.

Zuckerman, B.M. (1983). Hypotheses and possibilities of intervention in nematode chemoresponses. *Journal of Nematology, 15,* 173-183.

Zuckerman, B.M., and Jansson, H.B. (1984). Nematode chemotaxis and mechanisms of host/prey recognition. *Annual Review of Phytopathology, 22,* 95-113.

In: Nematodes
Editors: F. Boeri and J. A. Chung
ISBN: 978-1-61470-784-4

Chapter II

Resistance and Virulence in Plant-Nematode Interactions

Sergio Molinari
Institute for Plant Protection, National Research Council of Italy (CNR), Via G. Amendola 122/D – 70126 Bari, Italy.

Abstract

All cultivated plant species are subjected to biotic stresses that can endanger crop yields and cause relevant economic losses. Genetically-based resistance may be a simple and efficient solution to protect plants against pathogens and pests. Resistance to plant parasitic nematodes has been proved to be an effective management tool that improves crop yields, lowers pest population densities, and favors the developments of effective rotation systems. Most of the studies on resistance and virulence in plant-nematode interactions have been focused on those nematode families that are the most diffused and cause the major economic losses to worldwide agriculture, i.e. the sedentary endoparasites belonging to the cyst (*Globodera* and *Heterodera* spp.) and root-knot (*Meloidogyne* spp.) nematodes. Sedentary endoparasitic nematodes enter the roots as motile second-stage juveniles (J2) that do not kill parasitized cells and induce very specialized and complex relationships with their hosts. Cyst-forming J2 induce the formation of feeding sites, called syncytia, in which few cells merge by dissolving their cell walls, while root-knot nematodes induce the formation of few discrete giant or nurse cells; both types of feeding sites have the role of actively transferring solutes and nutrients toward the developing nematode. Once their feeding sites have been arranged, J2 become sedentary and begin to develop into enlarged adult females through subsequent molts to J3 and J4. Several genes conferring resistance to sedentary endoparasitic nematodes (*R*-genes) have been isolated from plants. They are generally expressed constitutively in proteins that are assigned to the recognition of the presence of nematode avirulent factors. The interaction between *R*-proteins and avirulent factors induces conformational changes in *R*-proteins that lead to signaling of defense response. Avirulence to specific *R*-genes is inherited as a single dominant trait in some plant parasitic nematode species. It seems that the loss of this trait determines the failure of pest recognition by resistant plants and produce the development of virulent populations.

Virulent populations are able to develop on resistant crop plants and can be selected in laboratory and in the field by repeated exposure to *R*-genes. This chance to break resistance by field populations can endanger resistance durability and the use of resistance as a valid management against nematodes for farmers. However, *R*-genes exert high selection pressures on nematode populations, thus leading to fitness costs, such as loss of competitiveness and lower reproduction potentials on susceptible hosts with respect to wild field populations. Moreover, virulent populations specifically develop only on plants that carry the gene on which selection occurred. In this chapter, the genetic bases of resistance and virulence, along with the most up-to-date studies on the molecular events and interactions leading to incompatible/compatible responses of the plants to nematodes, will be described, with particular attention to the role of some important hormones in plants, such as salicylic acid, and genetic variability of nematode populations.

Introduction

Plant-parasitic nematodes are among the most damaging and uncontrollable pests of cultivated crops causing severe economic losses in world agriculture, although detailed knowledge of exact economic damage on a worldwide basis is often inadequate. An estimation based on a 1987 international opinion survey, including the forty most important crops, mentioned 11% average loss of total production, which was evaluated as an economic loss of $118 billion for the year 2001 (McCarter, 2008). The highest economic losses concern rice and maize production, because of their overall predominance in world agriculture, as well as potato, wheat, tomato and grapes. Despite these important figures, nematodes are often considered to be pests of minor significance. However, the impact of these pests becomes apparent through the management strategies which have been employed thus far. The overall costs of the use of chemical nematicides can be in the range of $US 1 billion/year, as it occurred in the USA in 1982 (Landels, 1989). Management of plant-parasitic nematodes has always been difficult, and the most successful strategy for over 50 years has been the use of toxic fumigant nematicides, such as the most known methyl bromide. High-value crop production systems have strongly relied on the use of methyl bromide thus far, allowing growers to avoid the expense of crop rotations and the problem of scarcity of new appropriate land (McKenry, 1994). The phase out of methyl bromide under the Montreal Protocol is urging nematologists worldwide to look for alternative management strategies (Zasada et al., 2010). Although other tactics exist, such as crop rotation and other cultural manipulations, biological control and regulatory approaches, the use of resistant cultivars and rootstocks has been prioritized as a major goal for nematode management (Barker et al., 1994). Genetic resistance may provide an effective and economical management strategy in both high and low value cropping systems. Resistant plants can yield well on infested land, reduce nematode population densities to non-damaging levels for subsequent crops, do not require specialized applications and additional costs input and deficit, and, finally, their use can be integrated with other management tactics (Starr and Roberts, 2004). Although the benefits of using nematode resistance are generally recognized, the loss of methyl bromide cannot be replaced by simply planting resistant plants. Most of the plant resistance genes identified are effective against sedentary endoparasitic nematodes such as root-knot nematodes (*Meloidogyne* spp.) and cyst-forming nematodes (*Globodera* and *Heterodera* spp.). As cyst and root-knot are the most damaging groups of nematodes, most of the efforts to study parasitism and find sources of resistance and other control strategies have been focused on them. Nearly 90% of all reports dealing with nematode resistance involve sedentary endoparasitic nematodes because of the relative abundance of resistance to species of this group, which probably results from their intimate relationship with host plants. These nematodes establish biotrophic associations with their hosts, allowing the root cells at the feeding site to remain alive and go through deep modifications, so that they form syncytia or giant cells, by which developing nematodes feed during all of their life cycle. The mechanisms and genes involved in parasitism of sedentary endoparasitic nematodes have been extensively reviewed (Abad and Williamson, 2010; Davis et al., 2008).

Migratory ectoparasitic nematodes that have a significant economic impact in agriculture as they can transmit damaging plant viruses, show a very different feeding behavior. For instance, *Trichodorus* spp. feed only briefly on single cells at the surface of the root not allowing the plants to mount an effective resistant response. Conversely, species of another group belonging to the same family, *Xiphinema* spp., have a long stylet by which they feed deep into the roots from the outside. One of these species, *X. index*, is recognized as the primary vector for the transmission of grapevine fanleaf virus (GFLV), the causal agent of the fanleaf degeneration disease, which is considered to be one of the major threats to the grapevine industry (Andret-Link et al., 2004). Roots respond to this nematode by producing galls and modified cells similar to the giant cells produced by RKNs (Wyss et al., 1980). Probably because of this more complex interaction, many genetic determinants of resistance to *X. index* have been found in grape (Xu et al., 2008), whereas resistance to other species has not been reported. Migratory endoparasitic nematodes also do not induce feeding sites and feed while migrating into the roots. Some groups cause different types of tissue distortions, some other visible lesions and tissue necroses. *Pratylenchus, Radopholus,* and *Ditylenchus* spp. have a broad host range, whereas *Bursaphelenchus xylophilus* is known as the pine wilt nematode as it causes severe damage to a limited number of species of *Pinus*. All of these groups cause economically important losses in agriculture. Sources of resistance to migratory endoparasitic nematodes have recently been reported (Moens and Perry, 2009).

This chapter focuses on plant resistance to the most diffused genera of sedentary endoparasitic nematodes such as *Meloidogyne*, *Globodera*, and *Heterodera*, against which resistance is currently used as a valid management strategy for some crops. At present, cultivars with high-standard agricultural traits of tomato, tobacco, cotton, peanut resistant to *Meloidogyne* spp., of potato resistant to *Globodera* spp., and soybean resistant to *Heterodera glycines*, are commercially available for growers (Starr and Roberts, 2004). An extensive list of major annual and perennial crops carrying resistance to root-knot, cyst, and other families of nematodes commonly used for nematode pest management has been reported in Molinari (2011). The most serious concern about the use of resistant cultivars is the selection of virulent populations that break resistance and may make this strategy ineffective at specific locations. Therefore, selection for virulence and durability of resistance will be extensively treated herein. The genetics of resistance with the description of the *R*-genes mapped and cloned thus far has already been discussed in detail in many previous reviews, some of which are reported for reference (Williamson and Roberts, 2009; Fuller et al., 2008; Bakker et al., 2006). Major emphasis to the assessment of crop resistance, the molecular mechanisms underlying the incompatible plant-nematode interaction, and the involvement of salicylic acid in plant response to nematode attack will be given in this chapter. Finally, the results of genetic modification of susceptible plants by the introgression of cloned *R*-genes will be reported, as dedicated reviews are available in literature for a description of other types of approaches for engineering resistance and their possible use (see Atkinson et al., 2009; McCarter, 2008).

Assessment of Crop Resistance

Plants are defined as resistant to nematodes when they have reduced levels of reproduction (Trudgill, 1991). The general use of nematode reproduction in assessments of resistance reflects the complexity to quantitatively measure specific symptom development and damage levels compared with the relative ease and accuracy of reproduction measurements. Within the nematode groups to which resistance is a current strategy of control, only few nematode-host interactions produce specific symptoms that can easily be evaluated for a quantitative assessment of disease severity. The infestation of root-knot nematodes (RKNs, *Meloidogyne* spp.) induces specific and clearly visible symptoms on roots, such as the typical root-knots or galls, which result from the expansion of the cortical tissue surrounding the giant cells used by nematodes to feed at the infection site (Williamson and Hussey, 1996). Wheat roots inoculated with cereal cyst nematode (*H. avenae*) show distinct “knots” with swelling and lateral root proliferation at sites where females are developing (Cook and Noel, 2004). Root symptoms in the form of root-tip swelling can be scored in the parasitism of *X. index* to *Vitis* spp. (Harris, 1983). Evaluation of reproduction of sedentary endoparasitic nematodes usually relies on the measurement of the numbers of eggs per gram of root fresh weight, or per root system. Reproduction rates of RKNs are usually evaluated by a reproduction index (RI) which refers to the total number of eggs and J2 produced on the test roots and is expressed as the percentage of the total number of eggs on the roots of a fully susceptible cultivar of the same species, taken as a reference (Triantaphyllou, 1975). With respect to this RI, different degrees of resistance (DR) have been established to occur on tomato cultivars (Hadisoeganda and Sasser, 1982). Only when reproduction is as low as 50-25% of that on a fully susceptible cultivar, tomato plants are generally considered as slightly resistant to RKNs; full resistant plants must support a reproduction less than 10%. In field and greenhouse tests, the multiplication rates may also be calculated as P_f/P_i ratio, where P_f is the nematode population in the soil at the end of the crop cycle or of the test time, and P_i the initial population present at sowing/planting or the amount of J2/eggs artificially inoculated in the pots. Reproduction rates are also evaluated according to the numbers of the egg-containing formations characteristic of sedentary endoparasitic nematodes, i.e. the gelatinous egg masses of RKNs emerging from infested roots and the cysts of cyst nematodes mainly recovered from infested soils. An egg mass index (EI), ranging from 0 (no egg masses) to 5 (more than 100 egg mass/root), was used to screen host suitability to RKNs of tomato, southern, and garden pea cultivars (Hadisoeganda and Sasser, 1982). Standardized screening protocols for potato cyst nematodes (PCN) rely upon multiplication rates calculated from P_i and P_f expressed as the number of cysts per pot (McKenzie and Turner, 1987), although relative expressions and rankings of cyst numbers to internal standards has been reported to be more accurate than absolute assessments (Fleming, 1998).

Although the most common definition of nematode resistance is based on a measurable restriction of the pest reproduction, the primary aim of resistance used as a control strategy is to protect crop yield potential. The ability in reducing nematode population densities to levels that are non-damaging to subsequent crops, thereby enabling shorter and more manageable crop rotations, is only an additional benefit of resistance to nematodes (Starr and Roberts, 2004). Therefore, suitable resistance tests should rely also on parameters that can be related to actual disease severity and plant damage. The extent of galling in RKN-plant interactions determines root damage that impairs nutrient and water transports along the plant, thus causing poor shoot growth and yield loss. A 0-5 gall index (GI), associated with the percentage of galled roots, has long been used to detect disease severity (Ammati et al., 1985). However, gall index detection may be more suitable to "yes or no" statements of nematode infestation but seems a rather subjective indicator in distinguishing intermediate degrees of galling as its statement is made by a direct visual analysis of the roots. Therefore, a more quantitative factor of root damage obtained by extracting and counting the total number of developed sedentary individuals into the roots (SF) has been recently proposed, as galling is a reaction to motile individuals that enter the roots, establish a feeding site and turn into sedentary developmental stages, and is present also when such stages do not reproduce (Molinari, 2010). Furthermore, in some compatible interactions, there is little or no galling produced around the egg-laying RKN female. Actually, restriction of sedentary endoparasitic nematode disease may be exerted by host plant resistance (HPR) at different levels of pathogenesis, i.e. root penetration, post-penetration development and reproduction. Apparently, the motile infective juveniles of RKNs (J2) are attracted by both *Mi-1*-carrying resistant and susceptible tomato and, at least at the early stages of invasion, similar amounts of J2 penetrate resistant and susceptible roots. Tomato resistance conferred by *Mi-1* gene is expressed by a hypersensitive reaction (HR), leading to a rapid and localized cell death, whose earliest visible indications can be seen about 12 h after inoculation of roots with J2, while they attempt to establish a feeding site (Paulson and Webster, 1972). Therefore, host defense response is likely to be triggered upon cell penetration by nematode's stylet and feeding site formation, long after J2 penetration of host root tips and migration through the root tissues. Root penetration of *M. incognita* J2 was not inhibited in cotton-resistant genotypes with respect to susceptible controls, although the number of J2 per root system remained unchanged with increasing days after inoculation only in resistant genotypes (Faske and Starr, 2009). It is likely that resistance progressively inhibits the formation of the feeding sites and that additional invading J2 may have to regress back into the soil (Bendezu and Starr, 2003). However, a variable fraction of J2 normally succeed in developing on resistant roots, by a putative local failure of the resistance response or by their apparent ability to overcome resistance. According to the consistency of this fraction, resistance may be defined as full or partial. In both cases, the lower plant damage makes the resistant plants much more tolerant of nematode infection than comparable susceptible genotypes of the same crop plants (Trudgill, 1991). Nematode-resistant cultivars usually yield more than susceptible cultivars when planted in fields with population densities exceeding the damage threshold, thus indicating to be more tolerant across a range of initial nematode densities (Starr and Roberts, 2004). However, fewer nematodes developing into healthier roots reduces the competition for food, thus enhancing the reproduction potential of the developing nematodes. For instance, in the case of partial resistance, as that shown by some soybean cultivars to *M. incognita*, the increase of population densities may last for more days after planting with respect to susceptible cultivars, as a result of less damaged plants (Niblack et al., 1986). A sort of inverse correlation exists between the numbers of nematodes developing into the root and average female fertility of that population, also when no resistance factor restricts successful feeding site formation. Decreases of the multiplication rate are correlated to increases of initial population density because of the concomitant effects of higher competition for food and root damage (Greco and Di Vito, 2009). Accurate measurements of the relationship between initial population density and reproduction rates in *Meloidogyne*-tomato/pepper interactions were obtained in pot tests carried out in greenhouses (Molinari, 2010). In these lab tests, determined numbers of J2 are inoculated per plant, and infestation factors, such as reproduction potentials ($RP=P_f/P_i$), monitored at opportune intervals of time after inoculation (Figure 1).

In the figure shown, RP variations are analyzed over a range of inoculated J2 from 50 to 400, seven weeks after inoculation. After reaching a maximum of 100 J2 inoculated per plant, RP drastically decreases to remain constant at increasing inoculum densities, thus indicating a lowering of female fertility consequent to the increasing numbers of nematodes developing into the roots. Time after inoculation at which plants are harvested is very important for analyzing the infestation process occurring into the plant root. Tropical RKN species hatch and develop rapidly at 25-30 °C, have a short life cycle and complete several generations per growing season, under favorable environmental conditions (Van Gundy, 1985). In the reported experimental conditions, nematodes complete their life cycle in about one month, then, the eggs belonging to the second generation begin to hatch, and a high number of the second-generation juveniles start to penetrate the roots and establish a feeding site.

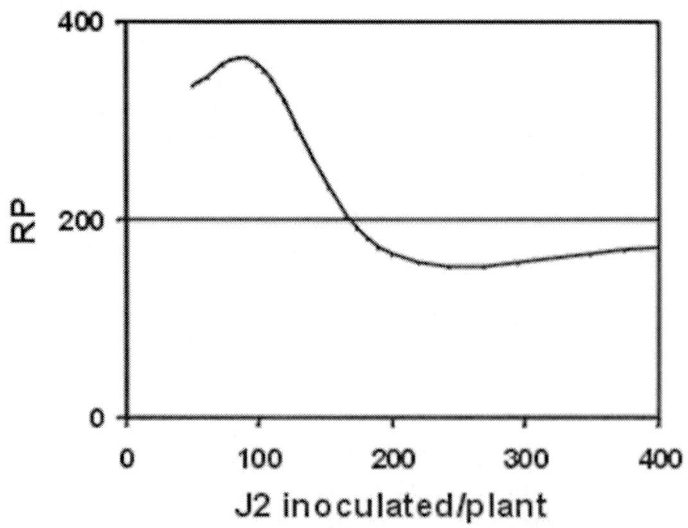

Figure 1. Reproductive Potential (RP) of RKNs as a function of increasing amounts of invading juveniles (J2) per plant.

For instance, in experiments in which 300 *M. incognita* J2 are inoculated in pots containing a single susceptible tomato plant harvested seven weeks after inoculation, the sedentary forms (developing individuals plus gravid females) extracted per root system (SF) are in the range of 1-2 thousands. Most of them belong to the nematode second generation but do not have time to reach the reproduction stage. It is evident that measuring this SF value may be a more accurate tool than gall index for an estimation of the potential plant damage occurring in a specific nematode-plant genotype interaction.

Mechanisms of Resistance and Selection for Virulence

Mi-1-mediated resistance is lost at temperatures above 28°C and the temperature sensitive period is limited to the first 24-48 h after infection (Dropkin et al., 1969). Once juveniles are permitted by the temperature-dependent failure of resistance to build their feeding site, resistance is not triggered any longer even at the permissive temperature. This means that HR-mediated resistance, like that conferred by *Mi-1* to tomato, may not be able to interfere with nematode development and reproduction of the individuals that succeed in forming their feeding sites. A detailed study, conducted on several near-isogenic lines of the susceptible Moneymaker-type and *Mi-1*-carrying tomato, measured the numbers of females and juveniles at the fourth stage (J4) per root, 40 days after the inoculation with 80 *M. incognita* J2 per plant (Zacheo et al., 1993). The individuals that developed into susceptible roots (in the range of about 100-250) were almost equally divided into J4 and females, and their total numbers exceeded those of inoculated J2, whilst the individuals that developed into resistant roots were almost exclusively females in the fraction of about one fourth of the inoculated J2. The number of galls detected per root were slightly lower than those of the total sedentary forms. Therefore, without any restriction to the viability of a full resistance response in incompatible interactions, a considerable fraction of the inoculated J2 succeeded in forming feeding sites and developed to the adult female stage; the absence of J4 probably indicates that roots, being "primed" by the first attack of avirulent nematodes, oppose a higher degree of resistance to the subsequent attack of the second generation, and/or that the second generation of invading J2 is delayed in resistant plants. Evidence exists that, when resistant tomato plants are "primed" by an inoculation with avirulent RKN populations, a subsequent inoculation with virulent populations produces less egg masses per root than inoculation with the sole virulent populations. Moreover, if production of egg masses is highly restricted on resistant tomato attacked by avirulent populations, fertility of the few gravid females is similar or even higher than that of females developed on susceptible roots (Molinari, personal communications). When seven different resistant genotypes of cotton were challenged with avirulent M. incognita, a delayed development of the nematodes was found, although the rate of delay differed among these genotypes (Faske and Starr, 2009).

In addition to resistance conferred by *Mi-1* in tomato against RKNs, HR-mediated resistance has been described in soybean to *H. glycines*, in wheat to *H. avenae*, and in pepper to *M. incognita*, although the resistance gene responsible for such a reaction has been characterized only in the latter interaction. Along with the HR-mediated resistance resulting in necrosis of the feeding sites within a couple of days after infection, a second type of resistance induces the degeneration of the feeding site about two weeks after infection with no apparent HR occurring. Resistance characterized by this delayed degeneration of the feeding site is observed for a wide variety of incompatible plant-nematode interactions, such as *H. avenae*-cereals, *H. schachtii*-sugar beet, *M. incognita*-pepper, *H. glycines*-common bean, and

PCN-potato (Bakker et al., 2006). It is interesting to note that pepper may carry two resistance genes, *Me-1* and *Me-3*, that induce, in the same crop, the two different mechanisms of resistance described above. A rapid HR early after inoculation characterizes the *Me-3* resistance, whilst *Me-1* resistance involves a delayed degradation of giant cells (Bleve-Zacheo et al., 1998). Interestingly, selection for virulence was possible only on *Me-3*-carrying pepper, whilst no virulent population has been collected on *Me-1*-carrying pepper thus far (Djian-Caporalino et al., 2010). The more delayed the degeneration of feeding cells is, the fewer the juveniles that are able to develop and reproduce on resistant hosts. Actually, a considerable number of resistance genes to RKNs have been reported to be located on the P9 chromosome of pepper (Djian-Caporalino et al., 2007). In pepper, the specificity of the resistance response appears to depend on the plant genotype and the specific R-gene-*Meloidogyne* species combination, indicating the likely presence of diverse mechanisms of resistance (Williamson and Roberts, 2009). Although *H1*-mediated resistance in potato against *G. rostochiensis* is characterized by HR and necrosis of tissue around the invading nematodes, the feeding sites begin to develop and nematodes become sedentary (Rice et al., 1987). However, in time, the feeding sites eventually collapse and the few nematodes that develop on *H1* potato plants are mostly male and reproduction is suppressed. This mechanism of resistance is probably responsible for the high durability of the H1 gene, at least in the UK where the G. rostochiensis originally introduced completely lacks the virulence allele. Where some populations already contained the virulence phenotype, like in the Netherlands, the selection for virulent *G. rostochiensis* on resistant cultivars has been much easier (Blok et al., 1997). On the other hand, the selection for virulence detected on the gene *Mi-1* has also been found to be low over 20 years of use of resistant tomato in the USA, with only some isolated cases of resistance breakdown (Kaloshian et al., 1996; Noling, 2000). As avirulence is usually dominantly expressed and gene flow relatively slow in undisturbed nematode populations, the selection for virulence in completely avirulent populations is likely to be an infrequent event. Conversely, selection for increased virulence will occur rapidly in populations which already contain virulent individuals, whether of the same or related species (Blok et al., 1997). Therefore, in agronomic systems, selection occurs on nematode populations that are already heterogeneous for virulence factors; in populations that lack virulent individuals, selection does not occur (Roberts et al., 1998). Human dissemination of infected plant material and soil is progressively increasing the chance of mixing together different gene pools, thus enhancing the potential for selection by high genetically variable nematode populations. Currently, occurrence of RKN virulent populations appears widespread and widely distributed geographically (Molinari, 2011).

Virulent populations are naturally selected from avirulent wild-type populations by repeated exposures to *R*-genes. Recently, in a study involving tens of *M. incognita* field populations, this "natural" selection has been mimicked in controlled greenhouse conditions by repeated inoculations of the progeny of those individuals which were able to develop and reproduce on *Mi-1*-carrying tomato (Molinari, 2010). It was possible to produce, from most of the avirulent field population tested, virulent isolates that reached their full reproductive potential within the second/third generation developed on resistant tomato (Figure 2).

Field RKN populations collected worldwide were very different in their capability to produce infestation on resistant tomato roots, and progenies from resistance-breaking isolates showed different degrees of virulence/aggressiveness when inoculated on the same tomato genotype. This survey indicated that selection for virulence on *Mi-1*-carrying tomato is highly probable, at least in greenhouse conditions, although not all of the selected isolates may represent a real threat to yields of resistant tomato crops. Experimentally, selection for virulence on tomato resulted in adverse fitness costs, detected as decreased reproduction and damage potentials of virulent populations with respect to field-starting populations (Figure 3).

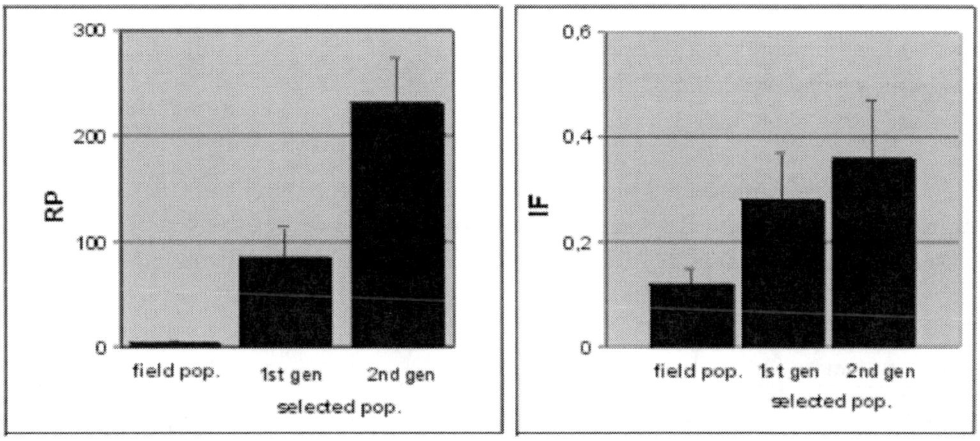

Figure 2. Reproductive Potential (RP) and Infestation Frequency (IF=egg masses per root/inoculated J2) on resistant tomato of an avirulent *M. incognita* field population (field pop.) and first/second generations of a virulent population (selected pop.) selected in a greenhouse from the field population.

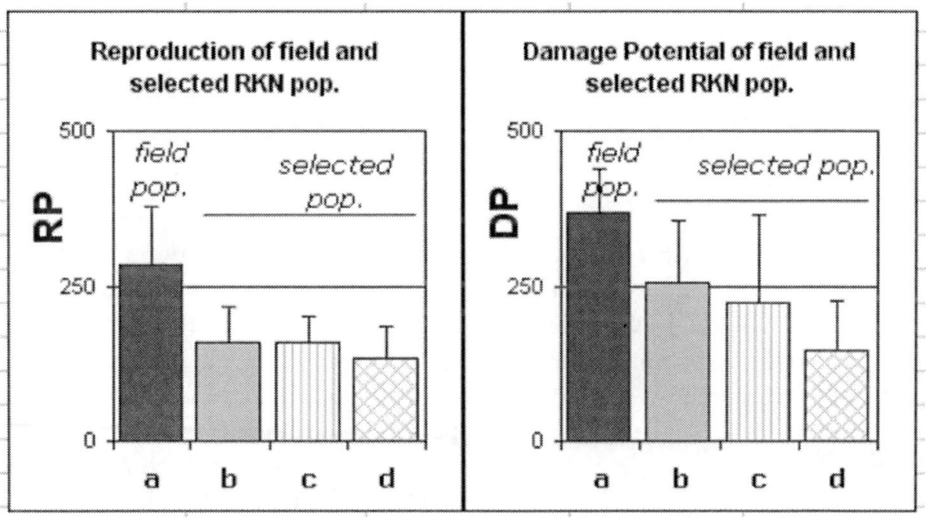

Figure 3. Average Reproductive Potential (RP) and Damage Potential (DP=SF/g root fresh weight) of RKN (a) virulent populations on tomato. a) avirulent field populations/susceptible tomato; b) selected virulent populations/susceptible tomato; c) selected virulent populations/heterozygotic resistant tomato; d) selected virulent populations/homozygotic resistant tomato.

In RKN-tomato interaction, a dosage effect of the *R*-gene can be presumed, as tomato carrying *Mi-1* in the homozygous form seems to be a poorer host with respect to hybrid tomato. Comparably, three isolates of *M. javanica* with low virulence consistently produced more egg masses on heterozygous *Mi-1*-resistant tomato than on two *Mi-1*-homozygous lines

(Tzortzakakis et al., 1998). It has been reported that field populations of *M. incognita* may comprise a mixture of virulent and avirulent lineages, in which the virulent forms had reduced reproductive potentials, indicating a trade-off between reproductive fitness and virulence (Petrillo and Roberts, 2005a). In such cases, reproduction on susceptible host plants in a rotation may result in a decline in the overall level of virulence in the field (Williamson and Roberts, 2009). Generally, *Mi*-virulent *M. incognita* isolates markedly decrease their aggressiveness on pepper with respect to the field populations from which they were selected (Table 1); in a single case, the ability to reproduce on pepper was found to be completely lost (Tzortzakakis and Blok, 2007).

Table 1. Egg masses produced per root (EM) and damage potential (DP) on susceptible pepper of a *M. incognita* field population (Mifield2V), used as the starting population for selection on *Mi-1*-carrying tomato, and the selected virulent isolate (SM2V). Values are expressed as averages (n = 12) ± SD

Infection factors	Mifield2V	SM2V
EM	199±47	115±28
DP	258±61	97±23

Reproductive fitness loss may result from the restriction of genetic variability produced by the high selection pressure exerted by plant genes conferring resistance to RKNs (Molinari, 2009). On the other hand, such a loss has not always been associated with virulence in some earlier reports (Castagnone-Sereno et al., 1994; Tzortzakakis and Gowen, 1996). However, it should be noted that those reports refer to nematode strains obtained from single egg mass (Tzortzakakis and Gowen, 1996), or to (a)-virulent strains obtained from isofemale lines (Castagnone-Sereno et al., 1994). In both cases, the avirulent isolates tested originate from a single or even part of a single egg mass, thus, they are probably characterized by a strongly reduced genetic variability with respect to the field population from which they were collected. When an isolate is artificially raised from a single egg mass, the restriction operated on the gene pool maintained by the parental population may be as drastic as that exerted by selective pressure of *R*-genes on virulent populations. Avirulent strains developed from a single egg mass can be experimentally proved to lose reproductive fitness with respect to the parental field population, which presumably carries a much higher genetic variability. Avirulent strains show reproductive potentials on susceptible tomato comparable with those of strains selected for virulence from the same parental population (Figure 4).

Isofemale lines from single egg masses were developed from both (a)-virulent parent isolates of *M. incognita* to *Rk*-carrying cowpea and cultured on cowpea plants with or without the *R*-gene *Rk* for up to 27 generations (Petrillo and Roberts, 2005b). Virulent progeny could also be selected from single egg masses of avirulent females, indicating that mechanisms exist by which these mitotically reproducing nematodes can produce progeny with different phenotypes. However, in these cases, the marked genetic rearrangements driven by selection pressure usually require over 20 generations to convert starting avirulent strains into stable virulent ones. It is possible that a gain of virulence may also be associated with genetic alterations impairing successful parasitism. However, it should be noted that genetic drawbacks and/or reduction of genetic variability resulting from the selection process may be recovered in time by means of numerous genetic processes, i.e. changes of ploidy of specific chromosomes, gene conversion, gene duplication, transposable elements, etc. In other words, if virulent populations are allowed to spread for many generations, their environmental fitness may increasingly improve due to the genetic rearrangements mentioned above. At least some of the *M. incognita* isolates, commonly referred to as "natural virulent", that are highly reproducible on resistant tomato at levels equivalent to those on susceptible plants (Roberts, 1995), may have had a previous and not documented exposure to *Mi-1* gene or analogues, and then allowed to spread in the agro-systems from which they were collected. It is intriguing that most of these "naturally virulent" isolates are collected from tropical or subtropical countries.

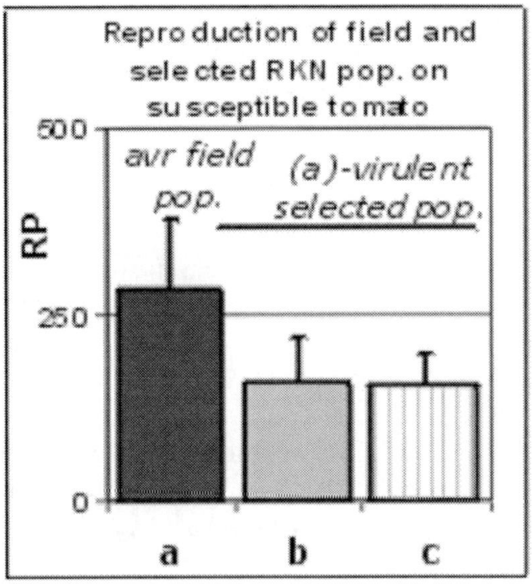

Figure 4. Reproductive Potential (RP) of a field population (a), an avirulent strain developed from a single egg mass collected from population a (b), and a virulent strain selected on *Mi-1*-carrying tomato from population a (c).

While tropical species of RKNs, such as *M. incognita*, *M. javanica*, and *M. arenaria*, reproduce by obligate mitotic parthenogenesis, other species adapted to milder climates, such as *M. hapla*, *M. chitwoodi*, and *M. fallax*, may reproduce by facultative meiotic parthenogenesis in which sexual crosses may occur in the presence of males. Conversely, cyst nematodes are mostly amphimictic. These species usually have a narrower host range than the asexual species, but sexual reproduction raises heterogeneity and adaptability among and within isolates. For instance, PCNs are very heterogeneous and populations often express several different alleles at a locus (Phillips et al., 1992). Although the use of resistant potato cultivars represents an effective method to enforce PCN control, a number of virulent populations have been identified and characterized as specific pathotypes, according to their ability to reproduce on a set of differential potato clones or cultivars carrying different resistance genes. European PCN populations are divided into eight pathotypes, five belonging to *G. rostochiensis* (Ro1-Ro5) and three to *G. pallida* (Pa1-Pa3). Recently, according to the elevated heterogeneity of PCN isolates, a very high polymorphism of superoxide oxidase (SOD) isozyme electrophoresis phenotypes allowed to discriminate among at least different groups of such pathotypes (Molinari et al., 2010). For the species in which sexual reproduction is possible, the occurrence of virulent populations may be enhanced due to within-population variability and the recombination of traits.

Probably, multiple genetic mechanisms may occur in transforming avirulent into virulent individuals. However, once acquired, this character is transmissible and stable over many generations. Virulence may be due to loss of the nematode gene product recognized by the plant resistance gene. In a study carried out on one near-isogenic line (NIL) of *M. javanica*, differing only for the gene(s) involved in the plant nematode interaction, virulence was determined by the loss of a single cDNA fragment, named *Cg-1*, present in the avirulent strain; soaking of nematode avirulent J2 in dsRNA corresponding to *Cg-1* transcript produced progeny that were virulent on resistant tomato (Gleason et al., 2008). Moreover, controlled crosses using inbred populations of *G. rostochiensis* or *H. glycines* have identified both dominant and recessive single traits that control the ability to reproduce on resistant plants (Dong and Oppermann, 1997; Janssen et al., 1991). Conversely, greenhouse studies on the development of virulence to *Mi-1* gene showed a progressive increase in virulence after prolonged selection on resistant plants, suggesting that several genes are involved (Castagnone-Sereno et al., 1994a). It is possible that the genetic determinants for virulence against the *Mi-1* gene may be different according to the origin and development of the different isolates tested. It has been reported that field and laboratory-selected virulent isolates exhibited differences in virulence and that the genetic changes induced by the selective pressure exerted by the *Mi-1* gene may differ in the "natural" selection of field populations or laboratory-driven selection of isofemale lines (Castagnone-Sereno et al., 1994b). Xu et al. (2001) divided virulent populations into those detected from a field where resistant tomatoes had been grown, those detected from fields where resistant tomatoes had not been grown ("natural virulent"), and those selected from avirulent populations in the laboratory. The conclusion was that the genetic events leading to the acquisition of virulence against *Mi*-gene are different between natural and selected virulent populations. Moreover, the process of selection for virulence could be associated with a gain of function enabling the nematode to circumvent the host response, for example, by enhancing antioxidant enzyme activities (Molinari, 2009b).

A Case Study: The *Mi* Multigene Family

Mi-1 was introgressed from the wild relative of tomato, *Solanum peruvianum*, into cultivated tomato, *Solanum lycopersicon,* using embryo rescue to obtain a hybrid of these normally incompatible species (Smith, 1944). All currently available fresh-market and processing tomato originate from the progeny of this single F1 plant and following extensive breeding. Moreover, *Mi-1* was the first root-knot nematode resistance gene to be cloned (Milligan et al., 1998; Vos et al., 1998), as well as the first plant *R*-gene proved to confer resistance to pests very different from nematodes, such as the potato aphid *Macrosiphum euphorbiae* (Rossi et al., 1998) and the white fly *Bemisia tabaci* (Nombela et al., 2003). It has been mapped to the short arm of chromosome 6, the same that in various other *Solanum* species carries an impressive collection of *R*-genes effective against different important pathogens. Like many other *R*-genes, *Mi-1* encodes a protein that carries a structural motif with a repeating pattern of 20-30 amino-acids (leucine-rich repeat, LRR), which function as sites of protein-protein interaction or protein-ligand binding, and a conserved region of about 320 amino acids containing a nucleotide binding site (NBS), which is generally involved in signal transduction associated with large conformational changes upon NTP hydrolysis. *Mi-1* includes a small gene family with seven highly homologous copies clustered together on the short arm of chromosome 6 on resistant tomato (Seah et al., 2007). Only one of these copies, *Mi-1.2*, was identified to be the functional resistance gene by the transformation of susceptible into resistant tomato (Milligan et al., 1998). Clustering of members of multi-gene families is common for plant *R*-genes, as well as the occurrence that a single gene in the cluster determines resistance. Several other *Mi* genes have been identified, different from *Mi-1* in functional characteristics, genetic locations and specificity (Ammiraju et al., 2003; Williamson, 1998). One of these genes, *Mi-9*, is located on the same chromosome as *Mi-1*, although it is heat-stable and confers resistance above 32 °C (Ammiraju et al., 2003), while *Mi-1* cannot (Dropkin et al., 1969). *Mi-9* has been indicated as a homologue of *Mi-1* by molecular studies using RNA interference (RNAi) (Jablonska et al., 2007). *M. incognita* populations virulent on *Mi-1*-carrying tomato may not be virulent on *Mi-3*, which is tightly linked to *Mi-5*; both of these genes are mapped on chromosome 12 and also confers heat-stable resistance to *M. javanica* (Williamson, 1998). The cloning of *Mi-3* and *Mi-9* is in progress, and fine mapping led to the construction of a DNA contig which spans the *Mi-3* locus with flanking markers within 0.25 cM of the gene (Yaghoobi et al., 2005).

The presence of *Mi-1.2* is necessary but not sufficient to make plants express resistance to RKNs, aphids, and white flies, as another gene, *Rme1*, is specifically required for the recognition/signaling pathways probably interacting with the *Mi-1* LRR (Martinez de Ilarduya et al., 2004). Other plant genes, such as *Hsp90, Sgt1,* and a glycosyltransferase have been shown to be required for *Mi-1.2* resistance (Bhattarai et al. 2007; Schaff et al. 2007), and that of many other plant *R*-genes. Chaperonins, like *Hsp90* (heat-shock protein 90), and *Sgt1* (suppressor of G_2 allele of SKP1) play a role in the assembly and stability of multi-subunit complexes and, ultimately, in regulating *R*-gene signaling (Kaloshian, 2004). Other strong epistatic gene interactions have also been found in resistance of cotton to *M. incognita* (Wang et al., 2008). Therefore, when resistance is conferred by a single dominant gene, it is evident that the resistant phenotypes may show quantitative and qualitative differences in contrasting nematode attack depending on the whole genetic background of the cultivar tested.

It is interesting that *Mi*-mediated resistance to aphids is not characterized by a HR; rather, *Mi-1* appears to modify aphid feeding behavior. Moreover, *Mi*-resistance to nematodes is inherited in a dominant fashion while resistant to aphids is semi-dominant and developmentally regulated, in that seedlings up to four-weeks old are susceptible to aphids, whereas six-week old plants are resistant (Bird and Kaloshian, 2003). How a single gene may be specific for resistance to such evolutionarily distant pests is an intriguing question. It is possible that *Mi-1* recognizes the avirulence effectors from each group of organisms or modifications of plant products altered by both pests. The fact that there are similarities in the feeding mechanisms of these piercing-sucking pests and in the defense responses that they induce in plants (Kaloshian, 2004), it is not convincing since cyst nematodes, for instance, are much more similar to RKNs than aphids, although completely not sensitive to *Mi-1*. Moreover, it has recently been reported that near-isogenic lines (NIL) of tomato harboring *Ol-4* and *Ol-6*, two genes for resistance to powdery mildew, are also resistant to nematodes and aphids (Seifi et al., 2011). In this case, the resistance to both nematodes and tomato powdery mildew is governed by linked, if not the same, *Mi-1* homologues in the *Mi-1* gene cluster. On the other hand, the *Am* gene-controlling resistance to Alfalfa mosaic virus in tomato is located in the same chromosome region (Parella et al., 2004). These findings indicate either that the *Mi-1* gene and/or *Rme1* probably recognizes different effector molecules from nematodes, aphids and white flies, which may share conserved signature motifs, or that the *Mi-1* region may produce a series of homologues of diverse resistance specificities in time.

The Role of Salicylic Acid in Nematode Resistance

Mi-mediated resistance seems to be regulated by a salicylic acid (SA)-dependent defense pathway (Branch et al., 2004; Molinari, 2007; Molinari and Loffredo, 2006), as it has generally been found in most *R*-gene-mediated defenses (Glazebrook, 2005). SA is undoubtedly involved in lesion formation and the contribution of SA to HR-associated resistance has been demonstrated to operate through the potentiation of Reactive Oxygen Species (ROS) generation and cell death (Shirasu et al., 1997). Overproduction of SA has been reported in resistant tomato attacked by *M. incognita*, although most of the produced SA was found in the leaves; however, a corresponding increase in SA root content was found in inoculated susceptible plants with respect to uninoculated controls (Vasyukova et al. 2003, 2007). Free SA was elevated to similar levels in both incompatible and compatible interactions between resistant *Hero A*-carrying tomato and *G. rostochiensis* (Uehera et al., 2010). No difference in total SA levels between root tissues of *Arabidopsis thaliana* plants uninfected and infected with the cyst nematode *H. schachtii* was observed, with an increase in total SA being present in shoots tissues of infected plants compared to the control (Wubben et al., 2008). High levels of SA may induce an accumulation of H_2O_2 (Molinari 2007). Accordingly, H_2O_2 accumulation was monitored in roots during the early stages of both compatible and incompatible *M. incognita*-tomato interactions, although the kinetics of H_2O_2 production was different (Melillo et al., 2006). In compatible interactions, H_2O_2 generation was transient and the amount of $CeCl_3$ precipitate, which was chosen as a quantitative measure for H_2O_2 concentration, turned back to the basal level within 48 h after inoculation, whereas in incompatible interactions, H_2O_2 concentration was still very high at this stage. It is likely that a difference exists in the detoxifying activity of ROS and H_2O_2 in particular. In fact, many studies confirmed that a specific and very early inhibition of H_2O_2-degrading enzymes, such as catalase, occurs in *M. incognita*-tomato incompatible interactions (Molinari, 2001; Molinari and Loffredo, 2006). This specific massive and prolonged H_2O_2 accumulation has been reported to be associated with HR and resistance response (Graham and Graham, 1999), as well as to be able to maintain elevated SA levels by stimulating its biosynthesis (Molinari, 2007).

Accumulation of SA is often produced in any abiotic and biotic stress in plants, but it tends to be transient because free SA concentration is normally lowered to non-phytotoxic levels by numerous biochemical pathways, the main of which is the conversion of SA, the active molecule, into SAG, an inactive storage glycosilated form (Molinari, 2007). Marked inhibition or even impairment of the enzymes responsible for maintaining oxidative/peroxidative balance may occur specifically and locally at the sites where *R*-gene encode/pest effector recognitions are realized.

The consequent anomalous increase of toxic compounds, that at lower concentrations normally operate as stress signals, may lead to cell death and tissue necroses. For instance, the impairing of electron transport chain and phosphorylation efficiency, as well as an increase in CN-resistant respiration, found in mitochondria isolated from resistant tomato roots challenged with *M. incognita* (Molinari et al., 1990), may be symptoms of uncontrolled spreading of SA in root cells (Lennon et al., 1997; Norman et al., 2004).

Contrasting results were obtained using *Mi-1.2*-carrying tomato transformed with a construct expressing *NahG*, which encodes salicylate hydroxylase, a bacterial enzyme that degrades SA into cathecol (Bhattarai et al., 2008). *NahG*-transformed plants did not sustain nematode development like untransformed plants, indicating that *Mi*-mediated resistance can be expressed in the absence, or in the presence, of low levels of SA. Actually, the levels of free SA and SAG monitored in roots of *NahG*-transformed plants inoculated by RKN J2 were very low.

Unfortunately, SA assays in untransformed and transformed *Mi-1* tomato roots were carried out only four weeks after nematode inoculation; therefore, the levels of SA produced at the early stages of J2 attack, determinant to trigger the resistance response, could not be predicted. Moreover, SA present in the leaves should also be taken into account when a defense response is expressed in the roots, as it has been proved that most of the free SA absorbed by the roots (and presumably that produced by the roots) is immediately translocated to the leaves (Molinari and Loffredo, 2006). Local and systemic SA accumulation following pathogen infection induces the expression of multiple pathogenesis-related (PR) protein genes (PR-1, PR-2, and PR-5) that are markers of the onset of SAR (Systemic Acquired Resistance, Durrant and Dong, 2004). SA-treated plants of *A. thaliana* had their susceptibility to *H. schachtii* significantly lowered compared with untreated plants (Wubben et al., 2008). Furthermore, PR-1 transcript induction was approximately 185-fold higher in shoots of SA-treated plants compared with root tissues. Thus, although responsiveness to exogenously added or endogenously produced SA may be higher in shoots and leaves than in roots, indirect effects may heighten defense reactions to soil pests in the roots.

On the other hand, the SA-dependent PR-genes were found to be induced early in roots of resistant tomato plants inoculated by *G. rostochiensis*, whereas it did not occur in *NahG*-transformed resistant tomato (Uehera et al., 2010). Recently, it has been reported that SA exogenously added to susceptible tomato, before inoculation with *M. incognita*, restricted the numbers of gravid females with respect to untreated plants (Molinari and Baser, 2010), thus confirming that SAR may also be activated against sedentary endoparasitic nematodes. Induction of SA *de-novo* synthesis is likely to occur during the early stages of both root-knot and cyst nematode infestation to resistant plants, as either RNA transcription or activity of the enzyme phenylalanine ammonia lyase (PAL) was enhanced only four days after nematode inoculation (Edens et al., 1995). PAL catalyzes desamination of phenylalanine with the formation of *trans*-cinnamic acid, which is the precursor of a high number of phenylpropanoids and phenolic compounds, SA included (Molinari, 2007). Finally, it can be concluded that SA may play a crucial role in signaling and expression of plant resistance to sedentary endoparasitic nematodes.

Genetic Modification of Plants by the Introgression of *R*-Genes

The identification and cloning of *R*-genes conferring resistance to nematodes enable their transfer to economically important crops which lack such a resistance. Moreover, cloned genes give the chance to transfer resistance directly into elite genotypes of the same crop, or into cultivars of different crops. The benefits of this biotechnology with respect to the classical genetic breeding are numerous and obvious. Major benefits are the avoidance of linkage drag and opportunity to transfer resistance in genotype backgrounds that prohibit introgression by cross-breeding. However, the susceptible plants that are chosen to be transformed should harbor suitable intracellular signaling pathways for being competent to raise a resistant response.

Recognition may occur by cooperation of *R*-genes with other complementary genes that may not be present in the transforming plant. Moreover, the avirulence targets that are guarded by the *R*-gene may be too divergent to be recognized by an *R*-gene ectopically expressed in a transgenic system (Atkinson et al., 2009). Transfer of *Mi-1* tomato gene to eggplant, which is taxonomically related to tomato, significantly lowered *Meloidogyne* reproduction, although aphid performance was not compromised (Goggin et al., 2006). Moreover, in a recent study, the full genomic region of *Mi-1* gene was cloned and transformed into lettuce that showed improved resistance to RKNs (Zhang et al., 2010). However, attempts to transfer the same gene to tobacco or *Arabidopsis* have been unsuccessful (Williamson and Kumar, 2006), as well as the transfer into potato of the tomato gene *Hero A*, conferring resistance to the potato cyst nematode *G. rostochiensis* (Sobczak et al., 2005).

If it is recognized that genotype backgrounds can have an effect on the efficacy of *Mi-1* in tomato (Jacquet et al., 2005), it is likely that the practical efficiency of *R*-genes operating in the heterologous genetic background of plants facing open field or green house conditions may be unpredictable. Furthermore, the process of transformation itself may affect the level of resistance of transformed plants as a result of the apparent dosage effect of *R*-gene copy number.

Also, it should be said that there is still much to understand about resistance gene function and expression in foreign genomes. However, even in the case of successful transfers of *R*-genes in heterologous species, it is reasonable to postulate that a more intense use of a monogenic resistance may lead to an augment of the nematode populations submitted to the relative selection pressure, with a major threat to the durability of such a resistance. Durability might be preserved only when the transgenic approach will be able to transfer multiple resistance genes into the same cultivar because of the extreme specificity of virulent populations to the *R*-gene on which they were selected (Molinari, 2010). Finally, the ultimate goal of transgenic biotechnology should be the introgression of resistance to multiple nematode species into cultivars characterized by resistance to different pests in order to develop a powerful pest management strategy able to speed up the end of the use of pesticides.

Conclusion

Genetically-based resistance may be a simple and efficient solution to protect plants against pathogens and pests. EU agricultural policy is focusing on pest management strategies alternative to pesticides. The current generally accepted perspective is that exploitation and management of plant genetic resistance must be the key elements of a durable Integrated Pest Management (IPM). A European network of excellence, named Endure®, has recently been constituted to the scope of introducing innovative strategies and reduce cropping system reliance on pesticides in EU farming systems (http://www.endure-network.eu/). In particular, host resistance to plant parasitic nematodes has generally been recognized as a management tactic that has much potential but needs to be more effectively utilized. One of the problems to a fitting use of natural host resistance is that it can be applied only after accurate investigations of the specificity of the resistance traits and their targets among the wide variety of nematode species and subspecies. An additional problem is the limited amount of germplasm resources that has been characterized for resistance to nematodes and the time-consuming procedures used to screen germplasm collections. The predictable increase of DNA-based markers (e.g. SSR, RFLP, and RAPDs) linked to resistance loci will determine sources of resistance more rapidly than traditional biological tests. For instance, through the use of this Marker-assisted Selection (MAS), it was possible to select soybean cyst nematode resistant lines based on genotype rather than phenotype (Young and Mudge, 2002), or, more recently, search for resistance genes to RKNs in cotton by means of SSR markers (Guitiérrez et al., 2010). Another challenge is to limit problems of linkage drag with undesirable horticultural traits and produce-resistant cultivars with optimal yield potential, as it has already been done with cotton, peanut, soybean, and tomato, for instance, although this may not be sufficient in the absence of effective grower-based education programs for a successful implementation of their use (Starr and Roberts, 2004).

Spread of virulent populations and resistance durability is an issue which should primarily be considered for an effective use of plant resistance to sedentary endoparasitic nematodes. Although, at least with RKN populations virulent on resistant tomato and pepper, it is becoming clear that selection, probably at its early stage, implies adverse fitness costs (Djian-Caporalino et al., 2011), actions to support resistance durability should be promptly or preventively applied, due to the presumed capacity of well-established virulent populations to compensate and restore fitness in time. These actions may involve spatial heterogeneity of resistant/susceptible varieties, random spatial patterning of differential monogenic resistances and, when they will be available, cultivars with multi-genic resistance. These strategies can significantly decrease plant damage, pest population density and the rate of spread of new virulent populations.

Major efforts should be spent to address all the above-mentioned drawbacks, since it is worth recalling that the use of resistance is extremely economically convenient for managing nematodes in both high and low value agricultural systems, as direct cost to the growers is minimal. Only the development of transgenic-resistant cultivars by private industry, and its high cost, may increase the price of seeds (Starr and Roberts, 2004). For instance, the growth of tomato resistant to *M. javanica* increased profits by 30,000 and 88,000 euros ha^{-1} in non-fumigated and methyl bromide-fumigated soils, respectively, compared with the growth of susceptible tomato (Sorribas et al., 2005). Comparably, although an amount of $1 million was spent to develop one soybean cultivar with resistance to *H. glycines*, benefits in the amount of $ 400 million resulted from this operation (Starr and Roberts, 2004). Currently, host resistance alone cannot be the solution of all problems caused by plant parasitic nematodes, but it should be the leading component of an integrated, multi-tactic approach to nematode management, that may involve, according to specific cases, the use of plant growth-promoting rhizobacteria, soil solarization, biological control, crop rotation, cover crops, etc. Although no stand-alone replacement for methyl bromide exists and no viable management practices will be identical to methyl bromide treatment, an integrated approach possibly using host resistance should effectively and economically manage the targeted nematode pest affecting diverse agricultural systems.

References

Abad, P. and Williamson, V.M. (2010). Plant Nematode Interaction: A Sophisticated Dialogue *Advances in Botanical Research, 53*, 147-192.

Ammati, M., Thomason, I.J. and Roberts PA (1985). Screening *Lycopersicon* spp. for new genes imparting resistance to root-knot nematodes (*Meloidogyne* spp.). *Plant Disease, 69,*112-115.

Ammiraju, J.S.S., Veremis, J.C., Huang, X., Roberts, P.A. and Kaloshian, I. (2003). The heat-stable, root-knot nematode resistance gene *Mi-9* from *Lycopersicum peruvianum* is localized on the short arm of chromosome 6. *Theoretical and Applied Genetics, 106*, 478-484.

Andret-Link, P., Laporte, C., Valat, L., Ritzenthaler, C., Demangeat, G., Vigne, E., Laval, V., PfeiVer, P., Stussi-Garaud, C. and Fuchs, M. (2004). Grapevine fanleaf virus: still a major threat to the grapevine industry. *Journal of Plant Pathology, 86*, 183–195.

Atkinson, H. J., Urwin, P. E., and Hussey, R. S. (2009). Plant biotechnology and control. In: R. N. Perry, M. Moens, and J. L. Starr (Eds.) *Root-knot Nematodes* (pp. 338-362). Wallingford, UK: CAB International.

Bakker, E., Dees, R., Bakker, J., and Goverse, A. (2006). Mechanisms involved in plant resistance to nematodes. In S. Tuzun, and E. Bent (Eds.), *Multi-genic and Induced Systemic Resistance in Plants* (pp. 314-334). USA: Springer Science + Business Media, Inc.

Barker, K. R., Hussey, R. S., Krusberg, L. R., Bird, G. W., Dunn, R. A. et al. (1984). Plant and soil nematodes: social impact and focus for the future. *Journal of Nematology, 26*, 127-137.

Bendezu, I. F. and Starr, J. L. (2003). Mechanism of resistance to *M. arenaria* in the peanut cultivar COAN. *Journal of Nematology, 35*, 115-118.

Bhattarai, K. K., Liu, Q., Liu, Y., Dinesh-Kumar, S. P. and Kaloshian, I. (2007). The *Mi-1* mediated pest resistant requires *Hsp90* and *Sgt1*. *Plant Physiology, 144*, 312-323.

Bhattarai, K. K., Xie, Q. G., Mantelin, S., Bishnoi, U., Girke, T., Navarre, D. A. and Kaloshian, I. (2008). Tomato susceptibility to root-knot nematodes requires an intact jasmonic acid signaling pathway. *Molecular Plant Microbe Interactions, 21*, 1205-1214.

Bird, D.McK. and Kaloshian, I. (2003). Are roots special? Nematodes have their say. *Physiological and Molecular Plant Pathology, 62*, 115-123.

Bleve-Zacheo, T., Bongiovanni, M., Melillo, M.T. and Castagnone-Sereno, P. (1998). The pepper resistance genes *Me1* and *Me3* induce differential penetration rates and temporal sequences of root cell ultrastructural changes upon nematode infection. *Plant Science, 133*, 79-90

Blok, V.C., Ehwaeti, M., Fargette, M., Kumar, A., Philipps, M.S., Robertson, W.M., and Trudgill, D.L. (1997). Evolution of resistance and virulence in relation to the management of nematodes with different biology, origins and reproductive strategies. *Nematologica, 43*, 1-13.

Branch, C., Hwang, C. F., Navarre, D. A. and Williamson, V. M. (2004). Salicylic acid is part of the *Mi-1*-mediated defense response to root-knot nematode in tomato. *Molecular Plant Microbe Interactions, 17*, 351-356.

Castagnone-Sereno, P., Bongiovanni, M. and Dalmasso, A. (1994a). Reproduction of virulent isolates of *Meloidogyne incognita* on susceptible and *Mi*-resistant tomato. *Journal of Nematology, 26*, 324-328

Castagnone-Sereno, P., Wajnberg, E., Bongiovanni, M., Leroy, F. and Dalmasso, A. (1994b). Genetic variation in *Meloidogyne incognita* virulence against the tomato *Mi* resistant gene: evidence from isofemale line selection studies. *Theoretical and Applied Genetics, 88,* 749-753.

Cook, R., and Noel, G.R. (2004). Cyst nematodes: *Globodera* and *Heterodera* species. In J. L. Starr, R. Cook, and J. Bridge (Eds.), *Plant Resistance to Parasitic Nematodes* (pp. 71-105). Wallingford, UK: CAB International.

Davis, E.L., Hussey, R. S., Mitchum, M. G. and Baum, T. J. (2008). Parasitism proteins in nematode-plant interactions. *Current Opinion in Plant Biology, 11*, 360-366.

Djian-Caporalino, C., Fazari, A., Arguel, M.J., Vernie, T., VandeCasteele C., Faure, I., Brunoud, G., Pijarowski, L., Palloix, A., Lefebvre, V., and Abad, P. (2007). Root-knot nematode (*Meloidogyne* spp.) *Me* resistance genes in pepper (*Capsicum annuum* L.) are clustered on the P9 chromosome. *Theoretical and Applied Genetics, 114*, 473-486.

Djian-Caporalino, C., Castagnone-Sereno, P., Palloix, A., Ciancio, A. and Molinari, S. (2010). The use of robust and durable resistances against root-knot nematodes: an efficient method for growers in vegetable cropping systems. *Endure International Conference, Integrated Pest Management in Europe*, 21-24 Nov, Paris, France.

Djian-Caporalino, C., Molinari, S., Palloix, A., Ciancio, A., Fazari, A., Marteu, N., and Castagnone-Sereno, P. (2011). Reproductive potential of root-knot nematodes as affected by major resistance genes from tomato and pepper. *European Journal of Plant Pathology*, *in press*.

Dong, K. and Oppermann, C. H. (1997). Genetic analysis of parasitism in the soybean cyst nematode *Heterodera glycines*. *Genetics, 146*, 1311-1318.

Dropkin, V. H., Helgeson, J. P. and Upper, C.D. (1969). The hypersensitive reaction of tomatoes resistant to *Meloidogyne incognita*: reversal by cytokinins. *Journal of Nematology, 1*, 55-60.

Durrant, W. E. and Dong, X. (2004). Systemic acquired resistance. *Annual Review of Phytopathology, 42*, 185-209.

Faske, T. R. and Starr, J. L. (2009). Mechanism of resistance to *Meloidogyne incognita* in resistant cotton genotypes. *Nematropica, 39*, 281-288.

Fleming, C. C. (1998). The evaluation and durability of potato cyst nematode resistance in the potato. In R.J. Marks, and B. B. Brodie (Eds.), *Potato Cyst Nematodes: Biology, Distribution and Control* (pp. 197-208). Wallingford, UK: CAB International.

Glazebrook, J. (2005). Contrasting mechanisms of defense against biotrophic and necrotrophic pathogens. *Annual Review of Phytopathology, 43*, 205-227.

Gleason, C. A., Liu, Q. L. and Williamson, V. M. (2008). Silencing a candidate nematode effector gene corresponding to the tomato resistance gene *Mi-1* leads to the acquisition of virulence. *Molecular Plant-Microbe Interactions, 21*, 576-585.

Goggin, F.L., Jia, L. L., Shah, G., Hebert, S., Williamson, V.M. and Ullman, D.E. (2006). Heterologous expression of the *Mi-1.2* gene from tomato confers resistance against nematodes but not aphids in egg plant. *Molecular Plant-Microbe Interactions, 19*, 383-388.

Graham, T.L. and Graham, M.Y. (1999). Role of the hypersensitive cell death in conditioning elicitation competency and defense potentiation. *Physiological and Molecular Plant Pathology, 55*, 13-20.

Greco, N. and Di Vito, M. (2009). Population dynamics and damage level. In: R. N. Perry, M. Moens, and J. L. Starr (Eds.) *Root-knot Nematodes* (pp. 246-274). Wallingford, UK: CAB International.

Gutiérrez, O. A., Jenkins, J. N., McCarty, J.C.,. Wubben, M.J., Hayes, R.W. and Callahan, F. E. (2010). SSR markers closely associated with genes for resistance to root-knot nematode on chromosomes 11 and 14 of Upland cotton. *Theoretical and Applied Genetics, 121*, 1323–1337.

Harris, A. R. (1983). Resistance of some *Vitis* rootstocks to *Xiphinema index*. *Journal of Nematology, 15*, 405-409.

Hoedisoganda, W. W. and Sasser, J. N. (1982). Resistance of tomato, bean, southern pea, and garden pea cultivars to root-knot nematodes based on host suitability. *Plant Disease, 66*, 145-150.

Jablonska, B., Ammiraju, J.S.S., Bhattarai, K.K., Mantelin, S., Martinez de Ilarduya, O., Roberts, P.A. and Kaloshian, I. (2007). The *Mi-9* gene from *Solanum arcanum* conferring heat-stable resistance to root-knot nematodes is a homologue of *Mi-1*. *Plant Physiology, 143*, 1044-1054.

Jacquet, M., Bongiovanni, M., Martinez, M., Verschave, P., Wajnberg, E. and Castagnone-Sereno, P. (2005). Variation in resistance to the root-knot nematode *Meloidogyne incognita* in tomato genotypes bearing the *Mi* gene. *Plant Pathology, 54*, 93-99.

Janssen, R., Bakker, J. and Gommers, F. (1991). Mendelian proof for a gene-for-generelationship between virulence of *Globodera. rostochiensis* and the *H1* resistance gene in *Solanum tuberosum* spp. *andigena* CPC 1673. *Revue de Nematologie*, 14, 213-219.

Kaloshian I, Williamson VM, Miyao G, Lawn DA, Westerdahl BB (1996). "Resistance-breaking" nematodes in California tomatoes. *California Agriculture, 50*, 18-19.

Kaloshian, I. (2004). Gene-for-gene disease resistance: bridging insect pest and pathogen defense. *Journal of Chemical Ecology, 30*, 2419-2438.

Landels, S. (1989). Fumigants and nematicides. In *Chemical Economics Handbook*. USA, California: Standford International Research Institute.

Lennon, A. M., Neuenschwander, U. H., Ribas-Carbo, M., Giles, L., Ryals, J. A. and Siedow, J. N. (1997). The effect of salicylic acid and tobacco mosaic virus infection on the alternative oxidase of tobacco. *Plant Physiology, 115*, 783–91.

Martinez de Ilarduya, O., Nombela, G., Hwang, C.F., Williamson, V.M., Muniz, M. and Kaloshian, I. (2004). *Rme1* is necessary for *Mi*-mediated resistance and acts early in the resistance pathway. *Molecular Plant-Microbe Interactions, 8*, 811-815.

McCarter, J.P. (2008). Molecular approaches toward resistance to plant-parasitic nematodes. In R.H. Berg, and C.G. Taylor (Eds.), *Cell Biology of Plant Nematode Parasitism-Plant Cell Monographs* (pp. 239-267). Berlin Heidelberg, Germany: Springer-Verlag.

McKenry, M.V. (1994). Nematicides. In C.J. Arntzen (Ed.), *Encyclopedia of Agricultural Science* (vol. 3, pp. 87-95). California, San Diego: Acad. Inc.

McKenzie, M. M. and Turner, S. J. (1987). Assessing reproduction of potato cyst nematodes (*G. rostochiensis* and *G. pallida*) on potato cultivars for National Listing. *EPPO Bulletin, 17*, 345-357.

Melillo, M. T., Leonetti, P., Bongiovanni, M., Castagnone-Sereno, P. and Bleve-Zacheo, T. (2006). Modulation of reactive oxygen species activities and H_2O_2 accumulation during compatible and incompatible tomato root-knot nematode interactions. *New Phytologist, 170*, 501-512.

Milligan, S. B., Bodeau, J., Yaghoobi, J., Kaloshian, I., Zabel, P. and Williamson, V.M. (1998). The root-knot nematode resistance gene Mi from tomato is a member of the leucine zipper, nucleotide binding, leucine-rich repeat family of plant genes. *Plant Cell, 10*, 1307–1319.

Moens, M. and Perry, R.N. (2009). Migratory endoparasitic nematodes: a group rich in contrasts and divergence. *Annual Review of Phytopathology, 47*, 313-332.

Molinari, S. (2001). Inhibition of H_2O_2-degrading enzymes in the response of *Mi*-bearing tomato to root-knot nematodes and salicylic acid treatment. *Nematologia mediterranea, 29*, 235-239.

Molinari, S. (2007). New developments in understanding the role of salicylic acid in plant defense. *CAB Reviews, 67*, 1-10. doi:10.1079/PAVSNNR20072067.

Molinari, S. (2009a). Selection exerted by host resistance, Interaction Pepper-Tomato x *Meloidogyne*. *Endure Annual Meeting*, 20-23 October, Wageningen, The Netherlands.

Molinari, S. (2009b). Antioxidant enzymes in (a)virulent populations of root-knot nematodes. *Nematology, 11*, 689-697.

Molinari, S. (2010). A new approach of rating pathogen-host suitability between (a)-virulent populations of root-knot nematodes and tomato. *Proceedings of the 3rd International Symposium on Tomato Diseases*, 25-30 July, Ischia, Italy.

Molinari, S. (2011). Natural genetic and induced plant resistance, as a control strategy to plant-parasitic nematodes alternative to pesticides. *Plant Cell Reports, 30*, 311-323.

Molinari, S, Zacheo, G. and Bleve-Zacheo, T. (1990). Effects of nematode infestation on mitochondria isolated from susceptible and resistant tomato roots. *Physiological and Molecular Plant Pathology, 37*, 27–37.

Molinari, S. and Loffredo, E. (2006). The role of salicylic acid in defense response of tomato to root-knot nematodes. *Physiological and Molecular Plant Pathology, 68*, 69-78.

Molinari, S. and Baser, N. (2010). Induction of resistance to root-knot nematodes by SAR elicitors in tomato. *Crop Protection, 29*, 1354-1362.

Molinari, S., Greco, N. and Zouhar, M. (2010). Superoxide dismutase isoelectric focusing patterns as a tool to differentiate pathotypes of *Globodera* spp. *Nematology, 12*, 751.758.

Niblack, T. L., Hussey, R. S. and Boerma, H. R. (1986). Effects of environments, *Meloidogyne incognita* inoculum levels, and *Glycine max* genotype on root-knot nematode-soybean interactions in field microplots. *Journal of Nematology, 18*, 338-346.

Noling, J. W. (2000). Effects of continuous culture of a resistant tomato cultivar on *Meloidogyne incognita* soil population and pathogenicity. *Journal of Nematology, 32*, 452 (abstr.)

Norman, C., Howell, K. A., Millar, A. H., Whelan, J. M. and Day, D. A. (2004). Salicylic acid is an uncoupler and inhibitor of mitochondrial electron transport. *Plant Physiology, 134*, 492–501.

Parella, G., Moretti, A., Gognalons, P., Lesage, M.L., Marchaux, G., Gebre-Selassie, K. and Caranta, G. (2004). The *Am* gene-controlling resistance to *Alfalfa mosaic virus* in tomato is located in the cluster of dominant resistance genes on chromosome 6. *Phytopathology, 94*, 345-350.

Paulson, R. E. and Webster, J. M. (1972). Ultrastructure of the hypersensitive reaction in roots of tomato *Lycopersicon esculentum* L., to infection by the root-knot nematode, *Meloidogyne incognita*. *Physiological Plant Pathology, 2*, 227-232.

Petrillo, M. D. and Roberts, P. A. (2005a). Fitness of virulent *Meloidogyne incognita* isolates on susceptible and resistant cowpea. *Journal of Nematology, 37*, 457-466.

Petrillo, M. D. and Roberts, P. A. (2005b). Isofemale line analysis of *Meloidogyne incognita* virulence to resistance genes *Rk* and Rk^2 in cowpea *Journal of Nematology, 37,* :448-456.

Phillips, M.S., Harrower, B.E., Trudgill, D.L., Catley, M.A. and Waugh, R. (1992). Genetic variation in British populations of *Globodera pallida* as revealed by isozymes and DNA analysis. *Nematologica, 38*, 304-319.

Rice, S. L., Stone, A. R., and Leadbeater, B. S. C. (1987). Changes in cell structure in roots of resistant potatoes parasitized by potato cyst nematodes. 2. Potatoes with resistance from *Solanum vernei*. *Physiological and Molecular Plant Pathology, 31*, 1-14.

Roberts, P. A., Matthews, W. C. and Veremis, J. C. (1998). Genetic mechanisms of host plant resistance to nematodes. In K.R. Barker, G. A. Pederson, and G. I. Windham (Eds.), *Plant-Nematode Interactions* (pp. 209-238). Madison, Wisconsin: American Society of Agronomy.

Roberts, P. (1995). Conceptual and practical aspects of variability in root-knot nematodes related to host plant resistance. *Annual Review of Phytopathology, 33*, 199-221.

Schaff, J. E., Nielsen, D. M., Smith, C. P., Scholl, E. H. and Bird, D. M. (2007). Comprehensive transcriptome profiling in tomato reveals a role for glycosyltransferase in *Mi*-mediated nematode resistance. *Plant Physiology, 144*, 1079-1092.

Seah, S., Telleen A.C. and Williamson, V.M. (2007). Introgressed and endogenous *Mi-1* gene clusters in tomato differ by complex rearrangements in flanking sequences and show sequence exchange and diversifying selection among homologues. *Theoretical and Applied Genetics, 114*, 1289-1302.

Seifi, A., Kaloshian, I., Vossen, J., Che, D., Bhattarai, K.K., Fan, J., Naher, Z., Goverse, A., Tjallingii, W. F., Lindhout, P., Visser, R. G. F., and Bai1, Y. (2011). Linked, if not the Same, *Mi-1* homologues confer resistance to tomato powdery mildew and root-knot nematodes. *Molecular Plant-Microbe Interactions, 24*, 441-450.

Shirasu, K., Nakajima, H., Rajasekhar, V. K., Dixon, R. A. and Lamb, C. (1997). Salicylic acid potentiates an agonist-dependent gain control that amplifies pathogen signals in the activation of defense mechanisms. *Plant Cell, 9*, 261–70.

Smith, P.G. (1944). Embryo culture of a tomato species hybrid. *Proceedings of the American Society of Horticultural Science, 44*, 413-416.

Sobczak, M., Avrova, A., Jupowicz, J., Phillips, M. S., Ernst, K. and Kumar, A. (2005). Characterization of susceptibility and resistance responses to potato cyst nematode (*Globodera* spp.) infection of tomato lines in the absence and presence of the broad-spectrum nematode-resistant *Hero* gene. *Molecular Plant-Microbe Interactions, 18*, 158-168.

Sorribas, F. J., Ornat, C., Verdejo-Lucas, S., Galeano, M. and Valero, J. (2005). Effectiveness and profitability of the *Mi*-resistant tomatoes to control root-knot nematodes. *European Journal of Plant Pathology, 111*, 29-38.

Starr, J. L. and Roberts, P. A. (2004). Resistance to plant-parasitic nematodes. In Z. X. Chen, S. Y. Chen, and D. W. Dickson (Eds.), *Nematology, Advances and Perspectives* (vol. 2. pp. 879-907). Wallingford, UK: CAB International.

Triantaphyllou, A. C. (1975). Genetic structure of the race of *Heterodera glycines* and inheritance and ability to reproduce on resistant soybeans. *Journal of Nematology, 7*, 356-364.

Trudgill, DL (1991). Resistance to and tolerance of plant parasitic nematodes in plants. *Annual Review of Phytopathology, 29*, 167-193.

Tzortzakakis, E.A. and Blok, V.C. (2007). Differentiation in two populations of *Meloidogyne incognita* from Greece in relation to reproduction on resistant tomato and pepper. *Journal of Plant Disease Protection, 114*, 276-277.

Tzortzakakis, E.A. and Gowen, S.R. (1996). Occurrence of a resistance-breaking pathotype of *Meloidogyne javanica* on tomatoes in Crete, Greece. *Fundamental Applied Nematology, 19*, 283-288.

Tzortzakakis, E. A., Trudgill, D.L. and Phillips, M.S. (1998). Evidence for a dosage effect of the *Mi* gene on partially virulent isolates of *Meloidogyne javanica*. *Journal of Nematology, 30*, 76-80.

Uehara, T., Sugiyama, S., Matsuura, H., Arie, T. and Masuta, C. (2010). Resistant and susceptible responses in tomato to cyst nematodes are differentially regulated by salicylic acid. *Plant and Cell Physiology, 51*, 1524-1536.

Van Gundy, S.D. (1985). Ecology of *Meloidogyne* spp. – emphasis on environmental factors affecting survival and pathogenicity. In J.N. Sasser, and C.C. Carte (Eds.), *An Advanced Treatise on* Meloidogyne (vol. 1 - Biology and Control, pp. 177-182). USA, North Carolina, Releigh: State University Graphics.

Vasyukova, N. I., Zinov'eva, S. V., Udalova, Zh. V., Panina, Ya. S., Ozeretskovskaya, O. L. and Sonin, M. D. (2003). The role of salicylic acid in systemic resistance of tomato to nematodes. *Doklady Biological Science, 391,* 343-345

Vasyukova, N.I., Pridvorova, S.M., Gerasimova, N.G., Chalenko, G.I., Ozeretskovskaya, O.L., Udalova, Zh.V. and Zinov'eva, S.V. (2007). The involvement of phenylalanine ammonia-lyase and salicylic acid in the induction of resistance of tomato plants infested with gall nematode *Meloidogyne incognita*. *Doklady Biological Science, 416*, 382-385.

Vos, P.,Simons, G., Jesse, T.,Wijbrandi, J.,Heinen, L.,Hogers, R.,Frijters, A.,Groenendijk, J., Diergaarde, P., Reijans, M., Fierens-Onstenk, J., De Both, M., Peleman, J., Liharska, T.,

Hontelez, J., and Zabeau, M. 1998. The tomato *Mi-1* gene confers resistance to both root-knot nematodes and potato aphids. *Nature Biotechnology, 16*, 1365–1369.

Wang, C., Ulloa, M. and Roberts, P. A. (2008). A transgressive segregation factor (*RKN2*) in *Gossypium barbadense* for nematode resistance clusters with gene *rkn1* in *G. hirsutum. Molecular Genetics and Genomics, 279*, 41-52.

Williamson, V. M. and Hussey, R. S. (1996). Nematode pathogenesis and resistance in plants. *Plant Cell, 8*, 1735-1745

Williamson, V. M. and Roberts, P. A. (2009). Mechanisms and genetics of resistance. In: R. N. Perry, M. Moens, and J. L. Starr (Eds.) *Root-knot Nematodes* (pp. 301-325). Wallingford, UK: CAB International.

Williamson, V.M. (1998). Root-knot nematodes resistance genes in tomato and their potential for future use. *Annual Review of Phytopathology, 36*, 277-293.

Wubben, M. J. E., Jin, J. and Baum, T. J. (2008). Cyst nematode parasitism of *Arabidopsis thaliana* is inhibited by salicylic acid (SA) and elicits uncoupled SA-independent pathogenesis-related gene expression in roots. *Molecular Plant Microbe Interactions, 21*, 424-432.

Wyss, U., Lehmann, H. and Jank-Ladwig, R. (1980). Ultrastructure of modified root-tip cells in *Ficus carica* induced by the ectoparasitic nematode *Xiphinema index. Nematologica, 28*, 178-179.

Xu, K., Riaz, S. Roncoroni, N. C., Jin, Y., Hu, R., Zhou, R. and Walker, M.A. (2008). Genetic and QTL analysis of resistance to *Xiphinema index* in a grapevine cross. *Theoretical and Applied Genetics, 116*, 305–311.

Young, N. D. and Mudge, J. (2002). Marker-assisted selection for soybean cyst nematode resistance. In J. L. Starr, R. Cook, and J. Bridge (Eds.), *Plant Resistance to Parasitic Nematodes* (pp. 241-252). Wallingford, UK: CAB International.

Zacheo, G., Orlando, C. and Bleve-Zacheo, T. (1993). Characterization of anionic peroxidases in tomato isolines infected by *Meloidogyne incognita*. *Journal of Nematology, 25*, 249-256.

Zasada, I.A., Halbrendt, J.M., Kokalis-Burelle, N. La Mondia, J. McKenry, M.V. and Noling, J.W. (2010). Managing nematodes without methyl bromide. *Annual Review of Phytopathology, 48*, 311-328.

Zhang, L.Y., Zhang, Y.Y., Chen, R.G., Zhang J.H., Wang, T.T., Li, H.X. and Ye, Z.B. (2010). Ectopic expression of the tomato *Mi-1* gene confers resistance to root knot nematodes in lettuce (*Lactuca sativa*). *Plant Molecular Biology Reporter, 28*, 204-211.

In: Nematodes
Editors: F. Boeri and J. A. Chung

ISBN: 978-1-61470-784-4

Chapter III

Filariasis: Biochemistry and Chemotherapeutic Targets

Jitendra K. Saxena[1], Anita[1], Manish K. Suthar[1], Pawan K. Doharey[1] and Alok R. Singh[2]
[1] Division of Biochemistry, CSIR-Central Drug Research Institute, Lucknow-226001, India.
[2] Moores Cancer Center, University of California, San Diego, U.S.A.

Abstract

The phylum Nematoda (roundworms), encompasses up to 500,000 species. Nematode infections in humans include ascariasis, trichuriasis, hookworm, enterobiasis, strongyloidiasis, filariasis, and trichinosis, etc. Recent data has demonstrated that approximately 60 species of roundworms parasitize humans. According to World Health Organization (WHO) report 2005, approximately 0.807-1.221 billion humans have ascariasis, 604-795 million have trichuriasis, 576-740 million have hookworm infections, and more than 120 million have lymphatic filariasis, worldwide. Lymphatic filariasis is one of the only six infectious diseases considered by WHO as eradicable with the currently available tools by the year 2020.

The main problem regarding the chemotherapy of filariasis is that no safe and effective drug is available to combat the adult human filarial worms. For a rational approach to antifilarial chemotherapy, knowledge of the biochemical composition and metabolic pathways operative in filarial parasite are of paramount importance, so that more potent antifilarial agents based on specific drug targets can be identified. In fact, even the most widely employed drug in the treatment of lymphatic filariasis, diethylcarbamazine (DEC), was discovered by a random screening of compounds fifty years ago and still does not have a defined mechanism of action. Further advancement of drug treatment and vaccine development would benefit greatly from an improved understanding of basic filarial biology. The drugs used for lymphatic filariasis are albendazole, diethylcarbamazine (DEC) and ivermectin. None of these is effective in killing the long-lived adult worms (macrofilariae) and the treatments are therefore aimed at reducing transmission and pathology. Although it is impractical, prolonged treatment with doxycycline significantly reduces the numbers of microfilaria in circulation, which is an important strategy to control transmission of filariasis by mosquito vectors.

The ~90 megabase genome of the human filarial parasite *Brugia malayi* consists of 5 chromosomes, including an XY sex-determination pair and predicts ~11,500 protein coding genes in 71 Mb of robustly assembled sequence. The filarial parasites are capable of assimilating oxygen under appropriate conditions, but none can completely oxidize substrates to carbon dioxide and water. All accumulate organic end products, indicating little or no terminal respiration. Energy yielding biochemical pathways, nerve transmission and neuromuscular conduction regulating parasitic motility represent the targets for action of many antifilarials. Folate, nucleic acid, polyamine metabolism and other biochemical pathways are being explored as alternate target for chemotherapy. Microfilariae have proved to be the most vulnerable target so far. However, microfilaricides yield only transient cure. Vulnerable targets of adult parasites are badly needed to design macrofilaricidal drugs for any lasting solution to filariasis.

The following review summarizes the biochemistry of filarial parasites and the metabolic difference with the respective hosts in the gross pathway as well as in molecular and biochemical properties. This information in turn can be meaningfully utilized for understanding the host invader interaction and managing it in favour of mammalian host.

Introduction

Helminth parasites represent major cause of human misery because ascarid, hookworm and filarial infections are ubiquitous in developing nations and cause malnutrition, disfiguration and disability. Filariasis represents a class of diseases caused by 'Thread like' worms of the super family Filarioidea of phylum nematode. According to World Health Organization (WHO) report 2005, approximately 0.807-1.221 billion humans have ascariasis, 604-795 million have trichuriasis, 576-740 million have hookworm infections, and more than 120 million people have lymphatic filariasis (by *Wuchereria bancrofti, Brugia malayi* and *Brugia timori*), onchocerciasis or 'river blindness' (by *Onchocerca volvulus*), and loiasis (by *loa loa*), worldwide [1-3]. Lymphatic filariasis, resulting from infection with the mosquito-borne filarial nematodes *Brugia malayi* and *Wuchereria bancrofti*, is a widespread and debilitating disease in many developing countries [4]. The Global Program to Eliminate Lymphatic Filariasis (GPELF), started in 1997, is active in 43 of the 80 countries worldwide in which lymphatic filariasis is endemic. The World Health Organization (WHO) has targeted the year 2020 for the eradication of lymphatic filariasis [5]. Although 91% of the cases reported worldwide are caused by *Wuchereria bancrofti*, a further 9% are attributed to brugian filariasis [6], and are restricted to the Asian region, particularly in India, Indonesia, Malaysia, Philippines, Sri Lanka and Thailand [7, 8]. So far, at least four physiological types have been declared, *ie* the nocturnally periodic, nocturnally subperiodic [9], diurnally subperiodic and non-periodic [10, 11]. Subperiodic *Brugia malayi* has been reported in domestic cats from southern Thailand [12].

The filarial infection is not a major cause of mortality in the world, but it is responsible for high rates of morbidity in endemic areas. Physical manifestations can range from acute fevers and lymphadenopathy to debilitating chronic edema, known commonly as elephantiasis [13]. Most clinical manifestations are due to damaged lymphatic vessels, believed to be the result of several additive factors, including actions of the parasite itself, inflammatory response to the parasite, and parallel opportunistic infections [14]. It is estimated that over 40 million individuals suffer from gross manifestations of lymphatic filariasis, most commonly limb lymphedema and urogenital disease. As a result, lymphatic filariasis is the second leading cause of disability in the world [15, 16].

There are three distinct phases in the life cycle of filarial parasites viz., microfilariae (mf), infective larvae (L3) and adult worms. The adult parasites reside in connective tissues, muscles, circulatory or lymphatic system of the host. The microfilariae (Sheathed/unsheathed) are released into peripheral blood of the host, have a life span of 14-70 days and exhibit nocturnal/diurnal periodicity. All the filarial parasites of man are vectorborne, transmitted by mosquitoes, biting midges, tabanids and black flies. Arthropods such as fleas, ticks, and mites transmit some filarial infections of animals. In India, *Mansonia* species of mosquito propagate *B. malayi,* while *Culex* species is responsible for transmission of *W. bancrofti*. While circulating in peripheral blood some mf are taken up through the bite of insect vector and they undergo further development to the infective larval stage within 15 days. When the blood sucking arthropode takes its next meal, the L3 larvae (3rd stage larvae) are transmitted to the recipient host through the skin and the cycle is completed.

Several reviews describing the relationship of mosquito vectors to transmission of human lymphatic filariasis have been published [17-19]. *W. bancrofti* and *B. malayi* microfilariae have unique circadian rhythms manifested as a variability of their number or concentration in the peripheral circulation [20-24].

Biochemistry of Filarial Parasites

The biochemistry of filarial nematodes has been difficult to study because of their complex life cycle. Each parasite stage is necessarily adapted to its specific environmental conditions such as temperature, pH, and oxygen tension. Elaborate mechanisms involving rapid changes from aerobic to anaerobic metabolism and invertebrate to vertebrate body temperature must therefore operate. The parasite cuticle, as well as the gut, is major route of entry for amino acids, monosaccharides, and dyes [25]. Cuticle, which forms the nematode surface, differs from the plasma membrane or tegumental surfaces of other helminthes. Glycoproteins have been identified as a structural component of cuticle in *D. immitis, B. pahangi* and *B. malayi* [26-28]. Complex oligosaccharides and their conjugates with protein and lipids present on the parasite's surface play a significant role in determining its antigenicity and host immune responses [29].

Mf are sheathed in *W. bancrofti* and *B. malayi.* Sheath acts as a barrier between the parasite and host defences. Cytochemical and lectin binding methods have unravelled the components of microfilarial sheath as well as location of carbohydrates, proteins and enzymes [26, 28, 30]. The sheath and the epicuticle of *B. pahangi* stain positively with concavalin A (Con A) and mf sheath showed activity of acid phosphatase, 5'-nucleotidase and peroxidases. Presence of N-acetylglucosamine, glucose and mannose on mature mf and sialic acid, galactose and N-acetylglucosamine on sheath of immature mf has been demonstrated [27, 31, 32]. *D. immitis* mf (unsheathed species) after treatment with proteases, neuraminidases, DEC or EDTA showed no lectin binding sites [26]. Infective larvae and adult worms of *B. malayi* did not bind any of the lectins tested but mf bound wheat germ agglutinin (WGA) in a specific and saturable marner giving evidence for the presence of exposed N-acetylglucosamine. The changes in the surface structures of mf related with developmental status as well habitat of mf may be involved in host recognition and immune response. The sheath carbohydrates have been suggested to play a crucial role in exsheathment, recognition by specific receptors in mosquito and also for the further development of parasite [29].

B. malayi microfilariae and adult worms incorporate exogenous long-chain fatty acids such as arachidonic, palmitic, oleic, and linoleic acids [33, 34]. Arachidonic acid is rapidly esterified into phospholipids including phosphatidylinositol. Esterification of other fatty acids into phospholipids is quantitatively less and involves incorporation into phosphatidylcholine and phosphatidylethanolamine. Microfilariae may metabolize host platelet eicosanoids in a manner which leads to production of biologically active lipids such as 12-hydroxyeicosatetraenoic acid. *B. pahangi* and *Dirofilaria immitis* microfilariae carry out purine synthesis de novo. These parasites also possess purine salvage pathways [18]. It is not clear why two separate routines for purine synthesis coexist in these organisms. *In vitro* studies demonstrated that adult filarial nematodes including *B. pahangi*, *O. volvulus*, and *A. viteae* rapidly take up glucose via a transcuticular route [35, 36]. These parasites rely heavily on glycolysis for energy production, and 10 to 20% of the glucose taken up is converted to glycogen. In *B. pahangi* and *A. viteae*, the primary product of metabolism is lactate, indicating that the worms are primarily homolactate fermenters [37]. The release of small amounts of succinate suggests that a partially reversed tricarboxylic acid cycle is also active in these filarial nematodes [35, 36].

Oxygen consumption and oxidative phosphorylation may represent important metabolic facets of filarial nematodes. Low oxygen tension is required for in vitro growth, molting, and differentiation of *B. pahangi* and *A. viteae* [38]. Measurable rates of exogenous oxygen uptake have been demonstrated for adult *A. viteae, Dirofilaria immitis*, and *B. pahangi*. However, cytochrome c and cytochrome c oxidase were not detected in these systems. Another aspect of filarial metabolism which needs further study is the toxic nature of oxygen for these organisms. Antioxidant viz., catalase, superoxidase dismutase, and glutathione peroxidase vitamins such as vitamin C and ascorbic acid are present in filarial parasites [39]. The mechanisms by which filarial worms penetrate and migrate within host tissues remain largely unknown. *O. volvulus* and *B. malayi* contain and release proteolytic enzymes that solubilize collagen molecules from native type I collagen fibers [40].

Carbohydrate Metabolism

Glycogen is the main reserve food in parasitic nematodes for providing energy under adverse conditions. Helminth parasites inhabiting the intestine and living in an oxygen deficient atmosphere usually have a high content of glycogen. Carbohydrates are the major energy source for helminth parasites and are used as essential constituents of the media employed to maintain them *in vitro* [41]. In some other parasites amino acids and lipids may also be important in providing energy as evident from their conversion to carbohydrate [42]. The anaerobic breakdown of glycogen to lactate via hexose phosphates and triose phosphates follows a course which is superficially similar to that in yeast and mammalian muscles [43]. Evidence for the functioning of different metabolic pathways in parasites has been adduced mainly by the demonstration of the enzymatic steps or the identification of the intermediates of the pathway. The operation of a full glycolytic pathway for conversion of glucose to lactic acid has been demonstrated in *Dracunculus insignis*, *D. uniformis, C. hawking, L. carinii*, and *D. Immitis* [41]. *S. cervi* appeared to be equipped with most of the enzymes of glycolytic and oxidative pathways [44]. The enzymes of Embden-Meyerhof scheme were localized mainly in soluble fraction, showing resemblance with the mammalian system. Hexokinase of *S. cervi* was found to be specific for glucose but the corresponding enzyme from *L. carinii, C. Hawking* and *A. galli* were of non-specific type. Hexokinase from adult *D. Immitis* phosphorylated glucose, fructose, mannose and glucosamine [45], while the hexokinase from *L. carinii* and *C. hawkingi* phosphorylated glucose, mannose, galactose and fructose [42]. Three isoenzymes of hexokinase have been observed in *D. immitis*; the enzyme was inhibited by Glucose-6-phosphate and it could also use glycerol as a glycolytic substrate [46]. *B. malayi* hexokinase exhibited a unique property of activation by G6P. All other hexokinases are either unaffected or inhibited by G6P [47]. Malate dehydrogenase (MDH) is the most active enzyme in *S. cervi* adults although significantly high activities of lactate dehydrogenase (LDH), fumarase, glucose phosphate isomerase (GPI) phosphoglucomutase, glyceraldehydes-3-phosphate dehydrogenase, FDPaldolase, phosphopyruvate hydratase, PEP carboxykinase (PEP-CK) and pyruvate kinase (PK) were also detected. Phosphofructokinase (PFK), glucokinase, malic enzyme and fructose diphosphatase are less active. Diaphorase activity was not detected in the system. Glucose-6-phosphate dehydrogenase and 6-phosphogluconate dehydrogenase, the enzymes of pentose phosphate pathways could not be detected in measurable amounts in *C. hawking* and *L. carinii* but *S. cervi* contained significant quantities of these enzymes [29]. Under aerobic conditions *L. carinii* produces acetate which is derived from decarboxylation of pyruvate, probably via pyruvate dehydrogenase. This enzyme is present in significant amount in *L. carinii* but low activity is observed in *B. pahangi* and *D. viteae* [37, 48]. Since the discovery of PEP-oxaloacetate pathway in invertebrates [49], its operation has been studied in several helminth parasites. Balance between PK and PEPCK and their affinity for the substrate (PEP) are the factors that determine whether metabolic products are channeled to succinate or lactate [41]. *S. cervi* which converts only 25% carbohydrate (glucose) to lactic acid and possessing low levels of PK and LDH and high activities of PEPCK and MDH has a functional PEP-succinate pathway. On the other hand typical lactic acid producers like *C. hawking* converting 80 to 90% glucose into lactic acid resemble vertebrate tissues in possessing high levels of PK and LDH and low levels of PEPCK and MDH. These parasites have a metabolic pathway leading to lactate accumulation while values less than one suggest the operation of carbon dioxide fixing pathway and probable production of succinate [50]. Both mf and adult forms of *S. cervi* have a PK/PEPCK ratio less than one [51]. However, in adult forms of *D. immitis*, *C. hawkingi* and *L. carinii* this ratio was higher than one supporting the preferential formation of lactate [29]. The occurrence of G-6-P dehydrogenase and 6-phosphogluconate dehydrogenase in mf indicates the possible utilization of pentose phosphate pathway which may be related to the need for large amount of ribose. Similarly the presence of glucose-6-phosphatase in mf suggests that gluconeogenic processes may also occur in this life stage [51]. Few enzymes of glycolytic and PEP–succinate pathway have been purified and characterized in order to elucidate their regulatory roles.

Lactate and malate dehydrogenases are associated with the key role of reoxidation of NADH permitting continuous operation of metabolic cycles. This process is critical for helminthes because they thrive in anaerobic habitats and these enzymes from filarial parasites have been used as target for chemotherapy. LDH has been purified from *D. immitis, O. volvulus* and *S. cervi*. MDH has also been purified and characterized from *S. cervi* and *O. Volvulus* [29, 52]. Both LDH and MDH needed functional SH groups for the enzyme activity. Polycations viz., protamine, histone and spermine were found to be strong activators of the *S. cervi* MDH [53]. These polycations also protected the enzyme from dilution inactivation effect. Suramin, aurin tricarboxylic acid and dextran sulfate strongly inhibited the soluble enzymes [53-55].

The TCA cycle is of great importance of many aerobic organisms as energy yield through this pathway is much greater than from glycolysis. Most of the helminthes have been found to be aerobic fermentors and sensitive to deficiency of oxygen. They consume oxygen and have been shown to survive longer when traces of oxygen are present. The levels of TCA cycle enzymes are considerably less in filarial parasites as compared to mammals. Fairly normal looking cristate mitochondria have been reported in *B. pahangi* and *L. carinii* adults as well as adults and mf of *D. immitis* [48, 56]. However, neither cytochrome nor cytochrome oxidase could be detected in *B. pahangi* or *D. viteae* [37]. Cytochrome C (C and C1) and low level of cytochrome a have also been reported in adult *D. immitis* [57]. It has been suggested that *D. immitis* contains a branched cytochrome chain and O-type cytochrome may serve as alternative oxidase. *S. cervi* adult females contain cytochrome b5 but inspite of several attempts presence of cytochrome *P450* could not be established in this filarial parasite [29].

Protein Metabolism

Protein constitutes over 63 and 57% of the dry weight of adult and mf of filarial parasites [51]. Most of amino acids are present in bound form and only traces of free amino acids could be detected in these filarial parasites. The serine hydroxymethyl transferase is present in adult *B. pahangi* and *D. immitis* while methionine synthase and betaine: homocysteine transmethylase could be detected in adult *B. pahangi* and exogenous methionine is required by these two parasites [58]. Glutamate dehydrogenase which plays an important role in deamination of amino acid and formation of α-amino nitrogen group from ammonia has been demonstrated in mf of *D. immitis* and adult of *S. cervi* [59, 60]. Alanine amino transferase (GOT), glutamate pyruvate transferase (GPT), serine dehydratase, threonine dehydratase and arginase are present in filarial parasite [60]. All the enzymes involved in the methionine to cysteine pathway have been identified in the filarial nematodes *B. pahangi* and *D. immitis* [61]. However, the very low cystathionine catabolizing activity in helminths may represent a significant bottleneck in this pathway [62-64].

Presence of two non-specific acid phosphomonoesterases had been reported in *S. cervi* [65]. Acid phosphatase activity has been demonstrated reproductive organs and body wall in adult female of *D. immitis*. The enzyme showing optimal activity at pH 3.8-5.8 was inhibited by tartarate. N-acetyl-β-D-glucosaminidase, β-galactosidase, β-glucosidase, β-glucuronidase, acid phosphatase, alkaline phosphatase, acid ribonuclease, acid deoxyribonuclease and cathepsin were found to be present in adult and microfilarial stages of *S. cervi* female worms [66]. The distribution pattern of these enzymes also differed in various body parts of the parasite [29].

Phosphorylated proteins play major role in the control of diverse biological processes [67]. Glycogen and energy metabolism are regulated by cAMP and also by protein kinases

[68]. The occurrence of cAMP dependent and independent protein kinases has been reported in *O. Volvulus*, *B. malayi, D. viteae, L. carinii* and *S. cervi* and in their developmental stages [69]. Protein kinases of *B. malayi, D. viteae, S. cervi* and *L. carinii* phosphorylated wide variety of exogenous proteins and peptides. The functions of these protein kinases in the metabolism of filarial worms and the possible involvement in differentiation and development have not yet been investigated. The endogenous substrates for these protein kinases are still unknown in the case of filariids. Beside the involvement of protein kinase in regulation of glycogen metabolism, they activate phosphofructokinase of *A. suum* while pyruvate dehydrogenase is inactivated by cAMP dependent protein kinase [29].

Glutathione Metabolism

Glutathione (GSH) constitutes the antioxidant system responsible for the long term existence of filarial worms in mammalian hosts by protecting them from the reactive oxygen species produced by normal metabolism and by immune cells of the host [70, 71]. Therefore, it is useful to develop drugs that could selectively deplete or distort glutathione stores in these parasites. The GSH either acts in solo or as a substrate to various enzymes like glutathione peroxidase (GPX), phospholipid hydroperoxide GPX and glutathione-S-transferase (GST) to quench the free radicals [72, 73]. After GSH has been oxidized to GSSG, the enzyme glutathione reductase (GR) accomplishes the recycling of GSSG back to GSH [74, 75]. The group of six enzymes that are responsible for the synthesis and breakdown of GSH constitutes the γ-glutamyl cycle. At present only two enzymes of the γ-glutamyl cycle viz. glutamate-cysteine ligase (GCL) and γ-glutamyl transpeptidase (γ-GT) have been characterized from filarial species [76, 77]. The whole of this cycle needs to be characterized in filarial worms, although, it has been well characterized in mammalian systems [78-80]. The sensitivity of filarial γ-glutamyl cycle enzymes and their counterparts from mammalian sources to known inhibitors needs to be compared. If marked differences in sensitivity exist this may provide targets for antifilarial drug development [81].

Folate Metabolism

The role of folic and folinic acids in growth and reproduction of nematodes has been demonstrated in axenic cultures. The folate antagonists aminopterin and amethopterin are inhibitory to nematodes in axenic cultures. There is indirect evidence that adult filariae do not synthesize dihydrofolate, but require a source of preformed folate [61]. Adult *B. pahangi* and *D. immitis* contain most of the enzymes involved in folate metabolism viz., N5,N10-methylenetetrahydrofolate reductase, serine hydroxymethyl transferase, N5N10-methylenetetrahydrofolate dehydrogenase, N10-formyltetrahydrofolate synthetase, N10-formyltetrahydrofolate dehydrogenase (NADP dependent/independent), N5N10-methylenetetrahydrofolate cyclohydrolase, N5-formyltetrahydrofolate cyclodehydrase, a complex containing N5-formiminotetrahydrofolate cyclodeaminase and formiminoglutamate, tetrahydrofolate N5-formiminotransferase, N5-formyl, N10-formyltetra-hydrofolate mutase [58, 82]. Adult *B. pahangi* and *D. immitis,* unlike mammalian cells, convert N5-methyl tetrahydrofolate directly to N5N10-methylenetetrahydrofolate and other folate cofactors indicating qualitative differences between the folate metabolism of the host and parasite. The enzyme responsible for this step N5, N10-methylene tetrahydrofolate reductase is a flavoprotein that operates in these parasites preferentially in the reverse direction. In vertebrates, this enzyme catalyses the irreversible formation of N5-methyltetrahydrofolate which serves as methyl donar in the synthesis of methionine from homocysteine [58].

Nucleic Acid Metabolism

Only limited data are available about nucleic acid and nucleotide composition and metabolism of filariids. DNA comprises 0.3 and 1.44% while RNA 0.6 and 1.29% of the dry weight of adult and microfilarial stage of filarial parasite [51]. Large uptake of adenosine was demonstrated *in vitro* by adult male and female *D. immitis* but no uptake of thymidine occurred under similar conditions [83]. Incorporation of uridine and uracil into nucleic acids has been reported in adult *D. immitis* while mf of *D. immitis* incorporated uridine uracil, adenine and adenosine into RNA [84]. Orotic acid derivatives have been reported to be incorporated into RNA by *D. immitis* [85]. These results suggest that mf could synthesise pyrimidines but not purines and filarial parasites possess salvage pathways for both purines and pyrimidines. The gut is probably nonfunctional in mf and 3rd stage larvae of filarial worms, while it becomes functional in 4th stage larvae [86]. Little information is available regarding the uptake of nutrients by the infective larvae of filarial worms. Autoradiographic studies have demonstrated the transport of adenine, amino acids and uridine by developing larvae of *B. pahangi* and *B. patei* [87].

DNA Topoisomerases

DNA topoisomerases are intricately involved in maintaining the topological structure of DNA, transcription and mitosis. This enzyme has been identified as a target for the development of antifilarial compounds [88]. The studies have shown that filarial parasites *Brugia malayi*, *Acanthocheilonema viteae* and *Setaria cervi* adults and microfilariae contained ATP-independent (topoisomerase I) and ATP-dependent (Topoisomerase II) activities. The activities were localized in nuclear fraction and distribution pattern differed between adults and mf stages of filarial parasites. *B. malayi* and *S. cervi* topoisomerase II differed significantly from its human homologue in its kinetic properties [89].

Lipid Metabolism

Few reports are available regarding the lipid composition of filarial parasites [90]. Total lipids accounted for 9 and 12% of dry weight in adult and mf of *S. cervi* [91, 92]. However, 31-34% lipids have been reported in *C. hawking*. *D. immitis* has been reported to contain 2.1% lipid on wet weight basis [93]. Phospholipids constituted the major portion of total lipids. Phosphatidylinositol (PI), diphosphatidylglycerol, cerebrosides, plasmalogens, lysophosphatidyl choline (LPC), sphingolipids and traces of phosphatidic acid have all been found in adult filarial parasites [94]. All major classes of neutrallipids viz., triacylglycerols, diacylglycerols, sterols, sterol esters, hydrocarbons and traces of free fatty acids have been demonstrated in various filarial parasites [94, 95]. Adult female *D. immitis* has been shown to synthesize PE by three pathways viz., (a) via phosphorylethanolamine, CDP ethanolamine, diacylglycerol, (b) decarboxylation of phosphatidyl serine and (c) exchange of ethanolamine for choline or serine in PC or PS [96]. Filarial parasites like other nematodes, are unable to synthesize sequalene or sterols *de novo*, however, ubiquinone 9 and short and long chain isoprenoid alcohols are formed from mevalonate [82]. In short chain isoprenoid alcohol fraction geranyl geraniol constituted the main fraction while, long chain isoprenoid alcohol (dolichol) consisted of dolichol 18(C90), 19(C-95), 20(C100), 21(C105) and dolichol 22(C110). These dolichols and other isoprenols are responsible for the enzymatic transfer of sugar groups of glycoproteins and proteoglycans.

Neurotransmitters and Their Metabolism

Neurotransmitters play an important role in the regulation of motility and metabolism of the parasites. Various biogenic amines viz., Norepinephrine (NE), dopamine (DA), 5-hydroxytryptamine (5-HT) and histamine (Hm) are present in mf and adults of *L. carinii*, *D. viteae* and *S. cervi* [97-99]. Acetylcholine, serotonin, epinephrine and dopamine have been implicated as putative neurohormonal transmitters in parasitic worms. *L. carinii* was found to possess both high and low affinity uptake mechanism for 5-HT and initially uptake mechanism operated against a concentration gradient [100]. Although biosynthesis of 5-HT in mammalian brain has been well established, evidence for *de novo* synthesis of 5-HT in *L. carinii* could not be presented [100]. Presence of low concentration of 5-HT and lack of tryptophan hydroxylase in *L. carinii* indicate that the parasite must obtain their 5-HT from the host. Monoamine oxidase (MAO), responsible for the catabolism of neuroamines was present in both mf and adults of *S. cervi*. The enzyme, mainly localized in mitochondria was found to be more active in female worms as compared to male worms and mf [92]. Presence of dopamine-β- hydroxylase in mf and adults of *S. cervi* has also been shown [101]. Acetylcholinesterase (AchE), having a role in neuromuscular transmission, was detected both in mf and adults of *S. cervi* [92]. Microfilariae contained ten times more activity of AchE as compared to adult. Adults and mf release significant amount of AchE during *in vitro* incubation at 37°C in a defined medium.

Polyamine Metabolism

The polyamines putrescine, spermidine and spermine are involved in growth, differentiation and macromolecular synthesis [102-104]. Octopamine (OA) plays an important role in the regulation of a number of key processes in nematodes, including pharyngeal pumping, locomotion and egg laying [105]. *O. volvulus, D. immitis, Brugia malayi, S. cervi* and *L. carinii* contain high levels of spermidine and spermine but low levels of putrescine and N-acetylated polyamines [106, 107]. The enzymes of polyamine biosynthesis viz. ornithine decarboxylase (ODC), S-adenosyl metheonine decarboxylase (SAMDC) and arginine decarboxylase (ADC) are either very low or absent in filarial parasites [107, 108]. Moreover, uptake of polyamines from the incubation medium as well as interconversion and excretion of putrescine and N1- acetylputrescine has been found in filariids [107, 109]. There is evidence for the existence of complete reverse pathway generating putrescine from spermidine and spermine respectively in filarial worms [107]. The presence of considerable levels of polyamine oxidase, an important enzyme of the reverse pathway of polyamines, indicates a strong point in favour of salvage pathway for polyamines in helminthes [107, 110].

Molecular Biology of Filariasis

Most filarial nematodes, including B. malayi, carry three genomes: nuclear, mitochondrial and that of an alphaproteobacterial endosymbiont, Wolbachia (available at GenBank, accession no. AF538716) [111]. The *B. malayi* nuclear genome is organized as five chromosomes [112] and has been sequenced to approximately ninefold redundancy with the use of whole-genome shotgun (WGS) sequencing [111]. The *B. malayi* genome is 90 to 95 Mb as compared to 100 Mb of *C. elegans* and 104 Mb of *Caenorhabditis briggsae* genome. The overall G + C content (30.5%) is lower than that of *C. elegans* (35.4%) or *C. briggsae* (37.4%) [113]. The repeat content of the *B. malayi* genome, estimated at ~15% [114].

Several groups of investigators have recently reported the use of DNA probes to distinguish between different species of filarial parasites. Three different probes for O. volvulus which show different degrees of cross-reactivity with genomic DNAs from other filarial species have been described [115, 116]. McReynolds *et al.* [117], have constructed two oligonucleotide probes from a highly repeated sequence of *B. malayi* DNA which were sensitive and species specific for *B. malayi* and *B. pahangi*. Species-specific DNAs may also be useful for detecting the presence of human filariae in vector populations [118, 119].

Recombinant DNA technology has provided an avenue for obtaining substantial quantities of defined parasite antigens to examine details of the immune response and for vaccine development. Nilsen and co-workers have identified and characterized a recombinant clone from an adult *B. malayi* complementary DNA expression library that corresponds to a potentially protective native antigen [120]. The full-length complementary DNA corresponding to this antigen has been engineered into a vaccinia construct and shown to lower the level of *B. malayi* microfilaremia in jirds immunized with this preparation [121]. Several *B. malayi* recombinant clones encoding antigens, some of which are specific for the genus, some are specific for *B. malayi* only, and some of which are cross-reactive among different species have been identified [122].

Immunology

B. malayi interacts with two hosts during its life cycle. The mechanisms by which they are able to suppress, subvert, or exploit host defense systems is unknown [123]. Comparison of *B. malayi* protein sequences with interleukins, chemokines, and other signalling molecules from humans showed two genes encoding members of the macrophage migration inhibition (MIF) family [124], growth factor beta (TGFb) homologs (including Bm-tgh-1) [125], and a member of the PDZ domain/interleukin 16 family [111]. These proteins may function as immune modulators that promote parasite survival or growth [111, 126].

Filarial antigens have been shown to decrease activation and cytokine expression of monocytes and dendritic cells and to desensitize B cells [127-129]. Individual filarial proteins have been shown to induce Th2 response [130], through direct cytokinetic activity [131]. The abundant larval transcript (ALT) family of proteins in *B. malayi* have been implicated as virulence factors through their ability to modulate macrophage function [132]. The *B. malayi* genome sequence revealed an unexpected diversity of 13 ALT genes, most of which are expressed by adult parasites [111].

The differences in genome content and organization between Caenorhabditis and *B. malayi* underscore the importance of obtaining additional genome data from representative species from across the diversity of the Nematoda [133]. The molecules and pathways unique to nematode development and parasitism can serve as the targets for the next generation of antinematode drugs and vaccines.

Chemotherapeutic Targets

The differences in the metabolism of parasites and their host can serve as useful tool for chemotherapy of parasitic diseases. Identification of parasitic specific molecular targets is a more rational approach for chemotherapy of parasitic disease. In filaria most of the antifilarials are effective against the mf stage and exert less effect on adult worms. The benzimidazole derivatives mebendazole and flubendazole have wide spectrum of anthelmintic action [134]. The mebendazole is effective against adults and mf of *O. Volvulus* and it inhibits microtubule assembly of parasites [135]. The enzymes of glycolytic pathway and PEP-Succinate pathway of helminth parasites are inhibited by benzimidazole derivatives. The fumarate reductase system which serves as a respiratory chain in many helminths appears to be a target for antihelmintics viz., cambendazole, thiabendazole, tetramisole etc. Phosphofructokinase of *B. malayi, L. carvini, D. immitis* and *D. viteae* are specifically inhibited by stibophen (a trivalent antimonials) as compared to host enzyme [136], cyanine-863 inhibits glucose utilization, lactic acid production and aldolase activity of *C. hawkingi, L. carinii* and *S. cervi* [137].

Diethylcarbamazine acts as microfilaricidal against *W. bancrofti, D. viteae, B. malayi* and *O. volvulus*. But have some macrofilaricidal effect against adults of *W. bancrofti* and *B. malayi* [135, 138]. The microfilaricidal action of DEC is mediated through immunological system, antibodies, complement and eosinophils [139]. The antifilarial enhances the cell adhesion and clump formation of entangled mf in presence of immune serum and leucocytes. DEC-N-oxide, a major metabolic of DEC, exhibited similar effect without clump formation [140]. Severe immunological side reactions of allergic nature are encountered in some individual on administration of DEC which might be due to release of histamine from the mf which have been shown to contain high amount of histamine [98]. In addition to above effects of DEC direct action on mf and adult parasites has been observed in *O. volvulus* [141]. Cellular damage has been observed in mf obtained from DEC treated animals [142].

Avermectins obtained from mycelia of *Streptomyces avermitilis* are effective against nematodes and destroy mf of *D. immitis*. It blocks the transmissions of command from venteral neuron to motorneuron [143] and enhances GABA release and its binding to post synaptic GABA receptors [144].

Levaunisole, a broad spectrum antihelmintic is effective against *B. malayi, B. pahangi, W. bancrofti, L. carinii* and *S. cervi.* The drug effects neuromuscular system and causes tetamic paralysis of adults and mf stages. It decreased the level of glycogen phosphorylase and increased the levels of glycogen synthetase in parasites [145].

Wolbachia, a bacterial endosymbiont is essential for the growth, development, fertility and survival of the causative agents of lymphatic filariasis. Doxycycline, either alone or in combination with antifilaria therapy, strongly reduced *Wolbachia* count in *B. malayi* and led to a decrease in microfilaremia that was sustained for at least 1 year after treatment. Treatment of bancroftian filariasis with a 4-week, 6-week or 8-week course of a 200mg/day dose of doxycycline results in long term sterility and eventual death of adult worms [146, 147]. At the individual level, doxycycline might be a suitable alternative for individual treatment. Doxycycline is not appropriate for mass drug administration [148].

Most anthelmintic act directly or indifectly by inhibiting either neuromuscular transmission or energy generation, hence for understanding the mode of action of drugs and design, knowledge about energy generating pathway is essential. LDH, MDH of *S. cervi O. volvulus* are strongly inhibited by suramin. NADP-dependent malic enzyme of *O. volvulus* and *D. immitis* are specifically inhibited by suramin [52, 54, 149]. Aurin tricarboxylic acid is a potent inhibitor of LDH and MDH of *S. cervi* [54]. The inhibition of LDH and MDH in filarial parasites causes inhibition of NADH reoxidation generated by glycraldehyde-3-phosphate dehydrogenase leading to inhibition of glycolysis in the parasites.

Protein kinases might serve as usual targets for chemotherapeutic attack [150]. Suramin inhibited AMP-dependent protein kinases in *O. volvulus*, while a stibophen sensitive protein kinase has been reported from *D. immitis* and *S. mansoni* [52, 151]. Protein kinases from adults and different developmental stages of *B. malayi, D. viteae* and *S. cervi* are sensitive to inhibition by suramin. Photvitin-phosphorylating protein kinases of *S. cervi* are selectively inhibited by suramin [69].

Polyamines play important role in growth and differentiation of all living organism. The enzymes of polyamine metabolism differ from the host enzyme in their properties and response to inhibitors, thereby providing a discriminative target for alternative chemotherapy [152, 153]. The inhibitors of polyamine metabolism may block differentiation and cellular development of helminth and act as possible agents for blocking the infections in the host. S-adenosyl –methionine decarboxylase (SAMDC) a key regulatory enzyme of the polyamine biosynthesis is considered as a potentially important target for chemotherapy of filarial infection. Various inhibitors of SAMDC like verenil and aromatic methyl glyoxal bis (guanil hydrazone) analogues might have potential as drug candidates against filarial worms [154]. The *in vitro* treatment of adult filariae with polyamine analogues and inhibitors of enzymes involved in the polyamine biosynthesis were effective in killing the parasites [155, 156].

Folate metabolism of filarial parasite (*B. pahangi, D. immitis*) also differ from their hosts but like of folate. The parasites can however catalyse direct conversion of 5 methyl FH4 to 5, 10 methyl FH4 unlike the hosts and they do not use methyl FH4 for synthesis of methiomine. Thymidylate synthase of these parasites can also act as alternate target [157]. In conformity with folate insufficiency of filariids, the growth and development of *B. malayi* L_3 in mosquitoes in enhanced by prior feeding of folic acid or p-aminobenzoic acid to *Aedes aegypti* [158].

Glutamate-cysteine ligase is the rate limiting enzyme of GSH biosynthesis which catalyses the ligation of the L-glutamate and L-cysteine to form L-γ-glutamyl-L-cysteine [159]. This enzyme has been well characterized in *O. volvulus* and *S. cervi* [76, 160]. The Ki values of *O. volvulus* GCL for buthionine sulfoximine (BSO) and cystamine were found to be lower as compared to host [76]. Hence, this enzyme represents a potential target for antifilarial drug development.

The mode of action of macrofilaricidal arsenicals is principally through the depletion of filarial GSH [161-163]. *L. carinii* glutathione reductase (GR) was found to be much more vulnerable to inhibition with arsenical melarsen oxide than the enzyme isolated from human erythrocytes [163]. Similarly, studies on the inhibition of GR by melarsen oxide revealed that human erythrocyte GR is less susceptible to inhibition by the arsenical than the enzyme from two cattle filariae viz. *S. digitata* and *Onchocerca gutturosa* along with differences in mechanisms of inhibition [164]. These differences between host and parasite enzyme might reflect differences in the primary and secondary structures of the proteins that might be exploitable for the design of new specific filaricidal drugs.

Glutathione-S-transferases are one of the major detoxification systems ubiquitous among eukaryotes and have been found in a wide range of parasitic helminths [165]. Helminths lack the important cytochrome P-450 dependent detoxification system, enhancing the requirement for phase II GST activity [100, 166]. They perform functions ranging from catalysing the detoxification of electrophilic compounds to protecting against peroxidative damage [167]. The compounds that can be detoxified by GSTs include the cytotoxic products of lipid peroxidation, such as lipid hydroperoxides and reactive carbonyls. Such compounds, in the absence of GSTs, cause membrane damage. GSTs are therefore postulated to protect the parasite against host-mediated lipid peroxidation of the membrane [168]. The inhibition of parasite GST(s) thus deprives the parasite of its major defence against oxidative stress and impairs its ability to survive. Filarial GST(s) is therefore important target for antifilarial drug design.

The Glutathione peroxidase (GPX) family is well characterized in some filariids. Recombinant *B. pahangi* gp29 was shown to reduce fatty acid and phospholipid hydroperoxides but was not active with H_2O_2 as the substrate [169]. The nucleotide sequences encoding homologues of this enzyme were also found in *B. malayi* and *W. bancrofti*. These enzymes belong to the group of selenium-independent GPXs. Selenium containing GPXs, a major family of enzymes in mammals appear to be absent from filarial species [170, 171]. GPXs protect the filarial worms from oxidative damage, and are thus important targets for novel chemotherapy.

Trehalose present in many nematode species but absent in mammals is involved in interaction with external environment, osmoregulation, resistance to desiccation, cryopreservation and egg-hatching. Trehalose also functions as a reserve carbohydrate fuel for energy metabolism. The combined action of two enzymes, i.e. trehalose 6-phosphate synthase and trehalose 6-phosphate phosphatase, catalyse the synthesis of trehalose in most organisms. These enzymes have been detected in adult *L. carinii*, *B. pahangi* and *D. viteae* but the processes regulating their activity are unknown [172]. Trehalose metabolism may provide new targets for attacking nematodes parasitic in mammals and studies indicate that trehalose is synthesized from glucose [35, 36, 173].

Chitin metabolism has also been proposed to be a parasite unique target, as the vertebrate host does not contain chitin [174, 175]. Chitinases have been identified in three separate stages in the filarial life cycle [174, 176, 177]. In adult females of *Onchocerca gibsoni* this enzyme plays a role in embryogenesis [174].

The nematode cuticle consists of a network of collagen molecules, which are primarily held together by intermolecular disulfide bonds [178]. The central enzyme involved in the biosynthesis of collagen is prolyl-4-hydroxylase [179] which catalyses the post-transcriptional oxidation of proline to 4-hydroxyproline in nascent collagen chains. This enzyme has been a subject of acute concern as a potential chemotherapeutic target, because molecules inhibiting its activity might be expected to be relatively specific inhibitors of collagen biosynthesis [180, 181].

Among the various enzymes identified for drug development against parasitic disease, DNA topoisomerase II has been picked up as a novel target for antifilarial drug development due to several reasons. Eukaryotic DNA topoisomerase I has been identified as the primary target for the antineoplastic alkaloid camptothecin, whereas DNA topoisomerase II is the target for many anticancer agents including both non intercalating (VM-26) and interacalating (m-AMSA) compounds. Antibacterial agents coumarins and quinolones, are inhibitors of DNA-gyrase. The DNA topoisomerase inhibitors exerted significant effect and antifilarial compounds suramin and ivermectin proved to be strong inhibitors of the parasitic enzyme suggesting the potentials of the enzyme as drug target and designing of novel compounds against adult parasite [89, 182]. Various compounds viz., glycoconjugates, amonosugars proved to be strong inhibitors of filarial DNA topoisomerases. Various compounds having different standard moiety were synthesized and found to inhibition the topo activity. The 7-O-acetamidyl-4-alkyl- 2H-1-benzopyran-2-ones, glycosylated beta-amino acid derivatives and Ru (II) poly-pyridyl hydridocarbonyl complexes were prepared and screened against DNA topoisomerase II of filarial parasite *S. cervi* [88, 182, 183].

Conclusions

The limited knowledge of biochemistry of human filarial parasites is a major impediment in identification of promising novel chemotherapeutic targets. All the parasites examined so far employ predominantly anaerobic metabolism of carbohydrate as major energy yielding pathway and do not completely oxidize the substrate to carbon dioxide due to rudimentary electron transport chain. Carbohydrate, folate and glutathione metabolism of filarial parasites have been studied in detail and represent the target for many antifilarials. No unique drug target for amino acid and lipid metabolism has been identified in helminthes. The amine and polyamine metabolism in filarial parasites are important chemotherapeutic targets but detailed studies of these pathways are required. DNA topoisomerase of human and other filarial parasites have been demonstrated as new target for nucleic acid metabolism.

With availability of complete genomic data of *Brugia malayi* many potential chemotherapeutic targets can be identified and validated in future. The reverse genetic technique of RNA interference (RNAi) is helpful in identifying genes that are involved in parasitism phenotypes and in understanding in *vivo* role of individual genes. However, there is still a large gap between target identification and validation on one hand and lead discovery and optimization on the other hand. The biochemical and structural studies of the targets and target inhibitor interaction might be useful in filling this gap. The understanding of immunopathologic and protective role of B and T cell responses to specific parasitic antigens will be helpful in development of a vaccine to eradicate or control the infection in endemic areas.

Acknowledgements

Financial assistance from Council for Scientific and Industrial Research, New Delhi (India) in the form of Senior Research Fellowship to SG is greatly acknowledged.

CDRI Manuscript number: 171/2011/JKS.

References

[1] WHO. Global Programme to Eliminate Lymphatic Filariasis. *Wkly Epidemiol Rec*, 2005, 80, 202-212.

[2] WHO. Global Programme to Eliminate Lymphatic Filariasis. *Wkly Epidemiol Rec,* 2009, 84, 437-444.

[3] WHO. Global programme to eliminate lymphatic filariasis (GPELF), Progress report 2000-2009 and strategic plan 2010-2020. 2010.

[4] Michael, E; Bundy, DAP; Grenfell, BT. Re-assessing the global prevalence and distribution of lymphatic filariasis. *Parasitology,* 1996a, 112, 409-428.

[5] Ottesen, EA; Hooper, PJ; Bradley, M; Biswas, G. The Global Programme to Eliminate Lymphatic Filariasis: health impact after 8 years. *PLoS Negl Trop Dis,* 2008, 2, e317.

[6] Bockarie, MJ; Deb, RM. Elimination of lymphatic filariasis: do we have the drugs to complete the job. *Current Opinion in Infectious Diseases*, 2010, 23, 617-620.

[7] Michael, M; Bundy, DAP. Global mapping of lymphatic filariasis. *Parasitol Today* 1997, 13, 472-476.

[8] WHO. Lymphatic filariasis infection and disease: control stratragies. *TDR/CTD/ FIL/PENANG/94.1*, 1994.

[9] Wilson, T; Edeson, JFB; Wharton, RH; Reid, JA; Turner, LH; Laing, ABG. The occurrence of two forms of *Wuchereria malayi* in man. *Trans R Soc Trop Med Hyg,* 1958, 52, 480-1.

[10] Kanjanopas, K; Choochote, W; Jitpakdi, A; Suvannadabba, S; Loymak, S; Chungpivat, S; Nithiuthai, S. *Brugia malayi* in a naturally infected cat from narathiwat province, southern thailand. *Southeast Asian J Trop Med Public Health*, 2001, 32, 585-587

[11] Sudjadi, FA; Soeyoko; Noerhajati, S. Diurnally subperiodic and non-periodic *Brugia* type in Balikpapan, East Kalimantan, Indonesia. *Southeast Asian J Trop Med Public Health,* 1984, 15, 425-6.

[12] Chansiri, G; Khawsak, P; Phantana, S; Sarataphan, N; Chansiri, K. The efficacy of a single-oral-dose administration of ivermectin and diethylcarbamazine on the treatment of feline *Brugia malayi*. *Southeast Asian J Trop Med Public Health,* 2005, 36, 1105-1109.

[13] Rajan, TV; Gundlapalli, AV. Lymphatic filariasis. *Chemical Immunology,* 1997, 66, 125-158.

[14] Taylor, MJ; Hoerauf, A; Bockarie, MJ. Lymphatic filariasis and onchocerciasis. *Lancet,* 2010, 375.

[15] Seim, AR; Dreyer, G; Addiss, D. Controlling morbidity and interrupting transmission: twin pillars of lymphatic filariasis elimination. *Revista da Sociedade Bras de Med. Trop,* 1999, 32(3), 325-328.

[16] Smith, HL. Investigating development of infective stage larvae of filarial nematodes. *Frontiers in Bioscience,* 2000, 5, e95-102.

[17] Mattingly, PE. The biology of mosquito-borne disease. George Allen and Unwin Ltd., London. 1969.

[18] Nanduri, J; Kazura, JM. Clinical and Laboratory Aspects of Filariasis. *Clinical Microbiology Reviews*, 1989, 2, 39-50.

[19] Wharton, RH. The biology of Mansonia mosquitoes in relation to the transmission of filariasis in Malaya. *Bull. Inst. Med. Res. Fed.* Malaya, 1962, 11, 114.

[20] Hawking, F. The 24-hour periodicity of microfilariae: biological mechanisms responsible for its production and control. *Proc. R. Soc. London Ser. B*, 1967, 169, 56-76.

[21] Hawking, F. The distribution of human filariasis throughout the world. II. Asia. W.H.O. publication WHO/FIL/73.114. World Health Organization, Geneva, 1973.

[22] Hawking, F; Gammage, K. The periodic migration of microfilariae of Brugia malayi and its response to various stimuli. *Am. J. Trop. Med. Hyg.*, 1968, 17, 724-729.

[23] Hawking, F; Pattaayak, S; Sharma, HL. The periodicity of microfilariae. 1. The effect of body temperature and other stimuli upon the cycles of Wuchereria bancrofti, Brugia malayi, B. ceylonensis and Dirofilaria repens. *Trans. R. Soc. Trop. Med. Hyg.*, 1966, 60, 497-513.

[24] Turner, LH; Edeson, JFB. Studies on filariasis in Malaya: the periodicity of the microfilariae of Wuchereria malayi. *Ann. Trop. Med. Parasitol.*, 1957, 51, 271-277.

[25] Chea, SN; Howells, RE. The uptake in vitro of dyes, monosaccharides, and amino acids by the filarial worm Brugia pahangi. *Parasitology*, 1979, 78, 343-354.

[26] Cherian, PV; Stromberg, BE; Weiner, OJ; Soulsby, EJL. Fine structure and cytochemical evidence for the presence of polysaccharide surfacecoat of *Dirofilaria immitis* microfilariae. *Intemational Journal of Parasitology,* 1980, 10, 227-233.

[27] Fuhrman, A; Ash, LR. Analysis of *Brugia pahangi* surface carbohydrates: Comparison of the binding of a panel of fluoresceinated lectins to mature *in vivo* derived and immature in utero-derived microfilariae. *Acta Tropica* (Basel), 1983, 40, 45-51.

[28] Kaushal, NA; Simpson, AJG; Hussain, R; Ottesen, EA. *Brugia malayi:* Stage-specific expression of carbohydrates containing N-acetyl-D-glucosamine on the sheathed surfaces of microfilariae. *Experimental Parasitology,* 1984, 58, 182-187.

[29] Ghatak, S; Saxena, JK; Kausal, NA; Shukla, OP. Biochemistry of filarial parasites in relation to antifiarial drug design. *Perspectives in Parasitology,* 1988, 2, 80-113.

[30] Simpson, MG; Laurence, HR. Histochemical studies on microfilariae. *Parasitology,* 1972, 64, 6 -68.

[31] Fuhrman, A; Ash, LR. Characterisation of the exposed carbohydrates on the sheath surface of *in vitro-derived Brugia pahangi* microfilariae by analysis of lectin binding. *Journal of Parasitology,* 1983, 69, 1043-1047.

[32] Fuhrman, A; Ash, LR. Analysis of *Brugia pahangi* surface carbohydrates: Comparison of the binding of a panel of fluoresceinated lectins to mature *in vivo* derived and immature in utero-derived microfilariae. *Acta Tropica* (Basel), 1983, 40, 45-51.
[33] Longworth, DL; King, DC; Weller, PF. Rapid uptake and esterification of arachidonic acid and other fatty acids by microfilariae of Brugia malayi. *Mol. Biochem. Parasitol.*, 1987, 23, 275-284.
[34] Longworth, DL; Monaham-Earley, RA; Dvorak, AM; Weller, PF. Brugia malayi: arachidonic acid uptake into lipid bodies of adult parasites. *Exp. Parasitol.*, 1988, 65, 251-257.
[35] Powell, JW; Stables, JN; Watt, RA. An investigation of the glucose metabolism of *Brugia pahangi* and *Dipetalonema viteae* by nuclear magnetic resonance spectroscopy. *Molecular and Biochemical Parasitology*, 1986, 18, 171-182.
[36] Powell, JW; Stables, JN; Watt, RA. An NMR study on the effect of glucose availability on carbohydrate metabolism in *Dipetalonema viteae* and *Brugia pahangi. Molecular and Biochemical Parasitology*, 1986, 19, 265-271.
[37] Wang, EJ; Saz, HJ. Comparative biochemical studies of *Litomosoides carinii, Dipetalonema viteae* and *Brugia pahangi* adults. *Journal of Parasitology,* 1974, 60, 316-321.
[38] Franke, ED; Weinstein, PO. In vitro cultivation of Dipetalonema viteae third-stage larvae: evaluation of culture media, serum and other supplements. *J. Parasitol,* 1984, 70, 618-628.
[39] Weinstein, PP. Filariasis: problems and challenges. *Am. J. Trop. Med. Hyg.* 1986, 35, 221-233.
[40] Petralanda, I; Yarzabal, L; Piessens, WF. Studies on a filarial antigen with collagenase activity. *Mol. Biochem. Parasitol.* 1986, 19, 51-59.
[41] von Brand, T. Biochemistry of parasites, 2nd edition. Academic Press, New York. 1973.
[42] Srivastava, VML; Ghatak, S. Utilization of mannose, galactose and fructose by *Chandlerella hawkingi. Indian Journal of Experimental Biology,* 1974, 12, 472-473.
[43] Lehninger, AL. Biochemistry. 4th edition, W.H. freeman and company, New York, 2006.
[44] Anwar, N; Ansari, AA; Ghatak, S; Krishna Murti, CR. *Setaria cerrvi :* Enzymes of glycolysis and PEP-succinate pathway. *Zeitschriftfur Parasitenkunde.* 1977, 51, 215-283.
[45] Hutchison, WF; Turner, AC; Oalshlegel, FJJr. Hexokinase of 'the adult dog heart worm, *Dirofilaria immitis. Comparative Biochemistry and Physiology,* 1977, 58B: 131-134.
[46] Hutchison, WF; Turner, AC. Glycerol metabolism in, the adult dog heartwvrm, *Dirofilaria immitis. Comparative Biochemistry and Physiology,* 1979, 64B, 399-401.
[47] Singh, AR; Joshi, S; Arya, R; *et al.* Molecular cloning and characterization of Brugia malayi hexokinase. Parasitol International, 2008.
[48] Middleton, KR; Saz, HJ. Comparative utilization of pyruvate by *Brugia pahangi, Dipetalonema viteae,* and *Litomosoides carinii. Joumal of Parasitology,* 1979, 65, 1-7.
[49] Simpson, JW; Awapara, J. The pathway of glucose degradation in some invertebrates. *Comparative Biochemistryand Physiology,* 1966, 18, 537-548.

[50] Bueding, E; Saz, UJ. Pyruvate kinase and phosphoenolpyruvate carboxykinase activities of ascaris muscle, *Hymenolepis diminuta* and *Schistosoma mansoni. Comparative Biochemistry and Physiology,* 1968, 24, 511-518.

[51] Rathaur, S; Anwar, N; Saxena, JK; Ghatak, S. *Setaria cerv!:* Enzymes in microfilariae and *in vitro* action of antifilarials. *Zeitschriftfur Parasitenk!lnde,* 1982, 68, 331-338.

[52] Walter, RD; Schulz-Key, H. *Onchocerca volvulus:* Effect of suramin on lactate dehydrogenase and malate dehydrogenase. *Tropenmedizin und Parasitology,* 1980, 31, 55-58.

[53] Bharadwaj, N; Saxena, JK; Ghatak, S; Singh, C. Kinetic and macroionic characteristics of soluble and immobilized malate dehydrogenase of *Setaria cervi. Biological Memoirs,* 1987, 13, 29-36.

[54] Saxena, JK; Bharadwaj, N; Kaushal, NA; Ghatak, S. Purification and characterization of malate dehydrogenase from *Setaria cerai. Indian Journal of Parasitology,* 1986, 10, 93-100.

[55] Saxena, JK; Bharadwaj, N; Rathaur, S; Ghatak, S. Purification and a characterization of lactate dehydrogenase from *Setaria cervi. Indian Journal of Medical Research,* 1986, 84, 264-269.

[56] Lee, CC; Miller, JH. Fine structure of the intestinal epithelium. of *Dirofilaria immitis* and changes occurring after vermicidal treatment with caparsolate sodium. *Journal of Parasitology,* 1969, 55, 1035-1045.

[57] Hayashi, H; Oya, H. Comparative study on the alternative electron .transport pathway in *Clonorchis sinensis, Dirofilaria immitis* and *Angiostrongylus cantonensis. 4th International Congress of Parasitology,* Warsaw F, 1978, 73-74.

[58] Jaffe, JJ; Chrin, LR. Involvement of tetrahydrofalte cofactor in De novo Purine ribonucleotide synthesis by adult *Brugia pahangi* and *Dirofilaria immitis. Molecular and Biochemical parasitology,* 1981, 2, 259-270.

[59] Langer, BW; Jiampermpoon, D. Some metabolic processes in the microfilarial form *ef Dirofilaria immitis. journal of Parasitology,* 1970, 56,144-145.

[60] Singh, RP; Saxena, JK; Saxena, KC; Ghatak, S. Amino acid metabolism of *Seteria cervi Indian Joumal of Parasitology,* 1988, 12, 21-28.

[61] Jaffe, JJ. Filarial folate-related metabolism as a potential target for selective inhibitors in The Host Invader Interplay, (H.Vanden Fossche ed.) pp. 605-614. Elsevier Biomedical Press, Amsterdam. 1980.

[62] Bankov, I; Timanova, A; Barrett, J. Methionine and cysteine metabolism in *Fasciola hepatica. International Journal for Parasitology*, 1996, 26, 1401-1404.

[63] Gomez-Bautista, M; Barrett, J. Cysteine metabolism in the cestode *Hymenolepis diminuta. Parasitology*, 1988, 97, 149-159.

[64] Walker, J; Barrett, J. Cystathionine β-synthase and γ-cystathionase in helminths. *Parasitology Research*, 1991, 77, 709-713.

[65] Singh, K; Sharma, S. Some biochemical studies on the non-specific acid phosphomonoesterase of the parasitic nematode *Setaria cervi.* (Rudolfi, 1819). *Revista de Parasitologia*, 1981, 42, 425-432.

[66] Saxena, JK; Singh, RP; Ghatak, S. Distribution pattern (hydrolytic enzymes in *Setaria cervi. Indian Journal of Parasitology,* 1986, 10, 21-26.

[67] Cohen, P. The role of protein phosphorylation in neural and hormonal control of cellular activity. *Nature,* 1982, 296, 613-620.

[68] Donahae, MJ; LaHoussaye, BA; Harris, BG; Masaracchia, RA. Regulation of glycogenolysis by adenosine-3', 5'-monophosphate in *Ascaris suun.* muscle. *Comparative Biochemistry and Physiology,* 1981, 69B, 693-699.

[69] Saxena, JK; Srivastava, AK; Murti, PK; Chatterjee, RK; Ghatak, S; Walter, RD. Protein kinases in different life stages of *Brugia malayi and* other filarial worms. *Tropenmedizin* and *Parasitology,* 1984, 35, 174-176.

[70] Brophy, PM; Pritchard, DI. Immunity to helminths: Ready to tip the biochemical balance? *Parasitology Today*, 1992, 8: 419-422.

[71] Callahan, HL; Crouch, RK; James, ER. Helminth antioxidant enzymes: A protective mechanism against host oxidants? *Parasitology Today*, 1988, 4, 218-225.

[72] Lomaestro, BM; Malone, M. Glutathione in health and disease: Pharmacotherapeutic issues. *Annals of Pharmacotherapy*, 1995, 29, 1263-1273.

[73] Zhang, LP; Maiorino, M; Roveri, A; Ursini, F. Phospholipid hydroperoxide glutathione peroxidase: specific activity of tissue of rats of different age and comparison with other glutathione peroxidases. *Biochimica et Biophysica Acta*, 1989, 1006, 140-143.

[74] Carlberg, I; Mannervik, B. Glutathione reductase. *Methods in Enzymology*, 1985, 13, 484-490.

[75] Schirmer, RH; Schollhammer, T; Eisenbrand, G; Krauth-Siegel, RL. Oxidative stress as a defence mechanism against parasitic infections. *Free Radical Research Communications*, 1987, 3, 3-12.

[76] Lüersen, K; Müller, S; Hussein, A; Liebau, E; Walter, RD. The γ-glutamylcysteine synthetase of *Onchocerca volvulus. Molecular and Biochemical Parasitology*, 2000, 111, 243-251.

[77] Singh, SN; Srivastava, AK; Chatterjee, RK. Gamma-glutamyl transpeptidase in adult *Setaria cervi* and *Acanthocheilonema viteae* and the effect of inhibitors. *Journal of Parasitic Diseases*, 1996, 20, 163-166.

[78] Kozak, EM; Tate, SS. Glutathione-degrading enzymes of the microvillus membranes. *Journal of Biological Chemistry*, 1982, 257, 6322-6327.

[79] Seelig, GF; Meister, A. Cystamine-sepharose: A probe for the active site of γ-glutamylcysteine synthetase. *Journal of Biological Chemistry*, 1982, 257, 5092-5096.

[80] Tate, SS; Meister, A. γ-Glutamyl transpeptidase from kidney. *Methods in Enzymology*, 1985, 113, 400-419.

[81] Gupta, S; Srivastava, AK. Biochemical targets in filarial worms for selective antifilarial drug design. *Acta Parasitologica,* 2005, 50(1), 1-18.

[82] Comley, JCW; Jaffe, JJ; Chrin, LR; Smith, RB. Synthesis of Ubiquinone 9 by adult *Brugia pahangi* and *Dirofilaria immitis:* evidence against its, involvement in the oxidation of 5-methyltetrahydrofolate. *Molecular and Biochemical Parasitology.* 1981, 2, 271-283.

[83] Chen, SN; Howells, RE. The uptake *in vitro* of monosaccharides, disaccharides, and nucleic acid precursors by adult *Dirofilaria immitis. Annals of Tropical.Medicine and Parasitology,* 1981, 75, 229-234.

[84] Jaffe, JJ; MeCormack, JJ; Meymarian, E. Comparative properties of systosomal and filarial dihydrofolate reductases. *Biochemical Pharmacology,* 1972, 21, 719-731.

[85] Jaffe, JJ; Doremus, HM. Metabolic patterns of *Dirofilaria immitis* microfilariae in vitro. *Journal of Parasitolgy,* 1970, 56, 254-260.

[86] Howells, RE; Chen, SN. *Brugia pahangi:* feeding and nutrient uptake *in vitro* and *in vivo. Experimental Parasitology,* 1981, 51, 42-58.

[87] Simpson, MG; Laurence, BR. Incorporation of radioactive precursors into filarial larvae of *Brugia* developing in susceptible and refractory mosquitoes. *Journal of Parasitology,* 1979, 65, 732-736.

[88] Tripathi, RP; Rastogi, SK; Kundu, B; Saxena, JK; Reddy, VJM; Srivastava, S; Chandra, S; Bhaduri, AP. Identification of inhibitors of DNA topoisomerases II from a synthetic library of glycoconjugates. *Combinatorial Chemistry and High Throughput Screening*, 2001, 4, 237-244.

[89] Reddy, VJM; Singh, AR; Joshi, S; Khan, MY; Saxena, JK. Topoisomerase II of Filarial parasite *Setaria cervi*: Purification and Characterization. *Helminthologia*, 2009, 46, 67-72.

[90] Frahya, GJ; Smith, JD. Lipid metabolism in parasitic helminths. *Advances in Parasitology,* 1983, 22, 309-387.

[91] Ansari, AA; Yusufi, ANK; Siddiqi, M. Lipid composition of the filarial parasite *Setaria cervi. Journal of Parasitology,* 1973, 59, 939-940.

[92] Rathaur, S; Anwar, N; Ghatak, S. Biochemical composition of nlicrofilarial and adult stages of *Setaria cervi. Zeitschriftfur Parasitenkunde,* 1980, 62, 85-93.

[93] Hutchlson, WF; Turner, AC; Grayson, DP; White, HB. Lipid analysis of the adult dog heartworm. *Dirofilaria immitis. Comparative Biochemistry and Physiology,* 1976, 53B, 495-497.

[94] Comley, JCW; Jaffe, JJ. .Isoprenoid biosynthesis in adult *Brugia pahangi and Dirofilaria immitis. Journal of Parasitology,* 1981, 67, 609-616.

[95] Turner, AC; Hutchison, WF. Lipid synthesis in the adult dog heartworm, *Dirofilaria immitis. Comparative Biochemistry and Physiology,* 1979, 64B, 403-405.

[96] Srivastava, AK; Jaffe, JJ; Lambert, RA. Phosphatidylethanolamine synthesis in adult *Dirofilaria immitis* females. *International Journal for Parasitology,* 1985, 15, 429-433.

[97] Rathaur, S; Kaushal, NA; Ghatak, S. Biogenic amines and their metabolites in microfilariae and adults of *Setaria cervi. Indian Journal of Parasitology,* 1985, 9, 203-205.

[98] Saxena, JK; Bose, SK; Sen, R; Chatterjee, RK; Sen, AB; Ghatak,.S. *Litomosoides carinii :* biogenic amines in microfilariae and adults.· *Experimental Parasitology,* 1977, 43, 239-243.

[99] Singh, DP; Rathaur, S; Misra, S; Chatterjee, RK; Ghatak, S; Sen, AB. Studies on the causation of adverse reactions in microfilariaemic host following diethylcarbamazine therapy *(Dipetalonema viteae* in *Mastomys natalensis). Tropenmedizin und Parasitology,* 1985, 36, 21-24.

[100] Saxena, JK; Sen, R; Chatterjee, RK; Ghatak, S. The uptake of 5-hydroxytryptamine by *Litomosoides carinii* and the effect of antifilarials. *Tropical Medicine* (Japan), 1987, 29, 187-196.

[101] Agarwal, A; Mishra, SK; Rathaur, S; Chatterjee, RK; Katiyar, JC; Ghatak, S. Dopamine-β-hydroxylase in parasitic nematodes. *Indian Journal of Parasitology,* 1982, 6, 223-225.

[102] Tabor, CW; Tabor, H. Polyamines. *Annual Review of Biochemistry*, 1984, 53, 749-790.

[103] Pegg A.E., Kameji T., Shirahata A., Stanley B., Madhubala R., Pajunen A. Regulation of mammalian S-adenosylmethionine decarboxylase. *Advances in Enzyme Regulation*, 1988, 27, 43-55.

[104] Svensson, F; Kockum, I; Persson, L. Diethylglyoxal bis(guanylhydrazone), a potent inhibitor of mammalian S-adenosylmethionine decarboxylase. Effects on cell proliferation and polyamine metabolism in L1210 leukemia cells. *Molecular and Cellular Biochemistry*, 1993, 124, 141-147.

[105] Brownlee, DA; Holden-Dye, L; Walker, R. The range and biological activity of FMRC amide-related peptides and classical neurotransmitters in nematodes. *Advances in Parasitology*, 2000, 45, 109-180.

[106] Srivastava, DK; Roy, TK; Shukla, OP. Polyamines of helminths. *Indian Journal of Parasitology*, 1980, 4, 187-189.

[107] Wittich, RM; Kilian, HD; Walter, RD. Polyamine metabolism in filarial worms. *Molecular and Biochemical Parasitology*, 1987, 24, 155-162.

[108] Walter, RD. Polyamine metabolism of filarial and allied parasites. *Parasitology Today*, 1988, 4, 18-20.

[109] Singh, RP; Saxena, JK; Ghatak, S; Shukla, OP; Wittich, RM; Walter, RD. Polyamine metabolism in *Setaria cervi*. *Parasitology Research*, 1989, 75, 311-315.

[110] Sharma, V; Tekwani, BL; Saxena, JK; Gupta, S; Katiyar, JC; Chatterjee, RK; Ghatak, S; Shukla, OP. Polyamine metabolism in some helminth parasites. *Experimental Parasitology*, 1991, 72, 15-23.

[111] Ghedin, E; *et al.* 2007. Draft Genome of the Filarial Nematode Parasite *Brugia malayi*. *Science,* 2007, 317, 1756-1760.

[112] Sakaguchi, Y; Tada, I; Ash, LR; Aoki, Y. *J. Parasitol,* 1983, 69, 1090.

[113] Stein, LD; *et al.* The genome sequence of *Caenorhabditis briggsae*: a platform for comparative genomics. *PLoS Biol.*, 2003, 1(2), E45.

[114] Ghedin, E; Wang, S; Foster, JM; Slatko, BE. First sequenced genome of a parasitic nematode. *Trends Parasitol,* 20, 2004, 151-153.

[115] Perler, FB; Karam, M. Cloning and characterization of two Onchocerca volvulus repeated DNA sequences. *Mol. Biochem. Parasitol.* 1986, 21, 171-178.

[116] Shah, JS; Karam, M; Piessens, WF; Wirth, DF. Characterization of an Onchocerca-specific DNA clone from Onchocerca volvulus. *Am. J. Trop. Med. Hyg.*, 1987, 37, 376-384.

[117] McReynolds, LA; DeSumone, SM; Williams, SA. Cloning and comparison of a repeated DNA sequence from the human filarial parasite Brugia malayi and the animal parasite Brugia pahangi. *Proc. Natl. Acad. Sci.* USA, 1986, 83, 797-801.

[118] Shah, JS; Lamontagne, L; Unnasch, TR; Wirth, DF; Piessens, WF. Characterization of a ribosomal DNA clone of B. malayi. *Mol. Biochem. Parasitol.*, 1986, 19, 67-75.

[119] Sim, BKL; Mak, JW; Cheong, WH; Sutanto, I; Kurniawan, L; Marwoto, HA; Franke, E; Campbell, JR; Wirth, DF; Identification of Brugia malayi in vectors with a species-specific DNA probe. *Am. J. Trop. Med. Hyg.* 1986, 35, 559-564.

[120] Nilsen, TW; Maroney, PA; Goodwin, RG; Perrine, KG; Denker, AJ; Nanduri, J; Kazura, JW. Cloning and characterization of a potentially protective antigen in lymphatic filariasis. *Proc. Natl. Acad. Sci.* USA, 1988, 85, 3606-3607.

[121] Kazura, J; Kroeze, W; Perrine, K; Langford, C; Nilsen, T. *Clin. Res.*, 1988, 36, 579.

[122] Arasu, P; Philipp, M; Perler, F. Brugia malayi. Recombinant antigens expressed by genomic DNA clones. *Exp. Parasitol,* 1987, 64, 281-291.

[123] Murray, J; Gregory, WF; Gomez-Escobar, N; Atmadja, AK; Maizels, RM. *Mol. Biochem. Parasitol.,* 2001, 118, 89.

[124] Pastrana, DV; Raghavan, N; FitzGerald, P; Eisinger, SW; Metz, C; Bucala, R; Schleimer, RP; Bickel, C; Scott, AL. Filarial nematode parasites secrete a homologue of the human cytokine macrophage migration inhibitory factor. *Infect. Immun,* 1998, 66, 5955-5963.

[125] Gomez-Escobar, N; Lewis, E; Maizels, RM. A novel member of the transforming growth factor-beta (TGF-beta) superfamily from the filarial nematodes Brugia malayi and B. pahangi. *Exp. Parasitol*, 1998, 88, 200-209.

[126] Foster, JM; Zhang, Y; Kumar, S; Carlow, CKS. Mining nematode genome data for novel drug targets. *Trends in Parasitology,* 2005, 21, 101-104.

[127] Deehan, MR; Harnett, W; Harnett, MM. A filarial nematode-secreted phosphorylcholine-containing glycoprotein uncouples the B cell antigen receptor from extracellular signal-regulated kinase-mitigen-activated protein kinase by promoting the surface Ig-mediated recruitment of Src homology 2 domain- containing tyrosine phosphatise-1 and Pac-1 mitogen-activated kinase-phosphatase. *J. Immunol* 2001, 166, 7462-7468.

[128] Schonemeyer, A; Lucius, R; Sconnenburg, B. *et al.* Modulation of human T cell responses and macrophage functions by *onchocerca vovulus*. *J. Immunol,* 2001, 167, 3207-3215.

[129] Semnani, RT; Sabzevari, H; Iyer, R; Nutman, B. Filarial antigens impair the function of dendritic cells during differentiation. *Infect Immun,* 2001, 69, 5813-5822.

[130] Counni, AS; Spanel-Borowski, K; Palacios, M; *et al.* Pulmonary inflammation induced by a recombinant *Brugia malayi* gamm-glutamyl transpeptidase homolog: involvement of humoral autoimmune response. *Mol Med* 2001, 7, 344-354.

[131] Falcone, FH; Loke, P; Zang, X; *et al.* A *Brugia malayi* homolog of macrophage migration inhibitory factor reveals an important link between macrophages and eosinophil recruitment during nematode infection. *J. Immunol,* 2001, 167, 5348-5354.

[132] Gomez-Escobar, N; Bennett, C; Prieto-Lafuente, L; Aebischer, T; Blackburn, CC; Maizels, RM. Heterologous expression of the filarial nematode alt gene products reveals their potential to inhibit immune function. *BMC Biol*, 2005, 3-8.

[133] Blaxter, ML. Nematoda: genes, genomes and the evolution of parasitism. *Adv. Parasitol*, 2003, 54, 101-195.

[134] Chusattayanod, W; Denham, DA. Chemoprophylactic activity flubendazole against *Brugia pahangi* in jirds. *Joumal of Parasitology,* 1984, 70, 191-192.

[135] Goodwin, LG. Recent advances in research on filariasis: Chemotherapy. *Transactions of the Royal Society of Tropical Medicine and Hygiene,* 1984, 78(Supp), 1-8.

[136] Saz, HJ; Dunbar, GA. The effects of stibophen on phosphofructokinases and aldolases of adult filariids. *Joumal of Parasitology,* 1975, 61, 794-801.

[137] Saxena, JK; Ansari, AA; Ghatak, S. Purification and characterization of FDP-aldolase from *Setaria rervi. Indian Joumal.of Parasitology,* 1982, 61, 27-35.

[138] Ottesen, EA. Immunological aspects of lymphatic filariasis and Onchocerciasis in man. *Transactions of the Royal Society of Tropical .Medicine and Hygiene,* 1984, 78, 9-18.

[139] Hammerberg, B. Sheathed and unsheathed microfilarial surfaces, interactions with host serum proteins in diethylcarbamazine. *Tropical Medicine and Parasitology,* 1985, 36(Supp), 5-9.

[140] Chandrasekaran, B; Ghirnikar, SN; Harinath, BC. Effect of diethylcarbamazine and diethylcarbamazine-N-oxide on microfilariae *in vitro* in the presence ef immune sera and leukocytes. *Indian Journal of Experimental Biology,* 1980, 18, 1179-1180.

[141] Langham, ME; Kramer, TR. The *in vitro* effect of DEC on the motility and survival of *Onchocerca volvulus* microfilariae. *Tropenmedizin und Parasitology,* 1980, 31, 59-66.

[142] Mehlhorn, H; Steinhorst, U; Lindenbatt, B; Gruentzig, J; Thomas, H. Electronmicroscopic studies on microfilariae of *Dipetalonnna viteae* and *Litomosoides carinii:* The occurrence of intracellular parasites after treatment with metrifonate or Diethylcarbamazine. *Zentral Bakteriol Mikrobiol Hg Abt* 1, 1981, 251, 263-272.

[143] Kass, IS; Wang, CC; Walrond, JP; Stretton, AOW Avermectin Bla, a paralysing anthelmintic that affects interneurons and inhibitory motorneurons in Ascaris. *Proceedings National Academy Science,* USA, 1980, 77, 6211-6215.

[144] Pong, SS; Wang, CC. Avermectin fia modulation of α-aminobutyric acid receptors in rat brain membranes. *Journal of Neurochemistry,* 1982, 38, 375-379.

[145] Komuniecki, PR; Saz, HJ. The effect of levamisole on glycogen synthase and metabolism of *Litomosoides carinii. Journal qf Parasitology,* 1982, 68, 221-227.

[146] Taylor, MJ; Bandi, C; Hoerauf, A. Wolbachia bacterial endosymbionts of filarial nematodes. *Adv Parasitol,* 2005, 60, 245-284.

[147] Debrah, AY; Mand, S; Toliat, MR; *et al.* Plasma vascular endothelial growth factor-A (VEGF-A) and VEGF-A gene polymorphism are associated with hydrocele development in lymphatic filariasis. *Am J Trop Med Hyg*, 2007, 77, 601-608.

[148] Supali, T; Djuardi, Y; Pfarr, KM; Wibowo, H; Taylor, MJ; Hoerauf, A; Houwing-Duistermaat, JJ; Yazdanbakhsh, M; Sartono, E. Doxycycline Treatment of *Brugia malayi*–Infected Persons Reduces Microfilaemia and Adverse Reactions after Diethylcarbamazine and Albendazole Treatment. Clinical Infectious Diseases, 2008, 46, 1385-93.

[149] Walter, RD; Albiez, EJ. Inhibition of NADP-linked malic enzyme from *Onchocerca volvulus* and *Dirofilaria immitis* by suramin. *Molecular and Biochemical Parasitology,* 1981, 4, 53-60.

[150] Greengard, P. Phosphorylated proteins as physiological effectors. *Science,* 1978, 199, 146-152.

[151] Walter, RD; Buttner, DW. Inhibition of protein kinase activity from *Schistosoma mansoni* and *Dirofilaria immitis* by Stibophen. *National Conference of Parasitology, Varna abstract,* 1983, 214-215.

[152] Bacchi, CJ. Content, synthesis and formation of polyamines in Trypanosomatids: Relationship to Chemotherapy. *Journal of Protozoology,*1981, 28, 20-27.

[153] Sjoerdsma, A; Schechter, PJ. Chemotherapeutic implications of polyamine biosynthesis inhibition. *Clinical Pharmacology and therapeutics,* 1984, 35, 287-300.

[154] Da'dara, AA; Mett, H; Walter, RD. MGBG analogues as potential inhibitors of S-adenosylmethionine decarboxylase of *Onchocerca volvulus*. *Molecular and Biochemical Parasitology*, 1998, 97, 13-19.

[155] Müller, S; Lüchow, A; McCann, PP; Walter, RD. Effect of bis(benzyl)polyamine derivatives on polyamine transport and survival of *Brugia pahangi*. *Parasitology Research*, 1991, 77, 612-615.
[156] Müller, S; Wittich, RM; Walter, RD. The polyamine metabolism of filarial worms as chemotherapeutic target. *Advances in Experimental Medicine and Biology*, 1988, 250, 737-743.
[157] Jaffe, JJ. Report of a scientific working group on the biochemistry of filarial and allied parasites. WHO/OCP/Oct/83.1. 1983.
[158] Rao, UR; ChandraShekhar, R; Parals, PB; Subrahmanyam, D. The effect of p·aminobenzoic acid and folic acid on the development of infective larvae of *Brugia malayi* in *Aedes aegypti. Acta Tropica,* 1984, 41, 61-67.
[159] Plummer, JL; Smith, BR; Sies, S; Bend, JR. Chemical depletion of glutathione *in vivo*. *Methods in Enzymology*, 1981, 77, 50-59.
[160] Tiwari, VK; Tewari, N; Katiyar, D; Tripathi, RP; Arora, K; Gupta, S; Ahmad, R; Srivastava, AK; Khan, MA; Murthy, PK; Walter, RD. Synthesis and antifilarial evaluation of N1,Nnxylofuranosylated diaminoalkanes. *Bioorganic and Medicinal Chemistry*, 2003, 11, 1789-1800.
[161] Worthington, DJ; Rosemeyer, MA. Human glutathione reductase: purification of the crystalline enzyme from erythrocytes. *European Journal of Biochemistry*, 1974, 48, 167-177.
[162] Krohne-Ehrich, G; Schirmer, RH; Untucht-Grau, R. Glutathione reductase from human erythrocytes. Isolation of the enzyme and sequence analysis of the redox-active peptide. *European Journal of Biochemistry*, 1977, 80, 65-71.
[163] Bhargava, KK; Le-Trang, N; Cerami, A; Eaton, JW. Effect of arsenical drugs on glutathione metabolism of *Litomosoides carinii*. *Molecular and Biochemical Parasitology*, 1983, 9, 29-35.
[164] Müller, S; Walter, RD; Fairlamb, AH. Differential susceptibility of filarial and human erythrocyte glutathione reductase to inhibition by the trivalent organic arsenical melarsen oxide. *Molecular and Biochemical Parasitology*, 1995, 71, 211-219.
[165] Brophy, PM; Barrett, J. Glutathione transferase in helminths. *Parasitology*, 1990, 100, 345-349.
[166] Precious, WY; Barrett, J. Xenobiotic metabolism in helminths. *Parasitology Today*, 1989, 5, 156-160.
[167] Armstrong, RN. Glutathione-S-transferases: reaction mechanism, structure and function. *Chemical Research in Toxicology*, 1991, 4, 131-140.
[168] Brophy, PM; Pritchard, DI. Parasitic helminth glutathione Stransferases: an update on their potential as targets for immuno and chemotherapy. *Experimental Parasitology*, 1994, 79, 89-96.
[169] Tang, L; Smith, VP; Gounaris, K; Selkirk, ME. *Brugia pahangi*: the cuticular glutathione peroxidase (gp29) protects heterologous membranes from lipid peroxidation. *Experimental Parasitology*, 1996, 82, 329-332.
[170] Cookson, E; Tang, L; Selkirk, ME. Conservation of primary sequence of gp29, the major soluble cuticular glycoprotein, in three species of lymphatic filariae. *Molecular and Biochemical Parasitology*, 1993, 58, 155-159.
[171] Tripp, C; Frank, RS; Selkirk, ME; Tang, L; Grieve, MM; Frank, GR; Grieve, RB. *Dirofilaria immitis*: molecular cloning and expression of a cDNA encoding a selenium-

independent secreted glutathione peroxidase. *Experimental Parasitology*, 1998, 88, 43.50.

[172] Behm, CA. The role of trehalose in the physiology of nematodes. *International Journal for Parasitology*, 1997, 27, 215-229.

[173] Fairbairn, D. Trehalose and glucose in helminths and other invertebrates. *Canadian Journal of Zoology*, 1958, 36, 787-795.

[174] Gooday, GW; Brydon, LJ; Chappell, LH. Chitinase in female *Onchocerca gibsoni* and its inhibition by allosamidin. *Molecular and Biochemical Parasitology*, 1988, 29, 223-225.

[175] Shahabuddin, M; Toyoshima, T; Aikawa, M; Kaslow, DC. Transmission-blocking activity of a chitinase inhibitor and activation of malarial parasite chitinase by mosquito protease. *Proceedings of National Academy of Sciences of the United States of America*, 1993, 90, 4266-4270.

[176] Fuhrman, JA; Lane, WS; Smith, RF; Piessens, WF; Perler, FB. Transmission-blocking antibodies recognize microfilarial chitinase in brugian lymphatic filariasis. *Proceedings of National Academy of Sciences of the United States of America*, 1992, 89, 1548-1552.

[177] Raghavan, N; Freedman, DO; Fitzgerald, PC; Unnasch, TR; Ottesen, EA; Nutman; TB. Cloning and characterization of a potentially protective chitinase-like recombinant antigen from *Wuchereria bancrofti*. *Infection and Immunity*, 1994, 62, 1901-1908.

[178] Cox, GN. Molecular and biochemical aspects of nematode collagens. *Journal of Parasitology*, 1992, 78, 1-15.

[179] Kivirikko, KI; Mallyla, R; Pihlajaniemi, T. Protein hydroxylation: Prolyl 4-hydroxylase, an enzyme with four cosubstrates and a multifunctional subunit. *Federation of American Societies for Experimental Biology Journal*, 1989, 3, 1609-1617.

[180] Hanauske-Abel, HM. Prolyl 4-hydroxylase, a target enzyme for the drug development: Design of suppressive agents and *in vitro* effects of inhibitors and proinhibitors. *Journal of Hepatology*, 1991, 13, Suppl.3, S8-S16.

[181] Wilson, WR; Tuan, RS; Shepley, KJ; Freedman, DO; Greene, BM; Awadzi, K; Unnasch, TR. The *Onchocerca volvulus* homologue of the multifunctional polypeptide protein disulfide isomerase. *Molecular and Biochemical Parasitology*, 1994, 68, 103-117.

[182] Pandya, U; Saxena, JK; Kaul, SM; Murthy, PK; Chatterjee, RK; Tripathi, RP; Bhaduri, AP; Shukla, OP. DNA topoisomerase of filarial parasites: Characterization and effect of antifilarial compounds. *Med. Sci. Res.*, 1999, 27, 103.

[183] Katiyar, D; Tiwari, VK; Tripathi, RP; Reddy, VM; Bhattacharya, SM; Saxena, JK. Synthesis and antifilarial evaluation of 7-O-acetamidyl-4-alkyl-2H-1-benzopyran-2-ones. *Arzneimittelforschung,* 2003, 53, 857-862.

[184] Chandra, M; Sahay, AN; Pandey, DS; Tripathi, RP; Saxena, JK; Reddy, VJM; Puerta, MC; Valerga, P. Potential inhibitors of DNA topoisomerases II: ruthenium (II) poly-pyridyl and pyridyl-azine complexes. *Journal of Organometallic Chemistry*, 2004, 689, 2256-2267.

In: Nematodes
Editors: F. Boeri and J. A. Chung
ISBN: 978-1-61470-784-4

Chapter IV

Key Role of Free-Living Nematodes in the Marine Ecosystem

F. Semprucci and M. Balsamo
Dipartimento di Scienze della Terra, della Vita e dell'Ambiente (DiSTEVA),
Università degli Studi di Urbino 'Carlo Bo', loc. Crocicchia, 61029 Urbino, Italy.

Abstract

Nematodes are the most abundant metazoans on Earth. Estimates of global nematode species diversity vary widely, but currently seem to agree that there are about one million species, a number that is comparable to what is reported for other very diverse animal phyla. Nematodes are the dominant taxon in marine sediments in terms of both abundance and biomass. However, the great importance of this group is not only related to these characteristics, but also to its close relationship with the other organisms playing a key role in trophic webs. Theeffects of nematodes on sediment stability are similarly crucial. The relative stability of populations, the short turnover rate, and the generallyhigh tolerance to ecosystem alterations make Nematoda particularly suitable as bioindicators, a benefit that is enhanced because collection methods and sample treatment for these species are relatively simple. For this reason this phylum has been used as an indicator for assessing the ecological quality of marine ecosystems according to the Water Framework Directive (WFD, Directive 2000/60/EC). In addition, the high taxonomic diversity of nematodes is also related to their large trophic spectrum and the variety of their buccal structures. The latter, and other traits such as body and tail shape and life strategy, are easily recognizable, meaning that nematodes can be distinguished on both a morphological and functional basis. Functional diversity is an important component of biodiversity, but if compared to taxonomic diversity, it is still poorly applied. Nevertheless it could be very promising when it comes to improving our knowledge of ecosystems and relationship between the diversity patterns of nematodes and the abiotic environment. Recent studies have suggested that biodiversity loss might impair the functioning and sustainability of ecosystems. Accordingly, monitoring the quality of the environment appears to be essential for devising effective protection strategies and appropriate forms of management. In this respect, the benthic environment is the best domain and nematodes the ideal candidates. The aim of this paper, therefore, is to review past and present advances in the use of nematodes as bioindicators for the monitoring of marine ecosystems, and to highlight future perspectives.

1. Introduction

Nematodes have colonized all climatic areas and all kinds of environments, from aquatic (marine to freshwater) ecosystems to soil, plant and animal tissues. They have also developed a multitude of parasitic life-styles, causing numerous human diseases and large financial losses for those involved in agriculture and livestock rearing (Manzanilla-Lópes et al., 2004). A set of peculiar traits (a thick, multi-layered cuticle enveloping a large inner body cavity filled with high-pressure liquid, and many sophisticated physiological adaptations) have enabled nematodes to achieve great evolutive and ecological success. Free-living marine nematodes are essentially fine and slender worms, generally one or a few millimeters long, and with a body diameter that is 20-40 times less than their body length. Although this is their general aspect, according to Platt and Warwick (1980): "contrary to the popular opinion, marine nematodes do not all look the same, and it is time that this myth was finally put to rest".

The morphology of nematodesmay significantly change in relation to habitat type (Raes et al., 2008;Giere, 2009 for review), and they display a wide range of morphological adaptations and even very distinctive body shapes (Figures 1, 2). For instance, the free-living marine species of the order Desmoscolecida are characterized by short, thick bodies, usually with conspicuous transverse rings (*desmen*) or rows of stout setae. On the other hand, the Draconematidae and Epsilonematidae species stand out due to, respectively, their peculiar "S" and "ε"-shaped body (Figure 1). These families also have rows of stout ventral bristles, which are used during locomotion (*ambulatory setae*), and, often, a swollen anterior region. Meanwhile, the species of the enigmatic sub-family Stilbonematinae usually present with a multi or single layered coat made of ectosymbiotic bacteria which cover the body cuticle. Some other species (e.g. *Eubostrichus parasitiferus*)have filamentous ectocommensals covering their entire body surface, so giving the animal a bushy appearance. In other words: "One, No one and One Hundred Thousand!" (Zullini, 1992).

Nematodes have a key role in the trophic chains of marine habitats, since they contribute significantly to benthic energy flows through the sediment. They are ubiquitous and an integral part of ecosystem functioning; they can substantially affect ecological processes such as the recycling of nutrients, the transfer of energy to higher levels in the benthic food webs, and the bioturbation of sediments (Giere, 2009). Furthermore, the importance of nematode diversity as a useful tool for providing information on ecosystem functioning has also recently been proved (Danovaro et al. 2008). As a consequence, research on marine nematode assemblages should be increased to aid our understanding of functional benthic biodiversity. This paper aims to review both current knowledge of free-living nematodes in marine ecosystems and, in particular, the advances and future perspectives of using marine nematodes as bioindicators.

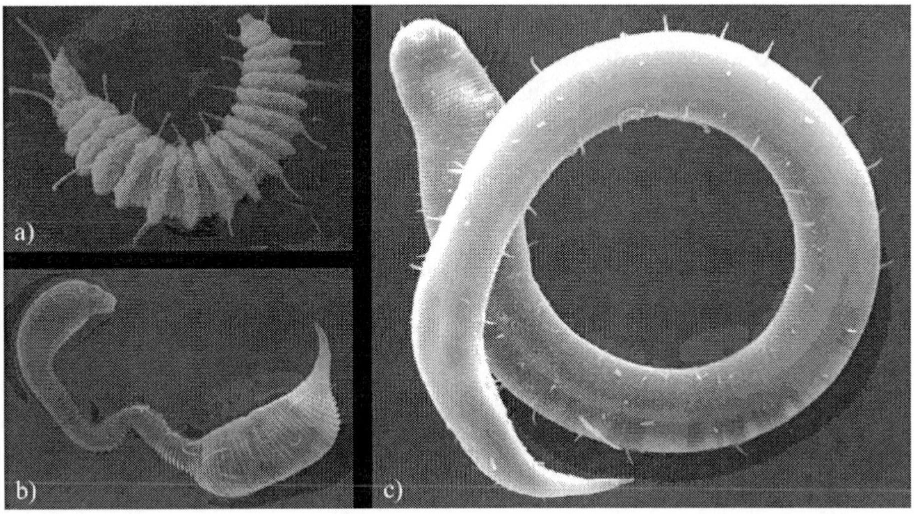

Figure 1. Body shapes of free-living marine nematodes: a) *Desmoscolex* sp., b) *Epsilonema* sp., and c) *Desmodora* sp.

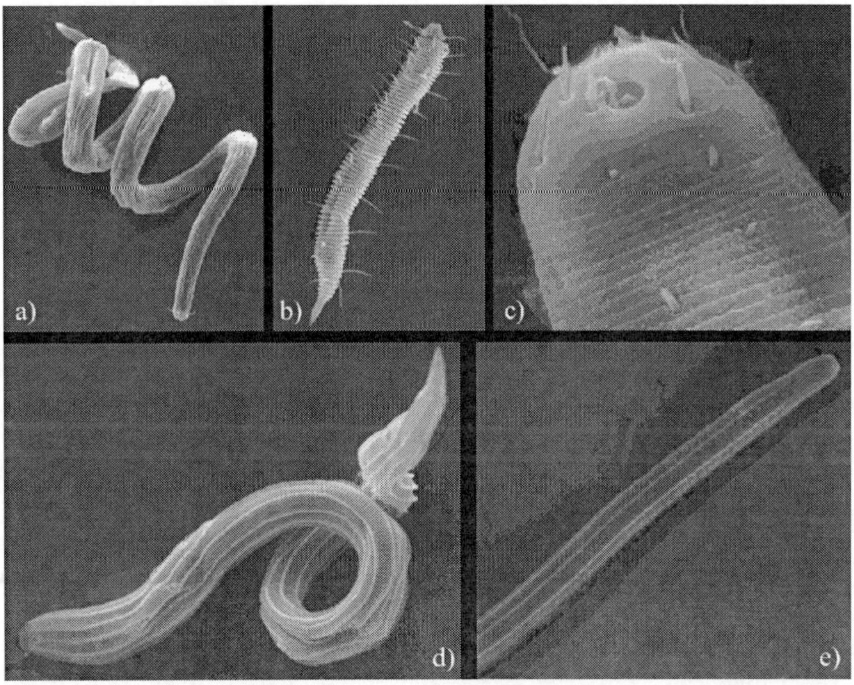

Figure 2. Cuticle morphologies of free-living marine nematodes: a) smooth; b) with desmens; c) striated; d) striated with longitudinal ridges; and e) annulated with longitudinal ridges.

2. Marine Nematode Biodiversity

Free-living nematodes are recognized as one of the most diverse metazoan taxa in marine ecosystems (Boucher and Lambshead, 1995). The total number of nematode species known is close to 27,000 (Hugot et al., 2001), while the number of described marine species is, thus far, in the range of 4000 – 5000 (Eyualem-Abebe et al., 2008). The value of the estimated global diversity of nematodes has varied widely in the past 15 years, e.g. between 10^5 (Andrássy, 1992) and 10^8 (Lambshead, 1993). The current conservative estimate is about one million species (Lambshead, 2004), a magnitude that is comparable to approximations of other very diverse animal taxa such as Insecta. However, more than 97% of the potential one million nematode species are currently unknown.

Among the difficulties hampering the assessment of the global biodiversity of the phylumare the small body size, the limited diagnostic taxonomical characters, and the low number of taxonomists unevenly distributed across the world. Numerous nematode species belonging to a single genus may be found in just one sample, but their identification at the specific level is often difficult. The reason for this is usually related to the high number of juveniles or females present, which in many cases lack all of the diagnostic features essential for an accurate classification. The identification of males also requires observations to be made on a large number of specimens. As a result of these difficulties, faunistic surveys have frequently been replaced with ecological studies that usually need a higher level of identification (i.e. genus level) or putative species on a morphological basis (genus sp.1, sp. 2, etc...). These accepted practices have notably contributed to the current standstill in the faunistic knowledge of marine nematodes, and, consequently, the lack of data about species distribution. Furthermore, recent studies have also suggested the existence of complexes of cryptic species that are currently included in a single morphological species in many meiobenthic groups, amongst which are nematodes. Species complexes are, in fact, especially numerous in microinvertebrates with few taxonomically diagnostic characters (Raes et al., 2008). As a result, putatively cosmopolitan species may actually have more limited geographical ranges (Schmidt and Westheide, 2000; Rocha-Olivares et al., 2001; Derycke et al., 2005, 2007, 2008; Meldal et al., 2007; Bhadury et al., 2008).

In the last few years, molecular techniques have been increasingly applied to make species identification easier and faster (see Bhaduryand Austen, 2010; Derycke et al., 2010 for review). In particular, polymerase chain reaction-based denaturing gradient gel electrophoresis has been evaluated as a tool for improving the assessment of nematode diversity in British coastal waters (Cook et al., 2005; Bhadury et al., 2006b), while DNA barcoding has been suggested as an alternative means of rapid nematode identification in biodiversity surveys (De Ley et al., 2005; Bhadury et al., 2006a; Bhaduryand Austen, 2010; Derycke et al., 2010; Pereira et al., 2010). Accordingly, the identification of nematodes could greatly benefit from the use of molecular tools, as these may provide a more reliable estimate of nematode diversity (Floyd et al., 2002; Bhadury et al., 2006a; Porazinska et al., 2009). Furthermore, the functional roles of nematodes may be highly species-specific, and their population dynamics can be affected by the presence of closely related species, often in the form of congeners (De Mesel et al., 2003, 2004, 2006; Postma-Blaauw et al., 2005; dos Santos et al., 2009). Consequently, diversity estimates that are only based on morphological criteria probably underestimate the real biodiversity of the phylum, suggesting that the 'ideal' taxonomic survey should be based on a combined approach between molecular and morphological analyses (Hebert et al., 2003; Will and Rubinoff, 2004; Hebert and Gregory, 2005).

Current data on the global distribution of marine taxa are available in important online open access databases, such as the Global Biodiversity Information Facility (GBIF, http://www.gbif.org) and the Ocean Biogeographic Information System (OBIS, http://www.iobis.org). Furthermore, the Nematoda and Mysida database (NeMys, http://nemys.ugent.be) is an essential tool for taxonomists and ecologists studying marine nematodes. Indeed, the most comprehensive record of species, identification keys and a wide range of taxonomic literature are all available in this database. Most records are from North-western Europe, followed by North America, since the vast majority of taxonomists of marine nematodes are active in the areas, which are where the most extensive faunistic investigations have been carried out (see the British Isles and the Belgian coasts in Europe). More limited information is available from other biogeographic regions (Figure 3).

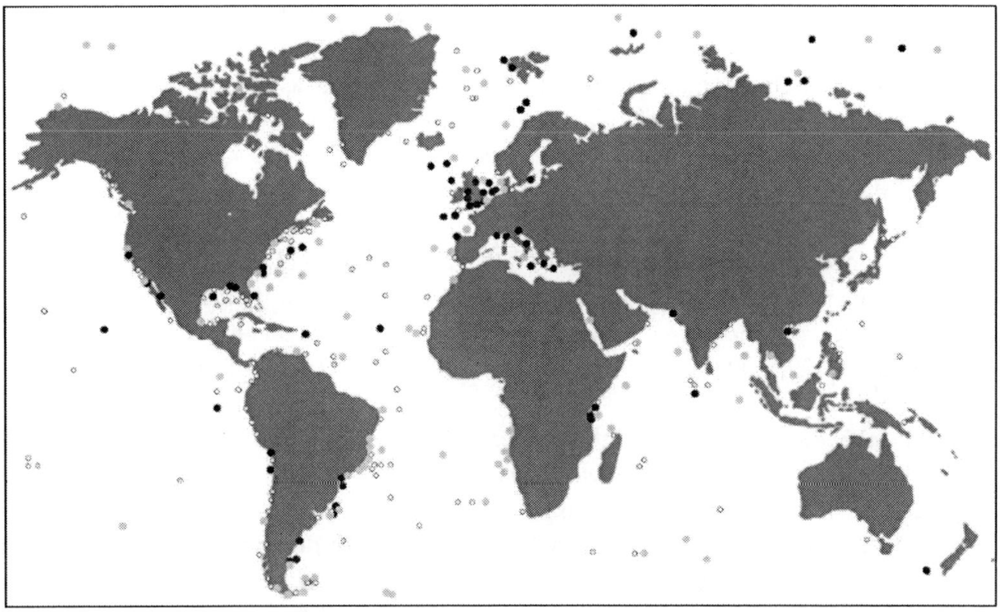

Figure 3. Global distribution of free-living marine nematodes. Black: range between > 101 to 51 species records; grey: 50 - 6 records; white: < 5 records (data source: OBIS, http://www.iobis.org).

The type of habitat is the main factor determining marine nematode diversity, and deep-sea biotopes (bathyal and abyssal) have proved to be the most diverse (Rex et al., 1993; Boucher and Lambshead, 1995; Mokievskiy et al., 2007).

The trend in species diversity towards an increase with the increasing latitude, which is well-known for several terrestrial and marine animal taxa, has not been confirmed for nematodes. Indeed, marine meiofauna show only weak latitudinal gradients, due to the presence of temporary meiofauna (Kotwick et al., 2005). In contrast, free-living nematodes are not more diverse in tropical regions than in subtropical or temperate areas, and in some cases they even show the opposite trend (Boucher, 1990; Boucher and Lambshead, 1995; Gobin and Warwick, 2006).

Future perspectives of the research on nematode biodiversity are aimed at increasing sampling efforts in habitats that are poorly studied, such as sediments in deep-sea, tropical and polar regions. In particular, the implementation of molecular approaches is planned, since these may produce promising results, especially if combined with morphological techniques.

3. Ecological Relevance of Free-Living Nematodes in Marine Systems

The density of free-living nematodes can reach several million ind./m^{-2} on the sea-floor in shallow subtidal sediments, representing a biomass of 0.2 - 0.5 g cm^{-2} (Heip et al., 1982). The top density values have been recorded in muddy, estuarine habitats and salt marshes, where ranges of 10-23 million ind./m^{-2} have been reported (Teal and Wieser, 1966; Warwick and Price, 1979).

The ecological value of nematodes is related to: their notable abundance (90 - 95%) and biomass (50 - 90%) in silty-sandy sediments; their interactions with bacteria (the population growth of which depends on nematodes); and their role within the trophic chains of aquatic ecosystems (see Balsamo et al., 2010 for review). Furthermore, as well as having a stabilizing effect (Cullen, 1973; Riemann, 1974; Riemann and Schrage, 1978), nematodes may also conduct bioturbation activity on sea-bottoms (Cullen, 1973), which suggests that they play a pivotal role in modifying the physical stability of shores (Platt and Warwick, 1980).

Sediment grain size appears to be the main environmental variable influencing the horizontal distribution of nematodes (Vanreusel, 1990; Vincx et al., 1990; Vanaverbeke et al., 2002). In contrast, detailed investigations of their vertical distribution have also highlighted the importance of sediment biogeochemistry, food availability and sediment oxygenation (Steyaert et al., 1999, 2003; Vanaverbeke et al., 2004a, b; Franco et al., 2008). Nematodes appear to be more sensitive than macrofauna, or other meiobenthic components such as harpacticoid copepods, to even slight changes in sediment composition (Heip et al., 1985; Vincx 1989; Raes et al., 2007). This feature can also be recognized at genus level (Semprucci et al., 2010b). Although nematode abundance is generally higher in muddy than in coarse sediments, their diversity trend appears to be the opposite; in fact, the ShannonDiversity (H') and Pielou evenness (J) indices have revealed the highest values in medium-coarse sands (Heip et al., 1985; Steyaert et al., 1999; Vanaverbeke et al., 2002; Semprucci et al., 2010b). This high diversity could be related to habitat heterogeneity; structurally complex habitats may provide more niches and diverse ways of exploiting environmental resources, and may consequently increase nematode species diversity (see Merckx et al., 2009 for details). Medium and coarse sediments are in fact rich in micro-niches because of the relatively large interstitial spaces that are useful for feeding and sheltering. Fine sands are strongly dominated by a few and often tolerant genera, which include species that are well adapted to anoxic and hypoxicconditions (Heip et al., 1985; Vanreusel, 1990; Steyaert et al., 1999; Vanaverbeke et al., 2002; Semprucci et al., 2010a). In contrast, very coarse sands, which are expected to be selective habitats because they are generally related to stronger hydrodynamic conditions, often host an intermediate diversity pattern, with a few genera that are equally represented (Semprucci et al., 2010b).

The high diversity of this phylum is related to its large trophic spectrum and the adaptive plasticity of its buccal cuticular pieces. The morphological type of the buccal cavity can be easily observed, meaning that nematodes can be distinguished in detritivorous, plant-feeding, omnivorous predators, as well as in species with various degrees of food specialization. Given the great number of nematode species, it is unsurprising that all trophic resources can be utilized, thus enhancing the importance of nematodes in marine ecosystem functioning.

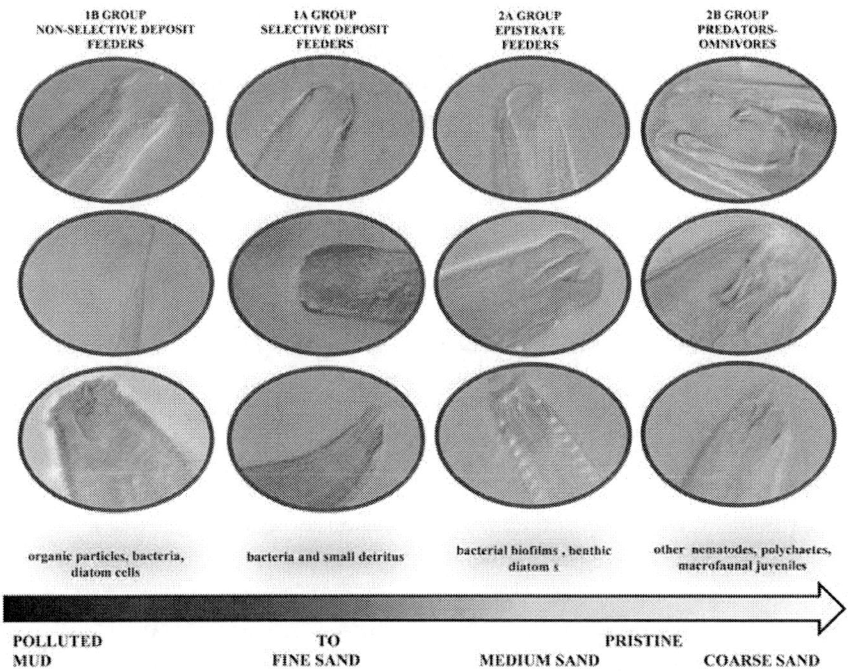

Figure 4. Nematode trophic groups in relation with the ecological status of sediment and grain size.

Wieser (1953) proposed the classification of nematodes into four groups according to the structure of the buccal cavity: 1A, selective deposit feeders; 1B, non-selective deposit feeders; 2A, epigrowth feeders; and 2B, predators and omnivores (Figure 4). Group 1A has a very small, non-cuticularized buccal cavity into which bacteria and small detritus, which are their main food sources, are pumped by means of the muscular pharynx. Non-selective deposit feeders (1B group), meanwhile, have a well-developed and weakly cuticularized buccal cavity, and take up food (organic particles, bacteria and diatom cells) mainly by means of the lips. Group 2A chiefly feeds on benthic diatoms. It scrapes off the (algal) surfaces of grains or pierces single cells by means of its buccal denticles and teeth. Finally, the cuticularized teeth and mandibles of group 2B are used for feeding on other nematodes, polychaetes and macrofaunal juveniles (Figure 4).

Heip (1996) reported a strong spatial segregation of nematode species in upper sediment layers; species with the same mouth structure, probably feeding on the same food (e.g. *Sabatieria*), were present at different depths in the sediment. In contrast, species with different buccal morphologies (e.g. *Acantholaimus*) co-existed in the same layer. This form of segregation suggests that nematofauna can be more different on a small, or even a micro-, spatial scale than on a large scale (Heip, 1996).

A limit of Wieser's scheme is the identification of each nematode species with a single trophic category, which does not correspond to the real trophic strategy of many of them. Furthermore, a number of nematode species are opportunistic, since they are actually able to change their feeding mode according to variations in the availability of food resources. On the other hand, Moens et al. (1999) have documented a high degree of feeding selectivity in favour of certain prey organisms, or even bacteria, due to the limited digestive capacities of some species. Consequently, the assignment of a nematode species to a particular feeding group is quite problematic in a number of cases. Wieser's classification has since been revised and modified (Jensen, 1987; Moens and Vincx, 1997). In particular, Moens and Vincx (1997) proposed an additional splitting of group 1A into two sub-groups (ciliate feeders and microvorous) and a separation between predators and facultative predators in group 2B. However, the use of trophic groups in field research is still hampered by the limited number of nematode genera considered by the Authors thereof, meaning that the Wieser's scheme is still in use.

4. Use of Nematodes as Biological Indicators

Free-living nematodes are essential for the functioning of estuarine and marine ecosystems, and could be used more widely as biological indicators of marine environmental pollution (Lambshead, 1986; Austen and McEvoy, 1997; Schratzberger et al., 2000; Austen, 2004; Danovaro et al., 2009; Moreno et al., 2011; Vanaverbeke et al., 2011).

Nematodes are the most diverse and numerically dominant metazoans in aquatic ecosystems, and, because of their rare ability to survive in extremely polluted conditions, they are usually the only persistent taxon in heavily polluted/stressed habitats (Coull and Chandler, 1992). Like all of the permanent members of the benthos, they are in direct contact with pollutants (Warwick, 1981; Heip et al., 1985; Vranken and Heip, 1986; Vranken et al., 1986). Their peculiar traits, as described above, along with their short biological cycles and the high stability of the populations, produce in this community a more rapid and unequivocal reaction to environmental changes than it is seen in macrobenthos (Platt and Warwick, 1980; Heip et al., 1985; Bongers and Van de Haar, 1990). Moreover, the high abundance of the phylum and the opportunity to study fixed specimens make sampling and analyzing operations very simple and quick (Platt and Warwick, 1980). The ecological and practical advantages associated with using nematodes in benthic studies are good reasons for utilizing them as an ideal indicator group for the assessment of sediment quality, as is the fact that doing so also follows the criteria reported by Kennedy and Jacoby (1999) and Goodsell et al. (2009). Nevertheless, nematodes are still poorly considered when it comes to the assessment of the ecological status of the benthic domain.

5. Measurement of Marine Nematode Taxonomic Diversity

5.1. Classic Diversity Estimations

Indices of biodiversity of selected taxa for environmental-impact assessment, or as a proxy for overall biodiversity, are generally expressed in terms of the number of taxa present (species richness), the distribution of relative abundances of taxa within samples (dominance or evenness), or as measures that aim to combine and balance the two dimensions (e.g. Shannon diversity).

Table 1. Main diversity and ecological indices reported for marine nematodes in the literature

Index	Equation	Application	Reference
Shannon Diversity	$H' = -\sum(pi \ln pi)$	This index is widely used in ecology. The Shannon index is generally more influenced by rare species than Simpson's one.	Shannon and Weaver (1949)
Hill Diversity	$N_1 = \exp[-\sum(pi \ln pi)] = \exp(H')$	N_1 is an exponential form of H', its value can be interpreted as the number of abundant taxa.	Hill (1973)
Margalef Diversity	$D = (S - 1) / \ln(N)$	D is unaffected by evenness or dominance, but it is sensitive only to species richness and sample size. Thus, its use should be restricted to comparing species richness among large communities.	Margalef (1958)
Simpson Index	$\lambda = \sum p_i^2$ $\lambda' = \{\sum_i N_i (N_i - 1)\} / \{N(N - 1)\}$	It measures the probability that two randomly chosen individuals of an infinite community belong to the same species. The revised form of Simpson index (λ') is appropriate when the total sample size (N) is small.	Simpson (1949)
Rarefation	$ES_n = \sum^{S}_{i=1}[1 - (N - N_i)!(N - n)!] / (N - N_i - n)!N!$	ES severely overestimates the species richness of small samples by making wrong assumptions concerning dominance and by assuming random rather than aggregated distributions of individuals (see Clarke and Warwick, 2001a).	Hurlbert (1971)
Maturity Index	$MI = \sum^{n}_{i=1} (v_i \times f_i)$	MI may be viewed as a measure of disturbance, with smaller values being indicative of a more disturbed environment and larger values characteristic of a less disturbed environment.	Bongers (1990)
Trophic Diversity	$ITD = \sum \theta^2$	The increase of ITD values is due to a marked dominance of an only trophic guild.	Heip et al. (1984)
Distinctness	$\Delta = [\sum\sum_{i<j} \omega_{ij} \chi_i \chi_j] / [N(N-1)/2]$ $\Delta^* = [\sum\sum_{i<j} \omega_{ij} \chi_i \chi_j] / [\sum\sum_{i<j} \chi_i \chi_j]$	Δ measures the average path length between any two individuals chosen at random (generalization of Simpson index), while Δ^* is the expected path length of any two randomly chosen individuals from the sample, provided that they belong to different species.	Clarke and Warwick (1995)

Whittaker (1960) suggested that species diversity can be measured in a range of scales, proposing the following measures: α-diversity (the basic unit of diversity) as the total number of species present in a sample or within a habitat; β-diversity as the turnover in species composition between spatial and/or temporal gradients; and γ-diversity as the species diversity measured at the largest scale. The latter measure does, however, appear to be redundant, since it equates to no more than evaluating the number of species in a region (Rosenzweig, 1995). An extensive review of the measurement of marine species biodiversity was proposed by Gray (2000) on Norwegian benthic fauna. However, measures of species richness, dominance, or evenness present problems, because the results of most diversity indices are sample-size or sampling-effort dependent. The limits of application of the most commonly used indices are reported in detail in Table 1.

5.2. Taxonomic Distinctness

More recently, Warwick and Clarke (1995) introduced a biodiversity measure based on the taxonomical relatedness of species (Table 1). This includes two indices: taxonomic diversity (Δ) and taxonomic distinctness (Δ^*). The former measures the average path length along the Linnean taxonomic classification between any two individuals chosen at random. This measure is a generalization of the Simpson diversity index, which is defined on the basis of the probability that any two specimens selected at random from a full set of individuals belong to the same species (Gray, 2000; Clarke and Warwick, 2001a). The second index, Δ^*, is the expected path length of any two randomly chosen individuals from the sample, provided that they belong to different species. This index has been created to remove the dominating effect of species abundance distribution. Clarke and Warwick (1998) examined the statistical sampling properties of measures based on both quantitative (Δ and Δ^*) and qualitative (presence/absence) data sets (Δ^+, AνTD). In particular, Δ^+ is simply calculated by adding the path lengths through a taxonomic tree connecting every pair of species on the list and dividing the result by the number of paths. The Authors demonstrated that the indices are independent of sample size or sampling effort, which is an important advantage over species richness measures. A further variation of this measure was then proposed, based on the taxonomic (or phylogenetic) relatedness of species, namely the 'variation in taxonomic distinctness' (VarTD, Λ^+) between every pair of species recorded in a study. This completes the previously defined 'average taxonomic distinctness' (AνTD, Δ^+). VarTD is simply the variance of these pairwise path lengths, and reflects the unevenness of the taxonomic tree (Clarke and Warwick, 2001b).

Although several Authors have documented the positive functioning of taxonomic distinctness, with a general decrease in the index from stressed to pristine areas (Warwick and Clarke, 1995; Clarke and Warwick, 1998; Barbuto and Zullini, 2005;Leonard et al., 2006), others have reported important limitations, especially in studies of the response of communities to several types of physical disturbance (Wlodarsaka-Kowalczuk et al. 2005; Somerfield et al., 2006; Vanavarbeke et al., 2007; Schratzberger et al., 2009). In particular, Vanaverbeke et al. (2007) statedthat the taxonomic relatedness indices (Δ^+ and Λ^+) do not reflect the changes in nematode community composition after dredging activities. They conclude that diversity indices based on taxonomic relatedness offer no valuable alternative to the species richness-based measures when it comes to assessing the possible impact of sand extraction on nematode communities. Instead, Schratzberger et al. (2009) suggested that patterns at genus level are more consistent thantaxonomic distinctness.

The potentialities of the taxonomic relatedness indices could be huge, but given their relatively recent history and the different results that have been reported, more data are necessary before they can be recommended as indicators of a wide range of types of disturbance. Furthermore, this approach does not seem to be helpful in assessing the ecological status of systems according to the Water Framework Directive (Salas et al., 2006; Prato et al., 2009).

6. Functional Diversity as an Alternative to Taxonomic Diversity?

Species in functional groups share morphological traits that are thought or known to be related toimportant ecological functions (Chalcraft and Resetarits, 2003). Functional groups of nematodes can be regarded as groups of species that have similar effects on ecosystem processes. The assessment of functional diversity is sometimes a practical necessity. This is because we are unable to determine in every case how each single species affects ecosystem processes, and nor can the measures based on a phylogenetic classification ever alone capture all of the important differences in nematode attributes (see Bongers and Bongers, 1998 for review). Furthermore, recent investigations have suggested that biodiversity loss might impair the functioning and sustainability of ecosystems (Danovaro et al., 2008).

6.1. Nematoda/Copepoda Ratio

The Nematoda/Copepoda ratio introduced by Raffaelli and Mason (1981) is among the methods of nematode community analysis that take account of the diverse biology and auto-ecology of the taxa. Nematodes and copepods have long been regarded as the most suitable taxa for studying the ecological conditions of the entire meiofaunal community (Heip et al., 1985; Moore and Bett, 1989).

The divergent auto-ecological characteristics of the two groups (the extreme tolerance of nematodes and the high sensitivity of copepods) led to the proposition that a ratio (total n° nematode ind./total n° copepod ind.) would be a useful index with which to assess variations in the sediment ecosystem. The Ne/Co ratio is easily measurable, but has unfortunately proved to be an excessive simplification, since it is influenced not only by pollution, but also by several environmental variables, such as horizontal distribution, sediment grain size and seasonal variability (Platt et al., 1984; Lambshead, 1986; Danovaro et al., 1995; Lee et al., 2001).

Furthermore, if copepods are generally sensitive and nematodes more tolerant to perturbations, it is also true that in some circumstances completely opposite trends have been observed (Travizi and Vidakovic, 1994;Danovaro et al., 1995). Consequently, despite a later revision of the index (Ne2A/Co, Warwick, 1981), the Ne/Co ratio has been generally abandoned.

6.2. Index of Trophic Diversity

Pollution may cause a modification in the trophic composition of the nematode community and, consequently, trophic diversity,due to changes in either the abundance of the benthic organisms or the organic supply, both representing trophic sources. The Index of Trophic Diversity (ITD; calculated on the basis of the percentage of each feeding type) was introduced by Heip et al. (1984) (Table 1). An increase of ITD values (due to a marked dominance of a single trophic guild) should be related to an increase of pollution. Some Authors have, nevertheless, reported an ambiguous influence on the ITD of an oil spill or the biodeposition impact of fish farms, suggesting that these types of stresses do not have a real impact on the trophic structure of the nematode community (Danovaro et al., 1995; Mirto et al. 2002). However, an increase or a decline of single trophic groups was usually observed. For example, group 1B decreased following contamination with oil, which could be explained by the ingestion of tar particles and oil emulsion with the sediment during feeding (see for details Danovaro et al., 2009). Mirto et al. (2002) documented that 1B species strongly increased in number in organic enriched sediments, which also contributed to raised, albeit not significantly so, ITD values. A lack of consistency of ITD values was reported by Schratzberger et al. (2009) in a study of the effects of physical disturbance on the seabed; the ITD trends were rather variable, appearing to depend on the nature and origin of the stress-generating factors. The limited efficiency of the ITD might plausibly be related to the confines of the Wieser classification, which did not perfectly reflect the trophic structure of nematode assemblages.

6.3. Maturity Index

The Maturity Index (MI), which is based on the ecological characteristics and reproductive strategies of nematodes, is computed as the weighted average of the individual colonizer-persistent (c-p) values in a sample (Bongers, 1990) (Table 1). It was initially proposed for the study of terrestrial and freshwater habitats, and was then also extended to marine and brackish ecosystems (Bongers et al., 1991). Bongers (1990) distinguished *r*-strategist species (colonizers or c-p1), which are more tolerant to environmental variations, and *k*-strategist species (persisters or c-p5), which are more sensitive (Tables 2 and 3). If the latter disappear after a stress is exerted on the ecosystem, colonizers occupy their niche. From an ecological point of view, the *r*-strategist nematodes (*Rhabditidae*and *Neodiplogasteridae*families) are small in size and have short biological cycles and voluminous gonads that produce small, but numerous, eggs. Consequently, colonizers have a great recovery capacity and respond rapidly to environmental changes. The *r*-strategists are generally bacterial-feeding, with high metabolic activity and almost constant movement. In contrast, the order Dorylaimida (Adenophorea) comprises species that are sensitive to pollutants and generally larger, with small gonads producing fewer but larger eggs. Other taxa of the Secernentea and Adenophorea occupy intermediate positions in this *continuum* of *r*- to *k*-selected (*sensulato*) features (Bongers and Ferris, 1999) (Tables 2 and 3). In particular, in the classification proposed by Bongers (1990), nematode families and genera are distributed in a c-p scale ranging from one (extreme colonizers) to five (extreme persisters) (Tables 2 and 3).

Table 2. Colonizer-persister (c-p) value assignments for select nematode families (Bongers et al., 1991)

c-p	Family
I	Neodiplogasteridae, Rhabditidae
II	Anoplostamidae, Anticomidae, Axonolaimidae, Comesomatidae, Linhomoeidae*, Thoracostomopsidae*, Tripyloididae, Xyalidae*,Monhysteridae[1]
III	Ceramonematidae, Chromadoridae, Cyatholaimidae, Desmodoridae*, Diplopeltidae, Ethmolaimidae*, Haliplectidae, Leptolaimidae*, Microlaimidae*, Monoposthiidae, Selachinematidae, Siphonolaimidae, Sphaerolaimidae, Tubolaimoididae
IV	Aegialoalaimidae, Desmoscolecidae, Draconematidae, Enchelidiidae, Epsilonematidae, Ironidae*, Oncholaimidae*, Oxystominidae, Phanodermatidae, Rhabdodemaniidae, Trefusiidae
V	Enoplidae, Leptosomatidae

Notes: * Heterogeneous families in which generic identification may provide additional precision.
[1] Monhysteridae c-p value was 1, but it was subsequently modified in 2 by Bongers et al. (1995).

According to Bongers's theory, a perturbation causing a numeric decrease in the *k*-strategy species leads to a decrease in MI value of the area. The MI seems to be useful for the study of organic enrichment, riverine inputs, hydrocarbon pollution and heavy-metals, but also of any perturbation that leads to an alteration of community structure (Bongers et al., 1991; Bongers and Ferris, 1999; Mirto et al., 2002; Gyedu-Ababio and Barid, 2006; Semprucci et al., 2010a). However, results were unsatisfactory in some cases (see Moreno et al., 2011 for review). Indeed, the lack of empirical evidence on the life histories of most marine nematode genera has led to the c-p being used conservatively and some limitations in the use of the MI in marine ecosystems. However, Semprucci et al. (2010a) underlined that the MI could be a good descriptor of stress, because it is influenced to a lesser extent by natural environmental variables (e.g. sediment granulometry) than the other available indices (i.e. H′ and J).

6.4. Biological Trait Analysis

Among the several morphological features widely discussed for nematodes there are not only the above discussed buccal structures and life history strategies, but also tail shape (Thistle and Sherman, 1985) and body size (length and width), the latter of which in particular is a function of physiological and ecological features (Soltwedel et al., 1996; Tita et al., 1999; Soetaert et al., 2002; Vanaverbeke et al., 2003, 2004a, b).

Schratzberger et al. (2007) were the first to propose a combined analysis of biological traits. In particular, an approach combining a number of functionality measures into a single one can provide interesting ecological information that it is impossible to discern from separate analyses (Schratzberger et al., 2007). The analysis of biological traits seems to offer more reliable correlative links with environmental factors than the taxonomic approach, and may reveal additional relationships within assemblages.

Table 3. Colonizer-persister (c-p) value assignments for select nematode genera (Bongers et al., 1991)

c-p	I	II	III	IV	V
	Diplolaimella	*Amphimonhystera*	*Antomicron*	*Actinonema*	*Cylicolaimus*
	Pellioditis	*Anoplostoma*	*Araeolaimus*	*Adoncholaimus*	*Enoplus*
		Anticoma	*Bolbolaimus*	*Aegialoalaimus*	*Leptosomatum*
		Ascolaimus	*Camacolaimus*	*Atrochromadora*	*Pseudocella*
		Axonolaimus	*Choniolaimus*	*Calyptronema*	*Synonchus*
		Bathylaimus	*Chromadora*	*Chromaspirina*	*Thoracostoma*
		Calomicrolaimus	*Chromadorella*	*Crenopharynx*	
		Chaetonema	*Chromadorina*	*Desmoscolex*	
		Daptonema	*Chromadorita*	*Dracognomus*	
		Desmodora	*Cobbia*	*Epsilonema*	
		Desmolaimus	*Cyartonema*	*Eurystomina*	
		Dichromadora	*Cyatholaimus*	*Gonionchus*	
		Dolicholaimus	*Dagda*	*Greeffiella*	
		Echinotheristus	*Dasynemoides*	*Halalaimus*	
		Eleutherolaimus	*Deontolaimus*	*Hypodontolaimus*	
		Enoploides	*Euchromadora*	*Ixonema*	
		Enoplolaimus	*Gammanema*	*Leptonemella*	
		Laimella	*Halaphanolaimus*	*Metaparoncholaimus*	
		Leptolaimus	*Haliplectus*	*Metepsilonema*	
		Linhomoeus	*Latronema*	*Metoncholaimus*	
		Metachromadora	*Mesacanthion*	*Nemanema*	
		Metadesmolaimus	*Molgolaimus*	*Oncholaimus*	
		Metalinhomoeus	*Monoposthia*	*Oxystomina*	
		Microlaimus	*Neotonchoides*	*Perepsilonema*	
		Monhystera	*Oncholaimellus*	*Polygrastrophora*	
		Neochromadora	*Onyx*	*Pomponema*	
		Neotonchus	*Paracyatholaimoides*	*Pontonema*	
		Odontophora	*Parallelocoilas*	*Praeacanthonchus*	
		Odontophoroides	*Paralongicyatholaimus*	*Prochaetosoma*	
		Oxyonchus	*Pselionema*	*Quadricoma*	
		Paracanthonchus	*Pseudonchus*	*Rhabdocoma*	
		Paracyatholaimus	*Ptycholaimellus*	*Rhabdodemania*	
		Paralinhomoeus	*Rhips*	*Stephanolaimus*	
		Paramesonchium	*Rhynchonema*	*Symplocostoma*	
		Paramonohystera	*Richtersia*	*Syringolaimus*	
		Prochromadorella	*Siphonolaimus*	*Thalassoalaimus*	
		Sabatieria	*Southerniella*	*Trefusia*	
		Spilophorella	*Sphaerolaimus*	*Tricoma*	
		Stylotheristus	*Terschellingia*	*Wieseria*	
		Theristus	*Tubolaimoides*		
		Trichotheristus	*Viscosia*		
		Trileptium	*Xyala*		
		Tripyloides			
		Valvaelaimus			

The number of biological traits and species seems to be highly correlated, suggesting that an increase in biodiversity may also lead to an increase in functional diversity (Schratzberger et al., 2007; Danovaro et al., 2008; Armenteros et al., 2009). At the present time, the biological traits approach seems to be no more powerful than the traditional taxonomic methods in detecting spatial patterns, but it can provide additional ecological information, even concerning a possible decrease in ecosystem functioning (Danovaro et al., 2008). Currently, one of the main problems hampering the application of functional traits is related to the approach's limits when it comes to defining biological traits, especially the feeding types thereof. The same problem had already been underlined in the ecology of terrestrial nematodes, for which feeding groups still provide inadequate information regarding the functioning of soil ecosystems (Bongers and Bongers, 1998). However, a good alternative appears to be based on the life history of nematodes or on the integration of both of the functional traits.

7. Free-living Marine Nematodes as Bioindicators in Accordance with the Water Framework Directive

The Water Framework Directive (WFD, Directive 2000/60/EC) highlights the importance of biological descriptors in evaluating and monitoring environmental conditions. Biological indicators are recognized as more appropriate tools for the assessment of ecological quality than physico-chemical or abiotic variables alone (GESAMP, 1995; Suter, 2001; Casazza et al., 2002; Goodsell et al., 2009). Indeed, environmental changes can significantly affect the density, diversity and structure of biological assemblages. In particular, sediments can be regarded as the most human-impacted domain, and they therefore represent the target zone for both studies and actions aimed at biodiversity conservation. According to the European Marine Strategy Framework Directive (2008/56/EC), seafloor integrity should be at a level that ensures the safeguarding of the structure and functions of the ecosystem, and benthic ecosystems should likewise not be negatively affected(Vanaverbeke et al., 2011). Furthermore, the European Water Framework Directive (WFD) has established the concept of Ecological Quality Status (EQS) as a way of assessing the biological quality of waters. Currently, ecological status is defined by a wide variety of biotic indices based on data on macrofauna (see Vanaverbeke et al., 2011 and references therein for details). This is justified by the fact that macrofauna are ecologically important, obviously widespread and easily identifiable, resulting in the wide availability of data on macrofaunal spatial distribution in the marine ecosystem. In contrast, few attempts have been made to date to investigate meiobenthic and nematode assemblages. However, Moreno et al. (2011) have, for the first time, carried out a meta-analysis to evaluate the use of nematodes as biological indicators of environmental quality in accordance with the WFD. Among the indices analyzed were those most frequently used in the ecological assessment of the nematode assemblage: the Shannon Diversity Index (H'); the Maturity Index (MI); the colonizer-persister class percentages (c-p %); the Index of Trophic Diversity (ITD); and the presence of sensitive/tolerant genera. Furthermore, specific thresholds with which to evaluate the EQS for each descriptor have been preliminarily proposed (Table 4). The results have highlighted that the taxonomic approach (multivariate analysis at genus level) reveals the best correspondence between environmental status and biological response. In particular, the presence/absence of specific indicator genera seemed to be very informative. Among the synthetic descriptors, the c-p % composition and the H′ measures can be used efficiently to evaluate EQS, although the ITD has produced very ambiguous results (Moreno et al., 2011).

Table 4. Proposed threshold values of the nematode indicators suggested by Moreno et al. (2011) in accordance with the Water Framework Directive

Ecological Quality Status	Maturity Index	Colonizer/ Persister percentage	Shannon Index	Index of Trophic Diversity	Sensitive/ Tollerant genera (> 10%)
High	MI > 2.8	c-p 2 < 50% c-p 4 > 10%	H' > 4.5	0 < ITD < 0.25	*Desmoscolecidae* *Microlaimus* *Richtersia* *Oncholaimus* *Pomponema* *Epacanthion*
Good	2.8 < MI < 2.6	c-p 2 > 50% c-p 4 > 10%	3.5 < H' < 4.5	0.25 < ITD < 0.4	*Halalaimus* *Setosabatieria* *Ptycholaimellus*
Moderate	2.6 < MI < 2.4	c-p 2 > 50% 3 < c-p 4 < 10%	2.5 < H' < 3.5	0.4 < ITD < 0.6	*Anticoma* *Desmodora* *Spirinia* *Marylynia* *Prochromadorella*
Poor	2.4 < MI < 2.2	c-p 2 > 60% c-p 4 < 3%	1 < H' ≤ 2.5	0.6 < ITD < 0.8	*Daptonema* *Theristus* *Paralongicyatholaimus* *Parodontophora* *Odontophora*
Bad	MI < 2.2	c-p 2 > 80%	0 < H' ≤ 1	0.8 < ITD < 1	*Paracomesoma* *Terschellingia* *Sabatieria*

Conclusion

Monitoring environmental quality is an essential element of both an effective protection strategy and the correct management of marine ecosystems. The ecological and practical advantages associated with using nematodes in benthic biological studies are good reasons to utilize them as a descriptor group in the assessment of the quality status of marine sediments.

A general limitation of ecological investigations of nematodes is related to the difficult taxonomic identification thereof. Indeed, several Authors point to the issue of the accuracy of this measure, because the functional roles of nematodes may be highly species-specific, and their identification at a specific level is therefore fundamental for understanding their ecological role. In this respect, a new and important challenge for the future is the implementation of molecular technique applications, which could be an important instrument for making nematode identification easier and faster. This method should be combined with the morphological approach to produce an 'ideal' taxonomic survey that could be used as an appropriate basis for estimations of nematode biodiversity in marine ecosystems.

However, the calibration and standardization of good molecular approaches to biodiversity research will take time, and other forms of analysis have to be explored. Consequently, on the other side of the coin are the improvements in the functional diversity approach and the development of new nematode indices based on taxonomic sufficiency.

One of the important challenges in science today is understanding and predicting the link between biodiversity and ecosystem functions. As discussed, the combination of taxonomic and functional diversity may actually be a powerful tool in ecological research, even if the relationships between them and with ecosystem functioning are still an explorative field. Although the analysis of biological traits seems currently to be no more powerful than the traditional taxonomic approach when it comes to detecting spatial patterns, this method does have more reliable links with environmental factors. In this regard, improving our knowledge of the functional role of nematode species will be the key to improving the sensitivity and interpretation of biological trait analyses of marine benthic communities. According to recommendations by the WFD, another perspective will be the development of new informative indices. This has already been comprehensively done for the macrofaunal community, but is still in its infancy for meiobenthos and nematodes. Indeed, if the responses to a change can be species-specific, this level of identification requires services that are too time-consuming and expensive to be suitable for monitoring programs. However, the creation of new nematode benthic indices based on taxonomic sufficiency (e.g. genus level) could, if adequately calibrated with a meta-analysis approach, give useful results to this end.

References

Andrássy, I. (1992). A short census of free-living nematodes. *Fundamental and applied Nematologie, vol. 15*, 187–188.

Armenteros, M., Ruiz-Abierno, A., Fernández-Garcés, R., Pérez-García, J.A., Díaz-Asencio, L., Vincx, M. and Decraemer, W. (2009). Biodiversity patterns of free-living marine nematodes in a tropical bay: Cienfuegos, Caribbean Sea. *Estuarine, Coastal and Shelf Science, vol. 85*, 179-189.

Austen, M. C. (2004). Natural nematode communities are useful tools to address ecological and applied questions. *Nematology Monographs and Perspectives, vol.2*, 1–17.

Austen, M. C. andMcEvoy, A.J. (1997). The use of offshore meiobenthic communities in laboratory microcosm experiments: response to heavy metal contamination. *Journal of Experimental Marine Biology and Ecology, vol.211*, 247-261.

Balsamo, M., Albertelli, G., Ceccherelli, V.U., Coccioni, R., Colangelo, M.A., Curini-Galletti, M., Danovaro, R., D'Addabbo, R., Leonardis, C., Fabiano, M., Frontalini, F., Gallo, M., Gambi, C., Guidi, L., Moreno, M., Pusceddu, A., Sandulli, R., Semprucci, F., Todaro, M.A. andTongiorgi, P. (2010). Meiofauna of the Adriatic Sea: current state of knowledge and future perspective. *Chemistry and Ecology, vol. 26*, 45 - 63.

Barbuto, M. andZullini, A. (2005). The nematode community of two Italian rivers (Taro and Ticino). *Nematology*, *vol. 7*, 667-675.

Bhadury, P. and Austen, M. C. (2010). Barcoding marine nematodes: an improved set of nematode 18S rRNA primers to overcome eukaryotic co-interference. *Hydrobiologia*, *vol. 641*, 245–251.

Bhadury, P., Austen, M. C., Bilton, D. T., Lambshead, P. J. D., Rogers, A. D. andSmerdon, G. R. (2006a). Development and evaluation of a DNA-barcoding approach for the rapid identification of nematodes. *Marine Ecology Progress Series*, *vol. 320*, 1–9.

Bhadury, P., Austen, M. C., Bilton, D. T., Lambshead, P. J. D., Rogers, A. D. andSmerdon, G. R. (2006b). Molecular detection of marine nematodes from environmental samples: overcoming eukaryotic interference. *Aquatic Microbial Ecology*, *vol. 44*, 97–103.

Bhadury, P., Austen, M. C., Bilton, D. T., Lambshead, P. J. D., Rogers, A. D. andSmerdon, G. R. (2008). Evaluation of combined morphological and molecular techniques for marine nematode (*Terschellingia* spp.) identification. *Marine Biology*, *vol. 154*, 509–518.

Bongers, T. (1990). The maturity index: an ecological measure of environmental disturbance based on nematode species composition. *Oecologia*. *vol. 83*, 14-19.

Bongers, T. and Bongers M. (1998). Functional diversity of nematodes. *Applied Soil Ecology,vol. 10*, 239-251.

Bongers, T. and Ferris, H. (1999). Nematode community structure as a bioindicator in environmental monitoring. *Tree*, *vol. 14*, 224-228.

Bongers, T. and Van de Haar, J. (1990). On the potential of basing an ecological typology of aquatic sediments on the nematode fauna: an example from the river Rhine. *Aquatic Ecology*, *vol. 24*, 37-45.

Bongers, T., Alkemade, R. andYeates, G.W. (1991). Interpretation of disturbance-induced maturity decrease in marine nematode assemblages by means of the Maturity Index. *Marine Ecology Progress Series*, *vol. 76*, 135-142.

Bongers, T., de Goede, R.G.M., Korthals, G.W. andYeates, G.W. (1995). An update to the cp-rating of nematode genera can be found in proposed changes of c-p classification for nematodes. *Russian Journal of Nematology*, *vol.3*, 61-62.

Boucher, G. (1990). Pattern of nematodes species diversity in temperate and tropical subtidal sediments. *PSZN I: Marine Ecology*, *vol. 11*, 133-146.

Boucher, G. andLambshead, P.J.D. (1995). Ecological biodiversity of marine nematodes in samples from temperate, tropical, and deep-sea regions. *Conservation Biology, vol. 9*, 1594-1604.

Casazza, G., Silvestri, C. andSpada, E. (2002). The use of bio-indicators for quality assessments of the marine environment: Example from the Mediterranean Sea. *Journal of Coastal Conservation*, *vol. 8*, 147-156.

Chalcraft, D.R. andResetarits, W.J. (2003). Mapping functional similarity of predators on the basis of trait similarities. *American Naturalist*, *vol.162*, 390–402.

Clarke, K. R. and Warwick, R. M. (1998). A taxonomic distinctness index and its statistical properties. *Journal of Applied Ecology*, *vol. 35*, 523–531.

Clarke, K. R. and Warwick, R. M. (2001a). *Changes in marine communities: an approach to statistical analysis and interpretation*. 2nd edition. Plymouth Marine Laboratory, Plymouth.

Clarke, K. R. and Warwick, R. M. (2001b). A further biodiversity index applicable to species lists: variation in taxonomic distinctness. *Marine Ecology Progress Series*, *vol. 216*, 265–278.

Cook, A. A., Bhadury, P., Debenham, N. J., Meldal, B. H. M., Blaxter, M. L., Smerdon, G. R., Austen, M. C., Lambshead, P. J. D. and Rogers, A. D. (2005). Denaturing gradient gel electrophoresis (DGGE) as a tool for identification of marine nematodes. *Marine Ecology Progress Series*, *vol. 291*, 103–113.

Coull, B.C. and Chandler, G.T. (1992). Pollution and meiofauna: field, laboratory, and mesocosms studies. *Oceanography and Marine Biology An Annual Review*, *vol. 30*, 191–271.

Cullen, D. J. (1973). Bioturbation of superficial marine sediments by interstitial meiobenthos. *Nature*, *vol. 242*, 323-324.

Danovaro, R., Fabiano, M. and Vincx, M. (1995).Meiofauna response to the AgipAbruzzooil spill in subtidal sediments of the Ligurian Sea. *Marine Pollution Bulletin*, *vol. 30*, 133–145.

Danovaro, R., Gambi, C., Dell'Anno, A., Corinaldesi, C., Fraschetti, S., Vanreusel, A., Vincx, M. and Gooday, A. J. (2008). Exponential Decline of Deep-Sea Ecosystem Functioning Linked to Benthic Biodiversity Loss.*Current Biology,vol. 18*, 1–8.

Danovaro, R., Gambi, C., Höss, S., Mirto, S., Traunspurger, W. andZullini, A. (2009). Case studies using nematode assemblage analysis in aquatic habitats. In M. J. Wilson, and T. K. Kakouli-Duarte, (Eds) *Nematodes as Environmental Indicators*. CAB Internationals. Wallingford, UK, 146-171.

De Ley, P., De Ley, I. T., Morris, K., Abebe, E., Mundo, M., Yoder, M., Heras, J., Waumann, D., Rocha-Olivares, A., Burr, J., Baldwin, J. G. and Thomas W. K. (2005). An integrated approach to fast and informative morphological vouchering of nematodes for applications in molecular barcoding. *Philosophical Transactions of the Royal Society B Biological Science*, *vol. 360*, 1945–1958.

De Mesel, I., Derycke, S., Moens, T., Van der Gucht, K., Vincx, M., et al. (2004). Topdown impact of bacterivorous nematodes on the bacterial community structure: a microcosm study. *Environmental Microbiology*, *vol. 6*, 733–744.

De Mesel, I., Derycke, S., Swings, J., Vincx, M. and Moens, T. (2003). Influence of bacterivorous nematodes on the decomposition of cordgrass. *Journal of Experimental Marine Biology and Ecology*, *vol. 296*, 227–242.

De Mesel, I., Derycke, S., Swings, J., Vincx, M. and Moens, T. (2006). Role of nematodes in decomposition processes: Does within-trophic group diversity matter? *Marine Ecology-Progress Series*, *vol. 321*, 157–166.

Derycke, S., Backeljau T., Vlaeminck C., Vierstraete A., Vanfleteren J., Vincx M. and Moens T. (2007). Spatiotemporal analysis of population genetic structure in Geomonhysteradisjuncta (Nematoda, Monhysteridae) reveals high levels of molecular diversity. *Marine Biology*, *vol. 151*, 1799–1812.

Derycke, S., Fonseca, G., Vierstraete, A., Vanfleteren, J., Vincx, M. and Moens, T. (2008).Disentangling taxonomy within the *Rhabditis* (*Pellioditis*) *marina* (Nematoda, Rhabditidae) species complex using molecular and morphological tools.*Zoological Journal of the Linnean Society, vol.* 152, 1-15.

Derycke, S., Remerie, T., Vierstraete, A., Backeljau, T., Vanfleteren, J., Vincx, M. and Moens, T. (2005).Mitochondrial DNA variation and cryptic speciation within the freeliving marine nematode *Pellioditis marina.Marine Ecology Progress Series*, *vol. 300*, 91–103.

Derycke, S., Vanaverbeke, J., Rigaux, A., Backeljau, T. and Moens, T. (2010). Exploring the Use of Cytochrome Oxidase c Subunit 1 (COI) for DNA Barcoding of Free-Living Marine Nematodes.*PlosOne*, *vol. 5*, 1-9.

dos Santos, G.A.P., Derycke, S., Genevois, V.G.F., Coelho, L., Correia, M.T.S., et al. (2009). Interactions among bacterial-feeding nematode species at different levels of food availability. *Marine Biology*, *vol. 156*, 629–640.

Eyualem-Abebe, E., Decraemer, W. and De Ley, P. (2008). Global diversity of nematodes (Nematoda) in freshwater. *Hydrobiologia, vol. 595*,67-78.

Floyd, R., Abebe, E., Papert, A. andBlaxter, M. (2002). Molecular barcodes for soil nematode identification. *Molecular Ecology*, *vol. 11*, 839–850.

Franco, M.A., Soetaert, K., Van Oevelen, D., Van Gansbeke, D., Costa, M.J., Vincx, M. and Vanaverbeke, J. (2008). Density, vertical distribution and trophic responses of metazoan meiobenthos to phytoplankton deposition in contrasting sediment types. *Marine Ecology Progress Series,vol. 358*, 51-62.

GESAMP Joint Group of Experts on the Scientific Aspects of Marine Environmental Protection, 1995.Biological indicators and their use in the measurement of the condition of the marine environment. *Reports and Studies GESAMP No. 55,* UNEP, Nairobi.

Giere, O., 2009. *Meiobenthology: the microscopic motile fauna of aquatic sediments.* 2ndedition. Springer, Berlin Heidelberg.

Gobin, J.F. and Warwick, R.M. (2006). Geographical variation in species diversity: a comparison of marine polychaetes and nematodes. *Journal of Experimental Marine Biology and Ecology*, *vol.330*, 234-244.

Goodsell, P.J., Underwood, A.J. and Chapman, M.G. (2009). Evidence necessary for taxa to be reliable indicators of environmental conditions or impacts. *Marine Pollution Bulletin*, *vol. 58*, 323–331.

Gray, J.S. (2000). The measurement of marine species diversity, with an application to the benthic fauna of the Norwegian continental shelf. *Journal of Experimental Marine Biology and Ecology*, *vol. 250*, 23-49.

Gyedu-Ababio, T.K. and Baird D. (2006). Response of meiofauna and nematode communities to increased levels of contaminants in a laboratory microcosm experiment. *Ecotoxicology and Environmental Safety,vol. 63*, 443-450.

Hebert, P.D.N. and Gregory, T.R. (2005). The promise of DNA barcoding for taxonomy. *Systematic Biology, vol.54*, 852 - 859.

Hebert, P.D.N., Cywinska, A., Ball, S.L. and de Waard, J.R. (2003). Biological identifications through DNA barcodes. *Proceedings of Royal Society London B*, *vol.270*, 313-321.

Heip, C. (1996). Biodiversity of marine sediments. In F. di Castri, and T. Younès, (Eds). *Biodiversity, Science and Development: Towards a New Partnership*. CAB International, UK, 139-148.

Heip, C., Herman, R. and Vincx, M. (1984). Variability and productivity of meiobenthos in the Southern Bight of the North Sea. *Rapport et process-verbaux des reunions. Conseil international pour l'Exploration de la Mer*, *vol. 183*, 507-521.

Heip, C., Vincx, M., Smol, N. and Vranken, G. (1982). The systematic and ecology of free-living marine nematodes. *Helminthological Abstracts*, *vol. 51*, 1-31.

Heip, C., Vincx, M. andVranken, G. (1985). The ecology of marine nematodes. *Oceanography and Marine Biology An Annual Review*, *vol. 23*, 399-489.

Hill, M.O. (1973). Diversity and evenness: a unifying notation and its consequences. *Ecology, vol. 54*, 427–432.

Hugot, J.P., Baujardand, P. andMorand, S. (2001). Biodiversity in helminths and nematodes as a field of study: an overview. *Nematology*, *vol. 3*, 199–208.

Hurlbert, S.H. (1971). The non-concept of species diversity: A critique and alternative parameters. *Ecology*, *vol. 52*, 577–586.

Jensen, P. (1987). Feeding ecology of free-living aquatic nematodes. *Marine Ecology Progress Series*, *vol. 35*, 187-196.

Kennedy, A.D. and Jacoby, C.A. (1999). Biological indicators of marine environmental health: meiofauna - a neglected benthic component? *Environmental Monitoring and Assessment*, *vol. 54*, 47-68.

Kotwicki, L., Szymelfenig, M., De Troch, M., Urban-Malinga, B. andWeslawski, J. M. (2005). Latitudinal biodiversity patterns of meiofauna from sandy littoral beaches. *Biodiversity and Conservation*, *vol. 14*, 461–474.

Lambshead, P. J. D. (1986). Sub-catastrophic sewage and industrial waste contamination as revealed by marine nematode faunal analysis. *Marine Ecology Progress Series*, *vol. 29*, 247–260.

Lambshead, P. J. D. (1993). Recent developments in marine benthic biodiversity research. *Oceanis*, *vol. 19*, 5–24.

Lambshead, P. J. D. (2004). Marine nematode biodiversity. In Z. X. Chen, W.Y., Chen, S.Y. Chen, and D. W. Dickson (Eds). *Nematology, Advances and Perspectives Vol 1: Nematode Morphology, Physiology and Ecology*. CABI Publishing, Wallingford, 436–467.

Lee, M. R., Correa, J. A. andCastilla J.C. (2001). An assessment of the potential use of the nematode to copepod ratio in the monitoring of metals pollution. The Chañaral Case. *Marine Pollution Bulletin*, *vol. 42*, 606-701.

Leonard, D. R. P., Clarke, K. R., Somerfield, P. J. and Warwick, R. M. (2006). The application of an indicator based on taxonomic distinctness for UK marine biodiversity assessments. *Journal of Environmental Management*, *vol. 78*, 52–62.

Manzanilla-Lópes, R. H., Evans, K. and Bridge, J. (2004). Plant diseases caused by nematodes. In Z.X. Chen, W.Y., Chen, S.Y., Chen and D.W. Dickson, (Eds.), *Nematology: Advances and Perspectives Vol 2: Nematode Management and Utilization*. CABI Publishing, Wallingford, 637–716.

Margalef, R. (1958). Information theory in ecology. *General Systems*, *vol. 3*, 36–71.

Meldal, B. H. M., Debenham, N. J., De Ley, P., De Ley, I. T., Vanfleteren, J. R., Vierstraete, A. R., Bert, W., Borgonie, G., Moens, T., Tyler, P. A., Austen, M. C., Blaxter, M., Rogers, A. D. andLambshead, P. J D. (2007). An improved molecular phylogeny of the Nematoda with special emphasis on marine taxa. *Molecular Phylogenetics and Evolution*, *vol. 42*, 622-636.

Merckx, B., Goethals, P., Steyaert, M., Vanreusel, A., Vincx, M. and Vanaverbeke, J. (2009). Predictability of marine nematode biodiversity. *EcologicalModelling*, *vol. 220*, 1449–1458.

Mirto, S., La Rosa, T., Gambi, C., Danovaro, R. and Mazzola, A. (2002). Nematode community response to fish-farm impact in the western Mediterranean. *Environmental Pollution*, *vol. 116*, 203-214.

Moens, T. and Vincx, M. (1997). Observations on the feeding ecology of estuarine nematodes. *Journal of Marine Biological Association of the United Kingdom*, *vol. 77*, 211-227.

Moens, T., Verbeeck, L., de Mayer, A., Swings, J., Vincx, M. (1999). Selective attraction of marine bacterivorous nematodes to their bacterial food *Marine Ecology Progress Series*, *vol. 176*, 165-178.

Mokievskiy, V. O., Udalov, A. A. andAzovskii, A. L. (2007). Quantitative distribution of meiobenthos in deep-water zones of the world ocean. *Oceanology*, *vol. 47*, 787-813.

Moore, C.G. andBett, B. J. (1989). The use of meiofauna in marine pollution impact assessment. *Zoological Journal of the Linnean Society*, *vol. 96*, 263–280.

Moreno, M., Semprucci, F., Vezzulli, L., Balsamo, M., Fabiano, M. andAlbertelli, G. (2011). The use of nematodes in assessing ecological quality status in the Mediterranean coastal ecosystems. *EcologicalIndicators*, *vol. 11*, 328-336.

Pereira, T. J., Fonseca, G., Mundo-Ocampo, M., Guilherme, B.C. and Rocha-Olivares, A. (2010). Diversity of free-living marine nematodes (Enoplida) from Baja California assessed by integrative taxonomy. *Marine Biology*, *vol. 157*, 1665–1678.

Platt, H. M. and Warwick, R. M. (1980). The significance of free-living nematodes to the littoral ecosystem. In: J.H. Price, D.E.G. Irvine and W.F. Farnham (Eds.). *The Shore Environment, Vol. 2: Ecosystems*. Academic Press, New York, 729-759.

Platt, H. M., Shaw, K. M. andLambshead, P. J. D. (1984).Nematode species abundance patterns and their use in the detection of environmental perturbations. *Hydrobiologia*, *vol. 118*, 59–66.

Porazinska, D. L., Giblin-Davis, R. M., Faller, L., Farmerie, W., Kanzaki, N., et al. (2009). Evaluating high-throughput sequencing as a method for metagenomic analysis of nematode diversity. *Molecular Ecology Resources*, *vol. 9*, 1439–1450.

Postma-Blaauw, M. B., de Vries, F. T., de Goede, R. G. M., Bloem, J., Faber, J. H. et al. (2005). Within-trophic group interactions of bacterivorous nematode species and their effects on the bacterial community and nitrogen mineralization. *Oecologia*, *vol. 142*, 428–439.

Prato, S., Morgana, J. G., La Valle, P., Finoia, M. G., Lattanzi, L., Nicoletti, L., Ardizzone G. D. and Izzo, G. (2009). Application of biotic and taxonomic distinctness indices in assessing the Ecological Quality Status of two coastal lakes: Caprolace and Fogliano lakes (Central Italy). *Ecological Indicators*, *vol. 9,* 568-583.

Raes, M., Decraemer, W. and Vanreusel, A. (2008) Walking with worms: coral-epifaunal nematodes. *Journal of Biogeography*, *vol. 35*, 2207–2222.

Raes, M., De Troch, M., Ndaro, S. G. M., Muthumbi, A., Guilini, K. and Vanreusel, A. (2007). The structuring role of microhabitat type in coral degradation zones: a case study with marine nematodes from Kenya and Zanzibar. *Coral Reefs*,*vol.* 26, 113-126.

Raffaelli, D. G. and Mason, D. F. (1981). Pollution monitoring with meiofauna, using the ratio of nematodes to copepods. *Marine Pollution Bulletin*, *vol. 12*, 158-163.

Rex, M. A., Stuart, C. T., Hessler, R. R., Allen, J. A., Sanders, H. L. and Wilson, G. D. F. (1993). Global scale latitudinal patterns of species diversity in the deep-sea benthos. *Nature*, *vol. 365*, 636–639.

Riemann, F. (1974). On hemisessile nematodes with flagelliform tails living in marine soft bottoms and on micro-tubes found in deep sea sediments. *MikrofaunaMeeresboden, vol. 40*, 1-15.

Riemann, F. and Schrage, M. (1978). The mucus-trap hypothesis on feeding of aquatic nematodes and implications for biodegradation and sediment texture. *Oecologia*, *vol. 34*, 75-88.

Rocha-Olivares, A., Fleeger, J. W. and Foltz, D. W. (2001). Decoupling of molecular and morphological evolution in deep lineages of a meiobenthic harpacticoid copepod. *Molecular Biology and Evolution*, *vol. 18*, 1088–1102.

Rosenzweig, M. L. (1995). *Species Diversity in Space and Time*. Cambridge University Press, Cambridge.

Salas, F., Patrício, J., Marcos, C., Pardal, M. A., Pérez-Ruzafaa, A. and Marques J.C. (2006). Are Taxonomic Distinctness measures compliant to other ecological indicators in assessing ecological status? *Marine Pollution Bulletin*, *vol. 52*, 817–829.

Schmidt, H. andWestheide, W. (2000). Are the meiofaunal polychaetes Hesionidesarenaria and Stygocapitellasubterranea true cosmopolitan species? Results of RAPD–PCR investigations. *ZoologicaScripta*, *vol. 29*, 17–27.

Schratzberger, M., Gee, J. M., Rees, H. L., Boyd, S. E. and Wall, C. M. (2000). The structure and taxonomic composition of sublittoral meiofauna assemblages as an indicator of the status of the marine environment. *Journal of Marine Biological Association of the United Kingdom*, *vol. 80*, 969– 980.

Schratzberger, M., Lampadariou, N., Somerfield, P. J., Vandepitte, L. andVandenBerghe, E. (2009). The impact of seabed disturbance on nematode communities: linking field and laboratory observations. *Marine Biology*, *vol. 156*, 709-724.

Schratzberger, M., Warr, K., Rogers, S. I. (2007). Functional diversity of nematode community in the south-western North Sea. *Marine Environmental Research*, *vol. 63*, 368-389.

Semprucci, F., Boi, P., Manti, A., Covazzi-Harriague, A., Rocchi, M., Colantoni, P., Papa, S. and Balsamo, M. (2010a).Benthic communities along a littoral of the Central Adriatic Sea (Italy). *Helgoland Marine Research*, *vol. 64*, 101–115.

Semprucci, F., Colantoni, P., Baldelli, G., Rocchi, M., Balsamo, M. (2010b). The distribution of meiofauna on back-reef sandy platforms in the Maldives (Indian Ocean). *Marine Ecology: An Evolutionary Perspective*, *vol. 31*, 592-607.

Shannon, C.E. and Weaver, W. (1949).*The Mathematical Theory of Communication*. Universityof Illinois, Urbana, Illinois.

Simpson, E.H. (1949). Measurement of diversity. *Nature, vol. 163*, 688.

Soetaert, K., Muthumbi, A. and Heip, C. (2002). Size and shape of ocean margin nematodes: morphological diversity and depth-related patterns. *Marine Ecology Progress Series*, *vol. 242*, 179–193.

Soltwedel, T., Pfannkuche, O. and Thiel, H. (1996). The size structure of deep-sea meiobenthos in the North-Eastern Atlantic: nematode size spectra in relation to environmental variables. *Journal of the Marine Biological Association of the United Kingdom*, *vol. 76*, 327–344.

Somerfield, P. J., Cochrane, S. J., Dahle, S. and Pearson, T. H. (2006). Free-living nematodes and macrobenthos in a high-latitude glacial fjord. *Journal of Experimental Marine Biology and Ecology*, *vol.330*, 284–296.

Steyaert, M., Garner, N., Gansbeke, D. and Vincx, M. (1999). Nematode communities from the North sea: environmental controls on species diversity and vertical distribution within the sediment. *Journal of the Marine Biological Association of the United Kingdom, vol. 79*, 253-264.

Steyaert, M., Vanaverbeke, J., Vanreusel, A., Barranguet, C., Lucas, C. and Vincx, M. (2003). The importance of fine-scale, vertical profiles in characterising nematode community structure. *Estuarine, Coastal and Shelf Science, vol. 58*, 353-366.

Suter, G. W. I. (2001). Applicability of indicator monitoring to ecological risk assessment. *Ecological Indicators, vol. 1*, 101–112.

Teal, J. M. and Wieser, W. (1966). The distribution and ecology of nematodes in a Georgia Salt Marsh. *Limnology and Oceanography, vol. 11*, 217-222.

Thistle, D. and Sherman, K. M. (1985). The nematode fauna of a deep-sea site exposed to strong near-bottom currents. *Deep-Sea Research, vol. 32*, 1077–1088.

Tita, G., Vincx, M. andDesrosiers, G. (1999). Size spectra, body width and morphotypes of intertidal nematodes: an ecological interpretation. *Journal of the Marine Biological Association of the United Kingdom, vol. 79*, 1007–1015.

Travizi, A. andVidakovic, J. (1994). An evaluation of eutrophication effects on northern Adriatic meio- and nematofauna communities. *PeriodicumBiologorum, vol. 96*, 469-473.

Vanaverbeke, J., Deprez, T. and Vincx, M. (2007). Changes in nematode communities at the long-term sand extraction site of the Kwintebank (Southern Bight of the North Sea). *Marine Pollution Bulletin, vol. 54*, 1351-1360.

Vanaverbeke, J., Gheskiere, T., Steyaert, M. and Vincx, M. (2002). Nematode assemblages from subtidal sandbanks in the Southern Bight of the North Sea: effect of small sedimentological differences.*Journal of Sea Research, vol. 48*, 197-207.

Vanaverbeke, J., Merckx, B., Degraer, S. and Vincx, M. (2011). Sediment-related distribution patterns of nematodes and macrofauna: Two sides of the benthic coin? *Marine Environmental Research, vol. 71*, 31-40.

Vanaverbeke, J., Soetaert, K. and Vincx, M. (2004a). Changes in morphometric characteristics of nematode communities during a spring phytoplankton bloom deposition. *Marine Ecology Progress Series,vol. 273*, 139-146.

Vanaverbeke, J., Steyaert, M., Soetaert, K., Rousseau, V., Van Gansbeke, D., Parent, J.Y. and Vincx, M. (2004b). Changes in structural and functional diversity of nematode communities during a spring phytoplankton bloom in the southern North Sea. *Journal of Sea Research, vol. 52*, 281-292.

Vanaverbeke, J., Steyaert, M., Vanreusel, A. and Vincx, M. (2003). Nematode biomass spectra as descriptors of functional changes to human and natural impact. *Marine Ecology Progress Series, vol. 249*, 17–157.

Vanreusel, A. (1990). Ecology of free-living marine nematodes in the Voordelta (southern Bight of the North Sea). I. Species composition and structure of the nematode communities. *Cahiers de Biologie Marine, vol. 31*, 439-462.

Vincx, M. (1989). Free-living marine nematodes from the Southern Bight of North Sea. Academia AnalectaMeded. K. Acad. Wet. Lett. Schone. Kunsten. Belg. K1. Wet. *vol.* 51, 39-70.

Vincx, M., Meire, P. and Heip, C. (1990). The distribution of nematode communities in the Southern Bight of the North Sea. *Cahiers de Biologie Marine, vol.* 31, 107-129.

Vranken, G. and Heip, C. H. R. (1986). Toxicity of copper, mercury and lead to a marine nematode. *Marine Pollution Bulletin, vol. 17*, 453-457.

Vranken, G., Herman, P. M. J., Vincx, M. and Heip, C. H. R. (1986). A re-evaluation of marine nematode productivity. *Hydrobiologia, vol.135*, 193-196.

Warwick, R. M. (1981). The nematode-copepod ratio and its use in pollution ecology. *Marine Pollution Bulletin, vol. 12*, 329-333.

Warwick, R. M. and Clarke, K. R. (1995). New "biodiversity" measures reveal a decrease in taxonomic distinctness with increasing stress. *Marine Ecology Progress Series, vol. 129,* 301–305.

Warwich, R. M. and Price, R. (1979). Ecological and metabolic studies on free-living nematodes from an estuarine mud-flat. *Estuarine, Coastal and Shelf Science, vol. 9*, 257-271.

Whittaker, R. H. (1960). Vegetation of the Siskiyou Mountains. Oregon and California. *Ecological Monographs, vol. 30*, 279–338.

Wieser, W. (1953). Die Beziehung zwischen Mundhöhlengestalt, Ernährungsweise und Vorkommen bei freilebenden marinen nematoden. Eine okologisch-morphologische studie. *Arkiv für Zoologie, vol. 4*, 439-484.

Will, K. W., Rubinoff, D. (2004). Myth of the molecule: DNA barcodes for species cannot replace morphology for identification and classification. *Cladistics, vol. 20*, 47 -55.

Wlodarsaka-Kowalczuk, M., Pearson, T. H. and Kendall, M. A. (2005). Benthic response to chronic natural physical disturbance by glacial sedimentation in an Arctic fjord. *Marine Ecology Progress Series, vol. 303*, 31–34.

Zullini, A. (1992). Nematodi. In M. Pandolfi, (Ed). *I rinogradi di HaraldStümpke e la zoologia fantastica*. ScienzeNaturaliTesti, Franco MuzzioEditore, Padova, 121.

In: Nematodes
Editors: F. Boeri and J. A. Chung

ISBN: 978-1-61470-784-4

Chapter V

Caenorhabditis elegans as a Research Tool to Unveil Bacterial Virulence Determinants: Lessons from the *Burkholderia cepacia* Complex

Christian G. Ramos and Jorge H. Leitão
Department of Bioengineering and Institute for Biotechnology and Bioengineering, Instituto Superior Técnico, Universidade Técnica de Lisboa, Av. Rovisco Pais, 1049-001, Lisboa, Portugal.

Abstract

Caenorhabditis elegans is a small, free-living soil nematode that survives by feeding on microbes, primarily bacteria. *C. elegans* has been used as a model organism for a variety of reasons, including economy, and the ease of maintaining a worm population in the laboratory. A useful feature of *C. elegans* is that it is relatively straightforward to test pathogenesis. As the worm feeds on bacteria, when feeding bacteria are replaced by pathogenic bacteria, the worm will suffer an infection. The number of surviving worms and other pathogenic traits can be easily monitored, helping in the deciphering the underlying pathogenic traits.

In this chapter, we will present and discuss some examples on the use of the nematode *C. elegans* as an infection model to study the pathogenic traits of several Gram-positive and Gram-negative pathogens. Results from our own experience in using the nematode as an infection model to unveil virulence determinants of bacteria of the *Burkholderia cepacia* complex are also described. The major advantages and pitfalls of the *C. elegans* infection model are also discussed.

Introduction

Caenorhabditis elegans is a small, free-living soil nematode (roundworm) that lives in many parts of the world and survives by feeding on microbes, primarily bacteria (Altun and Hall, 2005) (Figure 1). The nematode is an important model system in many biological research fields, including genetics and genomics, cell biology, neuroscience and aging. Among the many advantages of the nematode as a model system, are its short life cycle, compact genome, stereotypical development, easy maintenance and propagation, and small size. The adult bodyplan is anatomically simple, with about 1000 somatic cells. *C. elegans* is amenable to genetic crosses and produces a large number of progeny per adult. The life cycle of the animal is about 3 days under optimal conditions.

The nematode can be easily maintained in the laboratory where it grows on agar plates or liquid cultures with *E. coli* as the food source. It can be examined at the cellular level in live preparations by differential interference contrast (DIC) microscopy, since it is transparent throughout its entire life cycle (Altun and Hall 2005).

C. elegans individuals are sexually dimorphic. They can exist as self-fertilizing hermaphrodites (XX) or males (XO). Males arise infrequently (0.1%) by spontaneous non-disjunction in the hermaphrodite germ line and at higher frequency (up to 50%) through mating (hermaphrodites can also be induced to generate male progeny spontaneously at a higher rate by treatment at high temperature). Self-fertilization of the hermaphrodite allows homozygous worms to generate genetically identical progeny. Male mating facilitates the isolation and maintenance of mutant strains, as well as moving mutations between strains. Mutant worms can be readily obtained by chemical mutagenesis or exposure to ionizing radiation. The strains can be kept as frozen stocks for long periods of time. *C. elegans* can also endure harsh environmental conditions by switching to a facultative diapause stage called dauer larvae, which can survive up to 4 to 8 times longer than the normal 3-week life span of the worm. Despite its simple anatomy, the animal displays relatively simple behavior, including locomotion, foraging, feeding, defecation, egg laying, dauer larva formation, sensory response to touch, smell, taste and temperature. In addition, complex behaviors, including male mating, social behavior, learning and memory are also exhibited by the worm (Altun and Hall 2005).

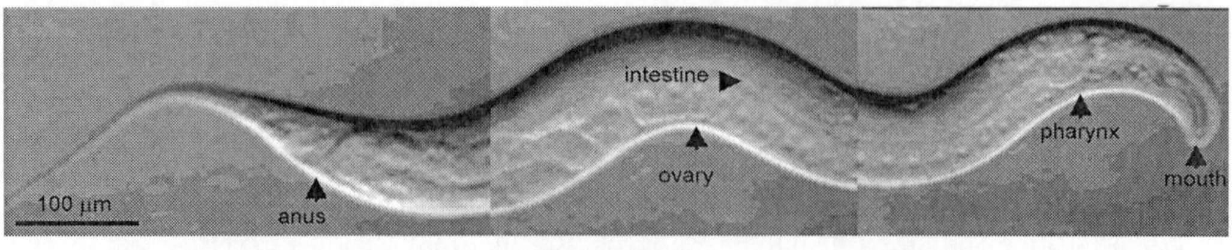

Figure 1. Photomicrograph of an adult *Caenorhabditis elegans*. The mouth, pharynx, ovary, intestine and anus (indicated with arrows) are clearly visible.

Similar to other nematodes, the life cycle of *C. elegans* comprises the embryonic stage, four larval stages (L1-L4) and adulthood (Figure 2). The end of each larval stage is marked by a molt where a new, stage-specific cuticle is synthesized and the old one is shed (Altun and Hall 2005).

C. elegans is used as a model organism for a variety of reasons, including economy and the easy of maintenance of a worm population in the laboratory. Worms can be frozen and subsequently thawed and remain viable, thus ensuring easy long-term storage of different worm strains (Hope 1999).

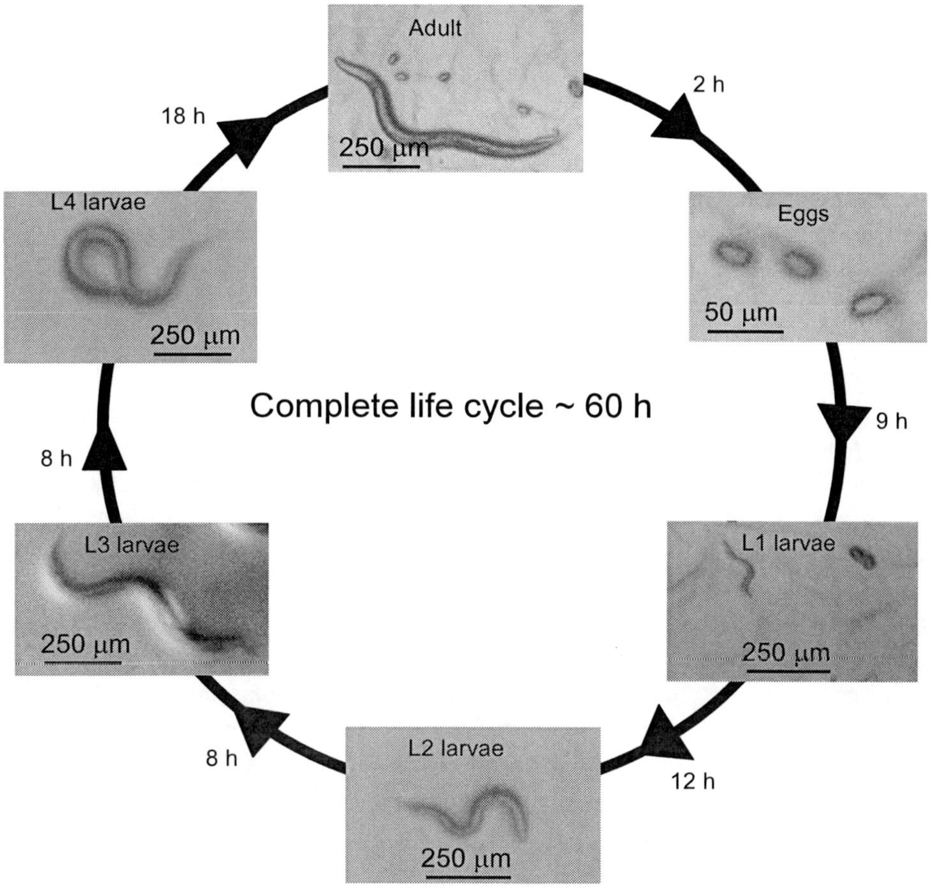

Figure 2. Photomicrographs of the representative stages of the *Caenorhabditis elegans* life cycle.

Another useful feature of *C. elegans* is that it is relatively straightforward to test pathogenesis. As the worm feeds on bacteria, when feeding bacteria are replaced by pathogenic bacteria, the worm will suffer an infection. The survival rate and other pathogenic traits related to infection can be easily monitored, helping in the deciphering the underlying pathogenic traits (Riddle et al. 1997).

Innate Immunity and Learned Behavior

The nematode *C. elegans* has provided, over the last decade, a simple experimental system to study host-pathogen interactions.

Bacteria are ubiquitous organisms, present in almost every square inch of our planet. All metazoans have already been exposed to a wide array of bacteria, eliciting the evolution of complex immune responses to cope with bacterial pathogens. With the exception of vertebrates, most metazoan organisms depend exclusively on their innate immune system to fight against bacterial pathogens. The innate immune system involves several autonomously-acting molecular factors, responsible for orchestrating the changes required to eradicate infection (Tan and Shapira 2011). This immune response relies largely on the recognition and elimination of the threatening pathogen, making use of evolutionary conserved signaling pathways (Gravato Nobre and Hodgkin 2005; Tan and Shapira 2011). In nature, *C. elegans* interacts with a wide range of soil bacteria, some of which are recognized human and nematode pathogens, as is the case of *Pseudomonas aeruginosa* (Tan and Shapira 2011) and *Burkholderia cepacia* complex bacteria (Cardona et al. 2005). Currently, over forty distinct microorganisms are recognized as *C. elegans* pathogens, including bacteria, fungi, viruses and microsporidia (Tan and Shapira 2011).

Compared to mammals, insects, and plants, relatively little is known about innate immune responses in the nematode *C. elegans*. Several distinct signal transduction cascades can be activated upon infection. This activation of specific cascades, in a pathogen-dependent fashion, ultimately leads to the activation of protective cellular responses and the production of molecules able to confine or destroy the pathogens (Gravato Nobre and Hodgkin 2005).

Initial genetic studies on the immunity of *C. elegans* have revealed that conserved innate immune signaling pathways also function in the nematode in response to bacterial infection.

C. elegans has at least four signaling pathways involved in response to bacterial infections. These are the p38 mitogen-activated protein (MAP) kinase pathway, the programmed cell death pathway, the tumor growth factor-β pathway, and the DAF-2 insulin/IGF-I like signaling pathway. As a result of the activation of these pathways, immune effector molecules such as lysozymes, lipases and saposin-like proteins are produced, which can act directly against the invading microorganisms (Nicholas and Hodgkin 2004). The effectors produced depend on the nature of the infecting agent as a result of the triggered signaling pathways. This suggests that the worm is able to detect and discriminate between different infecting microorganisms (Nicholas and Hodgkin 2004).

A study by Kim et al. (2002) using a large scale genetic screening for *C. elegans* mutants resistant to infection by *Pseudomonas aeruginosa* revealed two genes involved in the resistance to infection. These genes encode two distinct MAP kinases (*sek-1* and *nsy-1*). The study of Kim et al. (2002) has established that a p38 ortholog is required for pathogen defense. The authors suggest that the MAP kinase signaling pathway represents an ancient evolutionarily-conserved feature of innate immune responses among diverse species.

The critical role that this pathway plays in innate immune signaling pathways of mammals suggested that the study of the genetic determinants of pathogen resistance in *C. elegans* would illuminate evolutionarily conserved mechanisms of innate immunity (Ruiz-Diez et al. 2003; Speert 2002).

A major technological advance in the functional genomics analysis of *C. elegans* has been the development of methodologies enabling the systematic analysis of gene function using RNA interference (RNAi). For example, the effect of silencing, by RNAi technology, of each one of the more than 16,000 genes on the p38 MAPK signaling in the intact organism, under conditions of infection by the pathogen *P. aeruginosa,* was carried out by Vikstrom et al. (2005). This study was performed using a transgenic worm strain carrying a green fluorescent protein reporter gene fused to the promoter of a presumptive antimicrobial peptide identified from a genome-wide microarray analysis of p38 MAPK-regulated genes (Vikstrom et al. 2005).

The presence of structural homologues of components of the Toll pathway in *C. elegans*, suggests that a Toll-like pathway might function to trigger the *C. elegans* defense mechanisms towards bacterial pathogens (Pujol et al. 2001). Whereas studies of mammalian innate immunity have been focused largely on Toll-like receptor (TLR) signaling, TLR-independent signaling pathways have remained relatively unexplored in the nematode *C. elegans*. Recent work suggests that conserved TLR-independent signaling pathways play a key role in *C. elegans* immunity. Therefore, the identification of TLR-independent signaling pathways in innate immunity is of particular interest to elucidate the nematode response to infection (Kim et al. 2002; Riddle et al. 1997).

The major route of the nematode infections occurs via the digestive tract, although some pathogens are also able to infect the nematode via the vulva and the anal epithelia (Tan and Shapira 2011). The intestine of the nematode comprises roughly one third of all the animal somatic cells and is derived from a single endodermal cell. The two primary pathways that elicit immunity at the intestinal level encompass the p38 MAPK and the insulin/IGF1 signaling (IIS). These two pathways are responsible for regulating infection response genes, such as the lysozyme encoding gene *lys-2*, the ShK-like toxin encoding gene *T24B8.5*, the intestinal lysozymes LYS-7, or the saposin-like SPP-1 proteins (Tan and Shapira 2011). Studies on the role of the GATA transcription factor in innate immune response have shown that this transcription factor integrates external signals, being required for infection responses and survival when feeding on a pathogen. Moreover, the human homolog transcription factor GATA6 has a protective function in lung epithelial cells exposed to *P. aeruginosa* (Shapira et al. 2006).

In nature, *C. elegans* lives in the soil and feeds on bacteria, where it may encounter several pathogens. Thus, the ability to distinguish between pathogenic and "eatable" bacteria is of critical importance as a defense mechanism. This is achieved by a chemosensory system composed of 32 sensory neurons, located at both the head and the tail of the nematode (Tan and Shapira 2011). This system responds to several stimuli, including compounds produced by bacteria. The later can be used by the nematode to classify bacteria as "eatable" or noxious (Zhang et al. 2005). Specific changes in olfactory preferences are elicited by particular bacteria, inducing associative learning (Zhang et al. 2005). An increase in serotonin in the ADF chemosensory neurons is detected after exposure to pathogenic bacteria. Serotonin promotes aversive learning, representing a negative reinforcing stimulus in pathogenic infection (Zhang et al. 2005).

This chemosensory behavior has been attributed to the *tol-1* gene, from the Toll pathway, since *C. elegans tol-1* mutants are defective in their avoidance of pathogenic *Serratia marcescens* (Pujol et al. 2001). Nevertheless, wild-type worms avoid *S. marcescens* by sensing the repellent cyclic lipodepsipentapeptide serrawettin W2, a surfactant secreted by some *S. marcescens* strains (Tan and Shapira 2011). Interestingly, Beale et al. (2006) have showed that *C. elegans* is attracted by acylated homoserine lactones (AHLs). These molecules are used by quorum sensing systems of *P. aeruginosa* (Khan et al. 2005; Maddula et al. 2006; Riedel et al. 2001; van Delden et al. 2001; Vikstrom et al. 2005; Wopperer et al. 2006), *Burkholderia cepacia* complex bacteria (Aguilar et al. 2003; Baldwin et al. 2004; Eberl 2006; Lewenza et al. 1999; Lutter et al. 2001; Malott et al. 2005; Tomlin et al. 2005), as well as by other gram-negative bacteria. Apparently, *C. elegans* can distinguish AHLs and may use them to mediate aversive or attractive learning, showing a memory-like behavior (Tan and Shapira 2011). This "olfactory" system can be used by bacteria to elude the nematode, as is the case of *Bacillus nematocida*, which produces food-like volatile organic compounds that act as chemo-attractants. After their ingestion by the nematode, the bacterium secretes proteases able to kill *C. elegans* (Tan and Shapira 2011).

C. elegans as a Bacterial Infection Model

In several pathogens able to infect a broad range of hosts, it has become evident that universal virulence factors required for full pathogenicity are produced, regardless of the host. This is the case of the human opportunistic pathogens *P. aeruginosa* (Mahajan-Miklos et al. 1999) and of bacteria of the *Burkholderia cepacia* complex (Sousa et al. 2011). Non-mammalian infection models are therefore useful to assay universal virulence determinants, as is the case of *C. elegans*. One can test whether a given bacterium is pathogenic for *C. elegans* in a relatively straightforward way: as the worm feeds mainly on *E. coli*, the food source can be replaced by the pathogen under study, and the disease progress can be followed. Some pathogens produce lethal toxins which rapidly kill the nematode, such as *P. aeruginosa* (Mahajan-Miklos et al. 1999) and *B. cenocepacia* strain K56-2 (Ramos et al. 2010b). The production of toxins can be tested using special culture conditions (Cardona et al. 2005). Other pathogens are clearly virulent to *C. elegans*, such as certain strains of *P. aeruginosa*, *S. marcescens* and *Burkholderia*, which establish intestinal infection, provoking evident signs of illness, like locomotion defects, distended intestine, paralysis or erratic movement and cell lysis, which ultimately results in premature death of the worm (Ewbank 2002). When a particular bacterial strain clearly proves to be pathogenic to *C. elegans*, one approach that can be used to unveil the underlying virulence factors or determinants relies on the generation of collections of mutants and test the mutants for reduced virulence (Sousa et al. 2011). Problems may arise, when scoring the surviving worms, in particular when the worms are exposed to pathogenic bacteria that do not kill the nematodes before they are able to reproduce (Ewbank 2002). For example, the two clonally identical strains of *B. cenocepacia* J2315 and K56-2, kill the worm at distinct rates, with the later presenting a faster rate of killing, thus preventing reproduction of the worm (Ramos et al. 2010; Ramos et al. 2011). This problem can be circumvented by using a *C. elegans* mutant strain that cannot reproduce under certain circumstances, avoiding the scoring of the progeny.

C. elegans as an Infection Model to Study Gram-Positive Pathogens

The nematode *C. elegans* has been used as model host to study several Gram-positive human bacterial pathogens. Adult nematodes are killed by *Enterococcus faecalis, Streptococcus pneumoniae*, and *Staphylococcus aureus* (Garsin et al. 2001). Studies focused on the enterococcal species, found that both *E. faecalis* and *E. faecium* kill *C. elegans* eggs and hatchlings, but only *E. faecalis* kills the adults (Garsin et al. 2001).

The gram-positive bacterium *Streptococcus pyogenes*, a major human pathogen capable of causing a wide spectrum of diseases, has been used in *C. elegans* infection experiments to test pathogenesis (Jansen et al. 2002). This bacterium kills *C. elegans*, both on solid and in liquid medium. Unlike other pathogens, the killing of *C. elegans* by *S. pyogenes* is exclusively mediated by hydrogen peroxide, in a dose-dependent fashion (Jansen et al. 2002). Jansen et al. (2002) suggested that the killing of *C. elegans* by *S. pneumoniae* is also mediated by hydrogen peroxide.

Staphylococcus aureus is an important human pathogen, and multi-drug resistant strains have been increasingly recovered from patients worldwide. *S. aureus* has been shown to be capable of killing the nematode *C. elegans* (Sifri et al. 2003). Experiments performed with *C. elegans* feeding on *S. aureus* showed that the nematode dies over the course of several days, in a process correlated with the accumulation of bacteria within the nematode digestive tract (Sifri et al. 2003). Several *S. aureus* virulence determinants are required for full pathogenicity to the nematode, including the quorum-sensing global virulence regulatory system *agr* and the global virulence regulator *sarA*, the alternative sigma factor σ^B, alpha-hemolysin, and the V8 serine protease. These virulence determinants are believed to be also important in mammalian pathogenesis (Sifri et al. 2003). The ability of staphylococci to grow on biofilms is also recognized as an important virulence mechanism which allows bacteria both to adhere to the host cell and to resist to its immune system. Biofilms are also known to contribute to the bacterium resistance to antibiotics (Begun et al. 2007). The production of the biofilm-associated polysaccharide intercellular adhesin (PIA) in staphylococci, is required for the formation of a lethal *S. epidermidis* infection in the intestine of *C. elegans*. In fact, loss of PIA production renders the bacterium less virulent, suggesting that biofilms enhance virulence by playing an immunoprotective role during colonization of the *C. elegans* intestine (Begun et al. 2007). Mutation of the *C. elegans* PMK-1 p38 mitogen-activated protein (MAP) kinase or the DAF-2 insulin-signaling pathways rendered the worm more suspectible to *S. epidermidis* infection (Begun et al. 2007), highlighting the important role of the worm innate immune system in combating bacterial infections.

Recently, small colony variants (SCVs) of *S. aureus* have been isolated from patients. SCVs are slow-growing morphological variants that have been implicated in the persistence of infections, relapsing, and increased resistance to antibiotics (Sifri et al. 2006). SCVs are being described for an increasing number of bacterial pathogens, including both Gram-positive and Gram-negative bacteria. The altered phenotypes of SCVs has been attributed to several defects, including electron transport due to mutations in hemin or menadione biosynthesis (Sifri et al. 2006). Recently, Sifri et al. (2006) have evaluated the virulence mechanisms of *S. aureus* SCVs variants using the *C. elegans* infection model. These authors showed that SCVs clinical isolates, as well as *S. aureus hemB*- and *menD*-deficient mutants are greatly reduced in their virulence to the *C. elegans*. Another study using *C. elegans* as a infection model has revealed five previously unidentified virulence factors of *S. aureus* (Begun et al. 2005).

Yersinia pathogenicity determinants have also been investigated using *C. elegans*. Different strains of *Yersinia pseudotuberculosis* showed different levels of pathogenicity. Roughly 76 % of the *Yersinia* strains analysed caused no discernible effect; 5 % caused a weak infection, 9.5 % caused an intermediate infection, and 9.5 % caused a severe infection (Joshua et al. 2003). This study also revealed that *Y. pseudotuberculosis* mutant strains in genes encoding proteins involved in hemin storage or lipopolysaccharide biosynthesis were completely or partially attenuated in their virulence towards *C. elegans* (Joshua et al. 2003).

In addition, the study of interactions between the Gram-positive bacterial pathogen *Microbacterium nematophilum* and *Caenorhabditis elegans* provided a model to study the innate immune response in nematodes. *M. nematophilum* adheres to the rectal and post-anal cuticle of the worm, causing slower growth, constipation, and a defensive swelling response of rectal hypodermal cells. This response is triggered by genes encoding proteins containing C-type lectin domains, being important for innate immune defense in *C. elegans* (Sifri et al. 2005). Sifri et al. (2005) highlighted that the genes induced during infection by *M. nematophilum* appear to be largely distinct from genes induced by other pathogens, reinforcing the idea that *C. elegans* responses are pathogen-specific (Sifri et al. 2005).

C. elegans genes required for survival to *S. aureus* infection have also been identified. For example, enhanced susceptibility to killing by *S. aureus* was observed with loss-of-function mutations in the *C. elegans* genes *esp-2/sek-1* and *esp-8/nsy-1*, which encode components of a conserved p38 MAP kinase signaling pathway involved in nematode defense against multiple pathogens, suggesting that specific *C. elegans* factors can alter susceptibility to this gram-positive human pathogen (Sifri et al. 2003).

C. elegans as an Infection Model to Study Gram-Negative Pathogens

The enterobacterium *Salmonella typhimurium*, generally considered to be a highly adapted pathogen with a narrow range of target hosts, is also capable of infecting and killing *C. elegans*. Furthermore, *S. typhimurium* mutant strains exhibiting a reduced virulence to *C. elegans* were also attenuated in their virulence to mammals, showing that the nematode may constitute a useful model system to study this important human pathogen (Labrousse et al. 2000).

C. elegans has also been used as an infection model to evaluate the pathogenicity of several *Salmonella enterica* serovars, including *S. enteritidis* and *S. dublin* (Aballay et al. 2000). Killing of the nematode by *S. typhimurium* involves the accumulation of bacterial cells in the lumen of the worm intestine. The nematodes die over the course of several days. Killing requires contact with live bacterial cells (Aballay et al. 2000).

Pathogens such as *Salmonella enterica* are thought to co-evolved or to have had a long-standing association with their hosts. This pathogen appears to use highly host-specific strategies to establish a pathogenic relationship (Aballay and Ausubel 2002). The main ecological niche of Salmonella is the vertebrate intestinal lumen, where the pathogen responds to a number of relevant environmental conditions. As a consequence, the bacterium can produce several virulence factors. Despite the expectation that the narrow range of hosts of *S. enterica* would be limited to vertebrate species, several *S. enterica* serovars, including *Salmonella typhimurium*, are able to kill *C. elegans* (Aballay et al. 2000). This killing occurs over the course of several days, at a rate significantly slower than the killing of the nematode by *P. aeruginosa* strain PA14 (Mahajan-Miklos et al. 1999). Interestingly, *S. typhimurium*-mediated killing of *C. elegans* was initially underestimated, because the parental worms initially exposed to *S. typhimurium* produce large numbers of progeny that grow to full-sized adults before they are also killed, mitigating the killing of the parental worms (Aballay and Ausubel 2002).

The *C. elegans* infection model has shown to be particularly effective in unraveling both microbial and animal genes involved in toxin-mediated killing (Alegado et al. 2003). Additionally, research on worm infections by several bacterial pathogens has shown that a number of virulence regulatory genes mediate worm susceptibility, including regulators of the PhoP/Q two-component system of *Salmonella* and the LasR of *Pseudomonas aeruginosa*, which have also been implicated in virulence using mammalian models of infection. Again, these findings reinforce the use of the worm as a valid model to study virulence determinants and mechanisms in other bacteria (Alegado et al. 2003).

Work from the F. Ausubel Lab. has shown that the human opportunistic pathogen *Pseudomonas aeruginosa* strain PA14 (a clinical isolate from a cystic fibrosis patient), which is pathogenic to mice and plants, also kills *C. elegans*. Tan et al. (1999) showed that the rate of PA14-mediated killing of *C. elegans* depends on the composition of the agar medium on which the strain is grown. Killing of the worm can occur over the course of several days, when the bacterium is grown on minimal medium, and is referred to as "slow" killing. When killing is promoted by growing the bacterium on high-osmolarity medium, *C. elegans* dies within the course of several hours. This is known as the "fast" killing (Tan et al. 1999). Independent experimental evidences indicate that fast and slow killing occur through distinct mechanisms. The fact that heat-killed bacteria are still capable of causing fast, but not slow killing of *C. elegans*, contrasts with the slow killing process, which involves an infection-like process and correlates with the accumulation of live bacteria within the digestive tract of the worm (Tan et al. 1999). Tan et. al. (1999) also reported that six mutants of *P. aeruginosa* PA14 that had bccn shown previously to be affected in their pathogenicity to plants and mice, were also less effective in killing *C. elegans* under both fast- and slow-killing conditions (Tan et al. 1999). This suggests the presence of universal virulence determinants in *P. aeruginosa,* which are required for virulence towards distinct eukaryotic hosts.

The establishment of a *Caenorhabditis elegans–Pseudomonas aeruginosa* pathogenesis model (Tan and Ausubel 2000) has facilitated the systematic dissection of both host and pathogen genes involved in pathogenic interactions. The deep and thorough analysis of the *P. aeruginosa–C. elegans* interaction has opened new windows and opportunities to investigate host-pathogens interactions at the molecular level.

The use of *C. elegans* for the Identification of host factors required for the interaction with pathogens is crucial to the understanding of the infection process, since the direct screening for host mutations to aid in this task is not straightforward in mammals. Using a *P. aeruginosa-C. elegans* infection model, Darby et al. (1999) showed that *P. aeruginosa* excretes a toxin, only produced at high cell density under the control of the quorum-sensing regulators LasR and RhlR. This toxin rapidly and lethally paralyzes *C. elegans* (Darby et al. 1999). The data found in literature strongly contrasts *P. aeruginosa* infections with those mediated by *S. typhimurium*, where a small inoculum of the later can lead to proliferation in the *C. elegans* intestine and to the establishment of a persistent infection (Aballay et al. 2003). This probably derives from the fact that *Salmonellae* are more adapted to the intestinal environment than other pathogens.

C. elegans has also been used to study the pathogenicity of the human opportunistic pathogen *Serratia marcescens*. *C. elegans* is killed after intestinal colonization by *S. marcescens* (Kurz et al. 2003). Interestingly, nine *S. marcescens* mutants with an attenuated virulence phenotype in *C. elegans* also showed a reduced virulence in an insect model of infection (Kurz et al. 2003). Using the *C. elegans* as an infection model, the studies of Kurz et al. (2003) revealed that among the genes necessary for full *in vivo* virulence are those required for lipopolysaccharide (LPS) biosynthesis, iron uptake and hemolysin production, as well as conserved virulence factors also required for *Pseudomonas aeruginosa* pathogenicity.

The *Burkholderia cepacia* Complex

Burkholderia cepacia was initially described as *Pseudomonas cepacia* in 1950 as a phytopathogen responsible for the rot of onions (Burkholder 1950). Taxonomically, the species is one of the seventeen closely related species collectively known as the *Burkholderia cepacia* complex (Bcc) (Vanlaere et al. 2006; Vanlaere et al. 2009). The Bcc include plant, animal and human pathogens, being widely distributed in natural and man-made habitats (Mahenthiralingam et al. 2005). These bacteria are extraordinarily versatile, exhibiting an impressive metabolic plasticity endowed by a considerably large genome (~8 Mb), allowing their adaptation to a wide range of environments. Strains belonging to all Bcc species have been isolated from cystic fibrosis (CF) patients and from the environment. However, their frequency of isolation is uneven, while the majority of the isolates obtained from infected CF patients belong to the species *B. cenocepacia* and *B. multivorans* (Mahenthiralingam et al. 2008), the majority of the environmental isolates belong to the species *B. cepacia*, *B. ambifaria*, *B. cenocepacia*, and *B. pyrrocinia* (Chiarini et al. 2006).

Some evidences indicate that the environment is the reservoir for the acquisition of novel infections by Bcc strains. This is the case of the epidemic *B. cenocepacia* strain PHDC, which was recovered both from patients suffering from cystic fibrosis (CF) in the mid-Atlantic region of USA, and from agricultural soils (LiPuma et al. 2002).

Burkholderia cepacia H111, which was isolated from a cystic fibrosis patient, was shown to effectively kill *C. elegans,* both by the production of an extracellular toxin and by the accumulation of bacteria in the intestinal lumen of *C. elegans* (Kothe et al. 2003). The *cep* quorum-sensing system was demonstrated to be required for efficient killing of *C. elegans,* being involved in the entrance of the bacterium into the intestinal lumen of the worm, but dispensable for the colonization of the gut (Kothe et al. 2003).

The *cep* quorum-sensing system controls the expression of several virulence factors, as well as the formation of biofilms (Huber et al. 2004). The identification of a *cep*-regulated gene, encoding the protein AidA, was demonstrated to be essential for slow killing of *C. elegans* but has little effect on fast killing, suggesting that the protein plays a role in the accumulation of the strain in the nematode gut (Huber et al. 2004).

Lessons from the Use of *C. elegans* to Unveil Bcc Virulence Determinants

Over the last years, our research team has been involved in studies of bacteria of the Bcc, especially due to the involvement of these pathogens in severe and often lethal respiratory infections among patients suffering from cystic fibrosis. In order to identify genetic determinants that underlie their success as pathogens, an easy-to-maintain infection model was necessary to deepen our studies. In collaboration with Professor Leo Eberl from the Microbiology Department of the University of Zurich (Switzerland), we have started using the *C. elegans* wild-type strain Bristol N2 as an infection model.

In our initial studies, and depending on the Bcc strain under study, the scoring of surviving worms was not always a straightforward task. For highly virulent strains which were able to kill the nematode population within a time period of 3-4 days, the scoring of live worms was straightforward. However, when working with less virulent Bcc strains, the worms were still able to reproduce, leading to high numbers of progeny. In addition, the initial worms developed a sick appearance, very thin and with a small body size. As the progeny progressed throughout the larval stages of development, their appearance became indistinguishable from the progenitors, definitely compromising the scoring of live worms. These difficulties in scoring the surviving worms upon infection with less virulent Bcc strains were circumvented by using the *C. elegans* mutant strain DH26, which carries a *fer-15* mutation (Altun and Hall, 2005). This mutation affects the reproduction of the worm, rendering the animal unable to lay eggs at 25 ºC. Thus, we routinely use *C. elegans* DH26 at 25 ºC when assessing virulence of Bcc strains and mutants, regardless of the relative virulence of the Bcc strains under study.

The use of the *C. elegans* as a model of infection allowed us to identify several genetic determinants required for full virulence of Bcc bacteria. Our studies have been focused mainly on three strains: The *B. cepacia* IST408, a clinical isolate from a Portuguese cystic fibrosis patient (Richau et al., 2000), *B. cenocepacia* K56-2 (Mahenthiralingam et al., 2000) and *B. cenocepacia* J2315 (Vandamme et al., 2003), two clinical isolates obtained from CF patients and both belonging to the ET12 lineage.

Burkholderia cenocepacia J2315 is one of the most virulent and epidemic clinical isolates of the *B. cepacia* complex (Bcc). In a report by Sousa et al. (2008), a *B. cenocepacia* J2315 mutant on an acyl carrier protein encoding gene (*acp*), showed a reduced ability to colonize and kill *C. elegans*, indicating that ACP is involved in virulence (Sousa et al. 2008). Bacterial ACPs are the donors of the acyl moiety involved in the biosynthesis of fatty acids and several other biological processes. A complete conservation of ACP proteins in the *Burkholderia* strains sequenced so far was observed, together with the high structural similarity between the ACP proteins from *B. cenocepacia* J2315 and *E. coli*. This protein was therefore pointed out as a potential target to develop new antibacterial agents to combat infections caused not only by Bcc species, but also by other *Burkholderia* species, especially *B. pseudomallei* and *B. mallei,* the causative agents of melioidosis and glanders, respectively (Sousa et al. 2008). In fact, antisense phosphorodiamidate morpholino oligomers (PMOs) were tested for their ability to inhibit ACP expression in *Escherichia coli* by Geller et al. (2003). In their report, these authors showed that PMOs targeting the acyl carrier protein (*acpP*) mRNA significantly inhibited their targets in a dose-dependent response (Deere et al. 2005; Geller et al. 2003). Bacterial ACP are so promising targets to develop new antimicrobials that several patents have already been issued (Geller et al. 2009; Geller et al. 2006; Geller et al. 2010).

The transcriptional regulator Pbr from *B. cenocepacia* K56-2 was also found to be involved in virulence, using the nematode *C. elegans* as an infection model (Ramos et al. 2010b). The *pbr* mutant was found to be deficient in phenazine production. Phenazines are a large family of heterocyclic nitrogen-containing compounds with broad-spectrum antibiotic activity that can act as redox compounds, impairing the host oxidative stress defences (Mavrodi et al. 2006) Phenazines are widely produced by *Pseudomonas*, and have been showed to play an important role in the pathogenicity of *P. aeruginosa* to the nematode (Mahajan-Miklos et al., 1999).

A study involving a large screening of random mutants generated by plasposon mutagenesis and presenting an EPS-defective phenotype, revealed a *B. cepacia* IST408 mutant strain that also showed severely decreased ability to colonize and kill *C. elegans* (Sousa et al. 2010). The plasposon insertion in this mutant was mapped in a gene encoding the RNA chaperone Hfq. In addition, the *B. cepacia hfq* mutant was also found to be susceptible to osmotic, acidic and oxidative stresses (Sousa et al. 2010), conditions faced by bacteria when infecting the CF host. Moreover, this attenuated virulence phenotype was also observed for two *hfq* mutants derived from clinical isolates of the Bcc species *B. dolosa* and *B. ambifaria*. These results suggest that Hfq plays a major role on the pathoadaptation of Bcc, contributing to their success as pathogens for *C. elegans*, as well as for patients suffering from cystic fibrosis (Sousa et al. 2010).

Bcc members are almost unique among prokaryotes for harbouring two highly conserved and distinct copies of Hfq-like proteins in their genomes (Ramos et al. 2010a; Ramos et al. 2011). A recent study showed that the 188-amino acid residue Hfq2 is also required for *B. cenocepacia* J2315 full virulence, in particular to colonize and kill *C. elegans*, despite the presence of the other functional *hfq*-like gene encoding the small 79 amino acid residue Hfq protein (Ramos et al. 2011). The authors also presented results indicating a distinct pattern of regulation of the *hfq* and *hfq2* genes in *B. cenocepacia* J2315, suggesting that a tightly regulation of the levels of these RNA chaperones contributes to the fitness and virulence of the bacterium as a successful pathogen.

An interesting exercise involves the comparison of the survival rates of *C. elegans* when exposed to diverse bacterial pathogens, and then compare the data with their pathogenicity in other models of infection or correlate data with the outcome of the human infection. Results shown in Figure 3 summarize data from the literature, regarding the infectious potential of several bacterial pathogens towards the nematode *C. elegans*.

Interestingly, after 72-h of exposure to the indicated pathogens, remarkably distinct outcomes have been described. For instance, one would expect that *Salmonella* and *Vibrio* would cause a deadly infection in *C. elegans* shortly after colonization. However, this is not the case, and these pathogens, apparently very well equipped for establishing infections in the gastrointestinal tract of mammals, seem less pathogenic than *Staphylococcus, Pseudomonas* or *Burkholderia* strains.

It should be noted that, when comparing the *B. cenocepacia* J2315 and *B. cepacia* IST408, the later is considerably more pathogenic to *C. elegans* than the first, although this trend is the opposite to the observed when gp91$^{phox-/-}$ mice where challenged with these strains (Sousa et al. 2007). This was also been observed for humans. Indeed, *B. cenocepacia* J2315 caused several deaths among CF patients (Mahenthiralingam et al. 2005), while the death of the patient infected from whom *B. cepacia* IST408 was isolated, was not related to *B. cepacia* infection (Cunha et al. 2003).

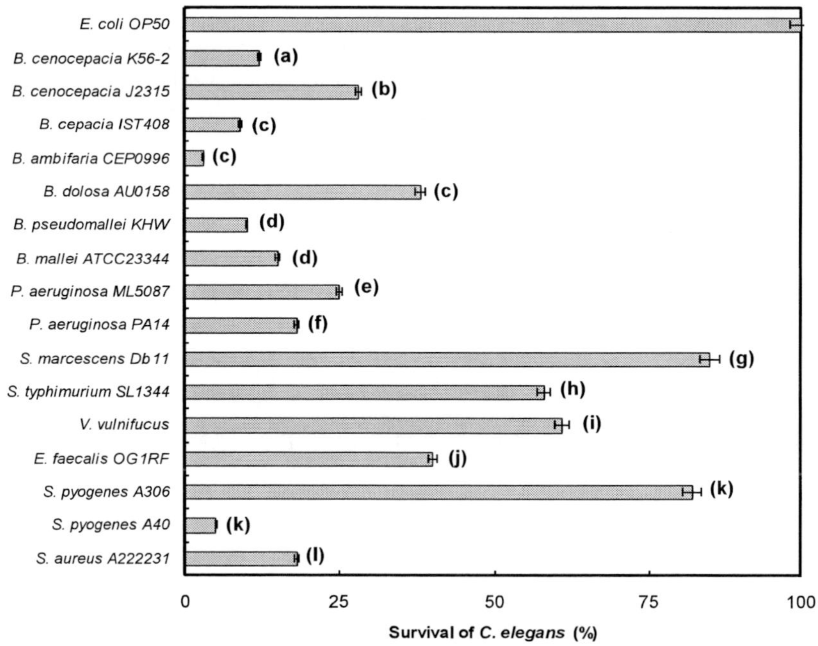

Figure 3. Comparison of the ability of the indicated bacterial pathogens, to kill *C. elegans*, after 72 h of infection. Data presented was obtained from (a), Ramos et al. (2010); (b), Ramos et al. (2011); (c), Sousa et al. (2010); (d), Gan et al. (2002); (e), Ruiz-Diez et al. (2003); (f), Mahajan-Miklos et al. (1999); (g), Kurz et al. (2003); (h), Aballay et al. (2000); (i), Dhakal et al. (2006); (j), Sifri et al. (2002); (k), Jansen et al. (2002) and (l), Sifri et al. (2006).

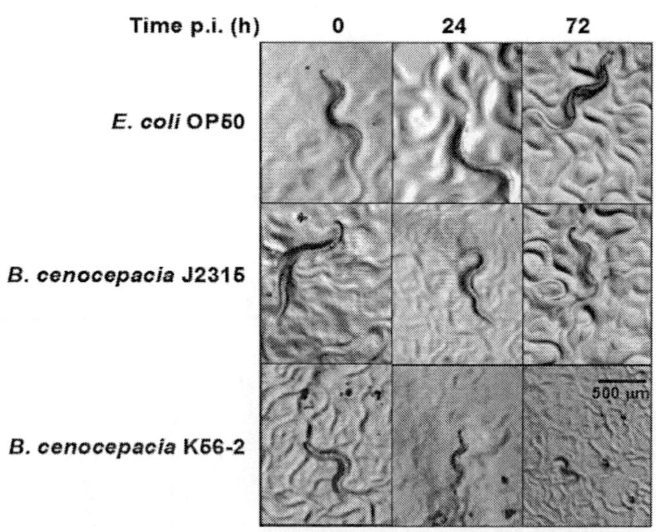

Figure 4. Microphotographs of *C. elegans* feeding on the indicated bacterial strains, for the indicated time. After 24 or 72 h of feeding on *B. cenocepacia* K56-2, the worms are clearly thinner and sicker than worms feeding on *B. cenocepacia* J2315.

Another interesting observation is the different killing rate of *C. elegans* by very closely related bacterial strains, such as *B. cenocepacia* strains J2315 and K56-2. As shown in Figure 4, the nematode appears much sicker after 72h of infection with the K56-2 strains than with the J2315 strain. The killing kinetics of *C. elegans* by each strain is slightly different, with the K56-2 strain killing 95% of the worm population within 4 days, while the J2315 strain kills about 90% of the worm initial population only after 5 days (Ramos et al. 2010; Ramos et al. 2011).

Altogether, the extensive list of studies has, undoubtedly, proved the utility of *C. elegans* as an *in vivo* model for the study of bacterial virulence and prompted several advances on the molecular understanding of bacterial pathogenicity.

Limitations of *C. elegans* as an Infection Model

The use of the nematode has been shown very useful for a variety of reasons to study host-pathogen interactions. However, like any other system, there are some pitfalls that should me mentioned.

The major drawback using the nematode as an infection model deals with temperature. Nematodes can cope with temperatures up to 25 ºC, which is 12 ºC below the human bacterial pathogens optimal growth temperature. At this low temperature, bacteria generally activate cold-shock responses, which *per se,* can lead to reduced virulence potential. In addition, caution has to be taken because those virulence factors that are expressed only at body temperatures will be missed when using the nematode as an infection model.

On the other hand, the lack of an adaptive immune system can "up-rate" some virulence determinants that would otherwise be easily depleted by the adaptive immune response.

It is worth mentioning that conditions found by bacteria when infecting the *C. elegans* can be quite distinct from those found in specialized niches in their hosts. This is the case of *B. cepacia*, which presents a marked preference for the lung of the human host.

Finally, the molecular mechanism involved in bacterial adhesion within the nematode's digestive tract might differ from those employed by the bacteria when colonizing its primary host, which will ultimately lead to distinct outcomes of the infection.

Conclusion

The nematode *C. elegans* is a practically and ethically attractive model system, increasingly recognized as relevant for the study of bacterial pathogenesis. In this chapter, we reviewed the use of the nematode *C. elegans* as an infection model for a surprisingly broad range of bacteria, and thus presented some descriptions undoubtedly indicating that it is a useful model for the study a variety of pathogens, as well as for the study of host responses to infection.

The evolutionarily conserved mechanisms involved in microbial pathogenesis and innate immunity have also been studied. In many cases, microbial genes known to be important for full virulence in mammalian models have also been shown to be similarly required for maximal pathogenicity in nematodes (Sifri et al. 2005). *C. elegans* has been used in mutation-based screening systems to identify novel virulence-related microbial genes and immune-related host genes (e.g., see Sousa et al. 2011), many of which have been validated in mammalian models of disease. *C. elegans*-based pathogenesis systems hold the potential to simultaneously explore the molecular genetic determinants of both pathogen virulence and host defence (Sifri et al. 2005).

Bacterial multi-drug resistance is increasing to an alarming level, pointing out the need for the development of novel therapeutic approaches to combat infections (Leitão et al. 2008; Leitão et al. 2010). To discover novel antimicrobials, Moy et al. (2006) have devised a simple screen to identify compounds that promoted the survival of *C. elegans* infected with the human opportunistic pathogen *Enterococcus faecalis* (Moy et al. 2006). In their assays, infected nematodes were rescued by antibiotic treatment in a dose-dependent manner, and antibiotic treatment markedly reduced the number of bacteria colonizing the nematode intestine (Moy et al. 2006). Using a high throughput screening of compound libraries, 6000 synthetic compounds and 1136 natural product extracts were screened. Sixteen compounds and 9 extracts that promoted nematode survival were identified (Moy et al. 2006). Those findings support that the whole-animal *C. elegans* screen is a very useful tool to identify not only traditional antibiotics, but also compounds that target bacterial virulence determinants or that stimulate the host defenses.

Bacterial pathogen–eukaryote host interactions are ubiquitous and have relevant medical and environmental implications. Peleg et al. (2008) have used *C. elegans* as a host to study the interactions between *Acinetobacter baumannii* and *Candida albicans,* two ecologically related and clinically difficult-to-treat pathogens (Peleg et al. 2008). In their study, these authors have co-infected *C. elegans* with both pathogens, and observed that *A. baumannii* inhibited filamentation, a key virulence determinant of *C. albicans* (Peleg et al. 2008). This antagonistic interaction led to attenuated virulence of *C. albicans* towards the nematode, when infected with both pathogens. These pioneer study findings raise interesting questions regarding the paradigm of studying mono-microbe pathogenesis in *C. elegans.* Moreover, it clearly shows that *C. elegans* can be a useful model system to study the complex dynamics of poly-microbial infections, as is the case of the respiratory infections that are often the cause of premature death of cystic fibrosis patients.

Acknowledgements

This work was partially supported by Fundação Ciência e Tecnologia, Portugal (contract PTDC/EBB-BIO/098352/2008) and a Doctoral grant to CGR from Fundação Calouste Gulbenkian.

References

Aballay, A. and F. M. Ausubel (2002). "*Caenorhabditis elegans* as a host for the study of host-pathogen interactions." *Current Opinion in Microbiology* 5(1): 97-101.

Aballay, A., E. Drenkard, L. R. Hilbun and F. M. Ausubel (2003). "*Caenorhabditis elegans* innate immune response triggered by *Salmonella enterica* requires intact LPS and is mediated by a MAPK signaling pathway." *Current Biology* 13(1): 47-52.

Aballay, A., P. Yorgey and F. M. Ausubel (2000). "*Salmonella typhimurium* proliferates and establishes a persistent infection in the intestine of *Caenorhabditis elegans*." *Current Biology* 10(23): 1539-1542.

Aguilar, C., A. Friscina, G. Devescovi, M. Kojic and V. Venturi (2003). "Identification of quorum-sensing-regulated genes of *Burkholderia cepacia*." *Journal of Bacteriology* 185(21): 6456-6462.

Alegado, R. A., M. C. Campbell, W. C. Chen, S. S. Slutz and M.-W. Tan (2003). "Characterization of mediators of microbial virulence and innate immunity using the *Caenorhabditis elegans* host–pathogen model." *Cellular Microbiology* 5(7): 435-444.

Altun, Z. F. and D. H. Hall (2005). *Handbook of C. elegans anatomy,* http://www.wormatlas.org/ver1/handbook/contents.htm

Baldwin, A., P. A. Sokol, J. Parkhill and E. Mahenthiralingam (2004). "The *Burkholderia cepacia* epidemic strain marker is part of a novel genomic island encoding both virulence and metabolism-associated genes in *Burkholderia cenocepacia*." *Infection and Immunity* 72(3): 1537-1547.

Beale, E., G. Li, M. W. Tan and K. P. Rumbaugh (2006). "*Caenorhabditis elegans* senses bacterial autoinducers." *Applied and Environmental Microbiology* 72(7): 5135.

Begun, J., J. M. Gaiani, H. Rohde, D. Mack, S. B. Calderwood, F. M. Ausubel and C. D. Sifri (2007). "Staphylococcal biofilm exopolysaccharide protects against *Caenorhabditis elegans* immune defenses." *PLoS Pathogens* 3(4): e57.

Begun, J., C. D. Sifri, S. Goldman, S. B. Calderwood and F. M. Ausubel (2005). "*Staphylococcus aureus* virulence factors identified by using a high-throughput *Caenorhabditis elegans*-killing model." *Infection and Immunity* 73(2): 872-877.

Burkholder, W. H. (1950). "Sour skin, a bacterial rot of onion bulbs." *Phytopathology* 40: 115 - 117.

Cardona, S. T., J. Wopperer, L. Eberl and M. A. Valvano (2005). "Diverse pathogenicity of *Burkholderia cepacia* complex strains in the *Caenorhabditis elegans* host model." *FEMS Microbiology Letters* 250(1): 97-104.

Chiarini, L., A. Bevivinoa, C. Dalmastria S. Tabacchioni and P. Visca (2006). "*Burkholderia cepacia* complex species: health hazards and biotechnological potential " *Trends in Microbiology* 14(6): 277-286.

Cunha, M. V., J. H. Leitão, E. Mahenthiralingam, P. Vandamme, L. Lito, C. Barreto, M. J. Salgado and I. Sá-Correia (2003). "Molecular Analysis of Burkholderia cepacia Complex Isolates from a Portuguese Cystic Fibrosis Center: a 7-Year Study." *Journal of Clinical Microbiology* 41(9): 4113-4120

Darby, C., C. L. Cosma, J. H. Thomas and C. Manoil (1999). "Lethal paralysis of *Caenorhabditis elegans* by *Pseudomonas aeruginosa*." *Proceedings of the National Academy of Sciences USA* 96(26): 15202-15207.

Deere, J., P. Iversen and B. L. Geller (2005). "Antisense phosphorodiamidate morpholino oligomer length and target position effects on gene-specific inhibition in *Escherichia coli*." *Antimicrobial Agents and Chemotherapy* 49(1): 249.

Dhakal, B. K., W. Lee, Y. R. Kim, H. E. Choy, J. Ahnn and J. H. Rhee (2006). "*Caenorhabditis elegans* as a simple model host for *Vibrio vulnificus* infection." *Biochemical and Biophysical Research Communications* 346(3): 751-757.

Eberl, L. (2006). "Quorum sensing in the genus *Burkholderia*." *International Journal of Medical Microbiology* 296(2-3): 103-110.

Ewbank, J. J. (2002). "Tackling both sides of the host-pathogen equation with *Caenorhabditis elegans*." *Microbes and Infection* 4(2): 247-256.

Gan, Y.-H., K. L. Chua, H. H. Chua, B. Liu, C. S. Hii, H. L. Chong and P. Tan (2002). "Characterization of *Burkholderia pseudomallei* infection and identification of novel virulence factors using a *Caenorhabditis elegans* host system." *Molecular Microbiology* 44(5): 1185-1197.

Garsin, D. A., C. D. Sifri, E. Mylonakis, X. Qin, K. V. Singh, B. E. Murray, S. B. Calderwood and F. M. Ausubel (2001). "A simple model host for identifying Gram-positive virulence factors." *Proceedings of the National Academy of Sciences USA* 98(19): 10892-10897.

Geller, B. L., J. D. Deere, P. L. Iversen and D. D. Weller (2009). Antisense antibacterial method and compound, *Google Patents*.

Geller, B. L., J. D. Deere, D. A. Stein, A. D. Kroeker, H. M. Moulton and P. L. Iversen (2003). "Inhibition of gene expression in *Escherichia coli* by antisense phosphorodiamidate morpholino oligomers." *Antimicrobial Agents and Chemotherapy* 47(10): 3233-3239.

Geller, B. L., P. L. Iversen and L. D. Tilley (2006). Antisense antibacterial method and compound, *Google Patents*.

Geller, B. L., P. L. Iversen and L. D. Tilley (2010). Antisense antibacterial method and compound, *Google Patents*.

Gravato Nobre, M. J. and J. Hodgkin (2005). "*Caenorhabditis elegans* as a model for innate immunity to pathogens." *Cellular Microbiology* 7(6): 741-751.

Hope, I. A. (1999). *C. elegans - A Practical Approach*, Oxford University Press.

Huber, B., F. Feldmann, M. Kothe, P. Vandamme, J. Wopperer, K. Riedel and L. Eberl (2004). "Identification of a novel virulence factor in *Burkholderia cenocepacia* H111 required for efficient slow killing of *Caenorhabditis elegans*." *Infection and Immunity* 72(12): 7220-7230.

Jansen, W. T. M., M. Bolm, R. Balling, G. S. Chhatwal and R. Schnabel (2002). "Hydrogen peroxide-mediated killing of *Caenorhabditis elegans* by *Streptococcus pyogenes*." *Infection and Immunity* 70(9): 5202-5207.

Joshua, G. W. P., A. V. Karlyshev, M. P. Smith, K. E. Isherwood, R. W. Titball and B. W. Wren (2003). "A *Caenorhabditis elegans* model of *Yersinia* infection: biofilm formation on a biotic surface." *Microbiology* 149(11): 3221-3229.

Khan, S. R., D. V. Mavrodi, G. J. Jog, H. Suga, L. S. Thomashow and S. K. Farrand (2005). "Activation of the *phz* operon of *Pseudomonas fluorescens* 2-79 requires the LuxR homolog PhzR, N-(3-OH-Hexanoyl)-L-homoserine lactone produced by the LuxI homolog PhzI, and a cis-acting *phz* box." *Journal of Bacteriology* 187(18): 6517-6527.

Kim, D. H., R. Feinbaum, G. Alloing, F. E. Emerson, D. A. Garsin, H. Inoue, M. Tanaka-Hino, N. Hisamoto, K. Matsumoto, M.-W. Tan and F. M. Ausubel (2002). "A conserved p38 MAP kinase pathway in *Caenorhabditis elegans* innate immunity." *Science* 297(5581): 623-626.

Kothe, M., M. Antl, B. Huber, K. Stoecker, D. Ebrecht, I. Steinmetz and L. Eberl (2003). "Killing of *Caenorhabditis elegans* by *Burkholderia cepacia* is controlled by the *cep* quorum-sensing system." *Cellular Microbiology* 5(5): 343-351.

Kurz, C. L., S. Chauvet, E. Andres, M. Aurouze, I. Vallet, G. P. F. Michel, M. Uh, J. Celli, A. Filloux, S. de Bentzmann, I. Steinmetz, J. A. Hoffmann, B. B. Finlay, J.-P. Gorvel, D. Ferrandon and J. J. Ewbank (2003). "Virulence factors of the human opportunistic pathogen *Serratia marcescens* identified by *in vivo* screening." *The EMBO Journal* 22(7): 1451-1460.

Labrousse, A., S. Chauvet, C. Couillault, C. Léopold Kurz and J. J. Ewbank (2000). "*Caenorhabditis elegans* is a model host for *Salmonella typhimurium*." *Current Biology* 10(23): 1543-1545.

Leitão, J. H., S. A. Sousa, M. V. Cunha, M. J. Salgado, J. Melo-Cristino, M. C. Barreto and I. Sá-Correia (2008). "Variation of the antimicrobial susceptibility profiles of *Burkholderia cepacia* complex clonal isolates obtained from chronically infected cystic fibrosis patients: a five-year survey in the major Portuguese treatment center." *European Journal of Clinical Microbiology and Infectious Diseases* 27: 1101-1111.

Leitão, J. H., S. A. Sousa, A. S. Ferreira, C. G. Ramos, I. N. Silva and L. M. Moreira (2010). "Pathogenicity, virulence factors, and strategies to fight against *Burkholderia cepacia* complex pathogens and related species." *Applied Microbiology and Biotechnology* 87(1): 31-40.

Lewenza, S., B. Conway, E. P. Greenberg and P. A. Sokol (1999). "Quorum Sensing in *Burkholderia cepacia*: Identification of the LuxRI homologs CepRI." *Journal of Bacteriology* 181(3): 748-756.

LiPuma, J. J., T. Spilker, T. Coenye and C. F. Gonzalez (2002). "An epidemic *Burkholderia cepacia* complex strain identified in soil." *The Lancet* 359: 2002 - 2003.

Lutter, E., S. Lewenza, J. J. Dennis, M. B. Visser and P. A. Sokol (2001). "Distribution of quorum-sensing genes in the *Burkholderia cepacia* complex." *Infection and Immunity* 69(7): 4661-4666.

Maddula, V., Z. Zhang, E. Pierson and L. Pierson (2006). "Quorum sensing and phenazines are involved in biofilm formation by *Pseudomonas chlororaphis* (*aureofaciens*) strain 30-84." *Microbial Ecology* 52(2): 289-301.

Mahajan-Miklos, S., M.-W. Tan, L. G. Rahme and F. M. Ausubel (1999). "Molecular mechanisms of bacterial virulence elucidated using a *Pseudomonas aeruginosa-Caenorhabditis elegans* pathogenesis model." *Cell* 96(1): 47-56.

Mahenthiralingam, E., T. Coenye, J.W. Chung, D.P. Speert, J.R. Govan, P. Taylor and P. Vandamme (2000). "Diagnostically and experimentally useful panel of strains from the *Burkholderia cepacia* complex." *Journal of Clinical Microbiology* 38(2): 910-913.

Mahenthiralingam, E., A. Baldwin and C. G. Dowson (2008). "*Burkholderia cepacia* complex bacteria: opportunistic pathogens with important natural biology." *Journal of Applied Microbiology* 104(6): 1539-1551.

Mahenthiralingam, E., T. A. Urban and J. B. Goldberg (2005). "The multifarious, multireplicon *Burkholderia cepacia* complex." *Nature Reviews Microbiology* 3(2): 144-156.

Malott, R. J., A. Baldwin, E. Mahenthiralingam and P. A. Sokol (2005). "Characterization of the *cci*IR quorum-sensing system in *Burkholderia cenocepacia*." *Infection and Immunity* 73(8): 4982-4992.

Mavrodi, D. V., W. Blankenfeldt and L. S. Thomashow (2006). "Phenazine compounds in fluorescent *Pseudomonas* spp. Biosynthesis and regulation." *Annual Review of Phytopathology* 44(1): 417-445.

Moy, T. I., A. R. Ball, Z. Anklesaria, G. Casadei, K. Lewis and F. M. Ausubel (2006). "Identification of novel antimicrobials using a live-animal infection model." *Proceedings of the National Academy of Sciences USA* 103(27): 10414-10419.

Nicholas, H. R. and J. Hodgkin (2004). "Responses to infection and possible recognition strategies in the innate immune system of *Caenorhabditis elegans*." *Molecular Immunology* 41(5): 479-493.

Peleg, A. Y., E. Tampakakis, B. B. Fuchs, G. M. Eliopoulos, R. C. Moellering and E. Mylonakis (2008). "Prokaryote–eukaryote interactions identified by using *Caenorhabditis elegans*." *Proceedings of the National Academy of Sciences USA* 105(38): 14585.

Pujol, N., E. M. Link, L. X. Liu, C. L. Kurz, G. Alloing, M.-W. Tan, K. P. Ray, R. Solari, C. D. Johnson and J. J. Ewbank (2001). "A reverse genetic analysis of components of the Toll signaling pathway in *Caenorhabditis elegans*." *Current Biology* 11(11): 809-821.

Ramos, C. G., A. M. Grilo and J. H. Leitão (2010a). *Small non-coding RNAs in Prokaryotes: Roles in Pathogenesis and Potential Therapeutic Targets*. Saarbrücken, Germany, *Lambert Academic Publishing*. ISBN: 987-3-8383-2116-5.

Ramos, C. G., S. A. Sousa, A. M. Grilo, L. Eberl and J. H. Leitão (2010b). "The *Burkholderia cenocepacia* K56-2 pleiotropic regulator Pbr, is required for stress resistance and virulence." *Microbial Pathogenesis* 48: 168-177.

Ramos, C. G., S. A. Sousa, A. M. Grilo, J. R. Feliciano and J. H. Leitão (2011). "The second RNA chaperone Hfq2, is also required for survival to stress and the full virulence of *Burkholderia cenocepacia* J2315." *Journal of Bacteriology* 193(7): 1515-1526.

Richau, J. A., J. H. Leitão, M. Correia, L. Lito, M. J. Salgado, C. Barreto, P. Cescutti and I. Sá-Correia (2000). "Molecular Typing and Exopolysaccharide Biosynthesis of Burkholderia cepacia Isolates from a Portuguese Cystic Fibrosis Center." *Journal of Clinical Microbiology* 38(4): 1651-1655.

Riddle, D. L., T. Blumenthal, B. J. Meyer and J. R. Priess (1997). *C. elegans II. Cold Spring Harbor Laboratory Press*, Plainview, New York 11803-2500, USA.

Riedel, K., M. Hentzer, O. Geisenberger, B. Huber, A. Steidle, H. Wu, N. Hoiby, M. Givskov, S. Molin and L. Eberl (2001). "N-Acylhomoserine-lactone-mediated communication between *Pseudomonas aeruginosa* and *Burkholderia cepacia* in mixed biofilms." *Microbiology* 147(12): 3249-3262.

Ruiz-Diez, B., P. Sanchez, F. Baquero, J. L. Martinez and A. Navas (2003). "Differential interactions within the *Caenorhabditis elegans-Pseudomonas aeruginosa* pathogenesis model." *Journal of Theoretical Biology* 225(4): 469-476.

Shapira, M., B. J. Hamlin, J. Rong, K. Chen, M. Ronen and M.-W. Tan (2006). "A conserved role for a GATA transcription factor in regulating epithelial innate immune responses." *Proceedings of the National Academy of Sciences USA* 103(38): 14086-14091.

Sifri, C. D., A. Baresch-Bernal, S. B. Calderwood and C. von Eiff (2006). "Virulence of *Staphylococcus aureus* Small Colony variants in the *Caenorhabditis elegans* infection model." *Infection and Immunity* 74(2): 1091-1096.

Sifri, C. D., J. Begun and F. M. Ausubel (2005). "The worm has turned-microbial virulence modeled in *Caenorhabditis elegans*." *Trends in Microbiology* 13(3): 119-127.

Sifri, C. D., J. Begun, F. M. Ausubel and S. B. Calderwood (2003). "*Caenorhabditis elegans* as a model host for *Staphylococcus aureus* pathogenesis." *Infection and Immunity* 71(4): 2208-2217.

Sifri, C. D., E. Mylonakis, K. V. Singh, X. Qin, D. A. Garsin, B. E. Murray, F. M. Ausubel and S. B. Calderwood (2002). "Virulence fffect of *Enterococcus faecalis* protease genes and the quorum-sensing locus *fsr* in *Caenorhabditis elegans* and mice." *Infection and Immunity* 70(10): 5647-5650.

Sousa, S. A., C. G. Ramos, F. Almeida, L. Meirinhos-Soares, J. Wopperer, S. Schwager, L. Eberl and J. H. Leitão (2008). "*Burkholderia cenocepacia* J2315 acyl carrier protein: a potential target for antimicrobials development?" *Microbial Pathogenesis* 45(5-6): 331-336.

Sousa, S. A., C. G. Ramos and J. H. Leitão (2011). "*Burkholderia cepacia* complex: emerging multi-host pathogens equipped with a wide range of virulence factors and determinants." *International Journal of Microbiology* 2011: Article ID 607575.

Sousa, S. A., C. G. Ramos, L. M. Moreira and J. H. Leitão (2010). "The *hfq* gene is required for stress resistance and full virulence of *Burkholderia cepacia* to the nematode *Caenorhabditis elegans*." *Microbiology* 156(3): 896 - 908.

Sousa, S. A., M. Ulrich, A. Bragonzi, M. Burke, D. Worlitzsch, J. H. Leitão, C. Meisner, L. Eberl, I. Sá-Correia and G. Doring (2007). "Virulence of *Burkholderia cepacia* complex strains in gp91$^{phox-/-}$ mice." *Cellular Microbiology* 9(12): 2817-2825.

Speert, D. P. (2002). "Advances in *Burkholderia cepacia* complex." *Paediatric Respiratory Reviews* 3(3): 230-235.

Tan, M.-W., S. Mahajan-Miklos and F. M. Ausubel (1999). "Killing of *Caenorhabditis elegans* by *Pseudomonas aeruginosa* used to model mammalian bacterial pathogenesis." *Proceedings of the National Academy of Sciences USA* 96(2): 715-720.

Tan, M.-W. and M. Shapira (2011). "Genetic and molecular analysis of nematode–microbe interactions." *Cellular Microbiology* 13(4): 497-507.

Tan, M. -W. and F. M. Ausubel (2000). "*Caenorhabditis elegans*: a model genetic host to study *Pseudomonas aeruginosa* pathogenesis." *Current Opinion in Microbiology* 3(1): 29-34.

Tomlin, K. L., R. J. Malott, G. Ramage, D. G. Storey, P. A. Sokol and H. Ceri (2005). "Quorum-sensing mutations affect attachment and stability of *Burkholderia cenocepacia* biofilms." *Applied and Environmental Microbiology* 71(9): 5208-5218.

Vandamme P, B. Holmes, T. Coenye, J. Goris, E. Mahenthiralingam, J. J. LiPuma and J. R. Govan (2003) "*Burkholderia cenocepacia* sp. nov.--a new twist to an old story" *Research in Microbiology* 154(2):91-96.

van Delden, C., R. Comte and A. M. Bally (2001). "Stringent response activates quorum sensing and modulates cell density-dependent gene expression in *Pseudomonas aeruginosa*." *Journal of Bacteriology* 183(18): 5376-5384.

Vanlaere, E., A. Baldwin, E. Mahenthiralingam, C. G. Dowson, G. W. Payne and P. Vandamme (2006). Proposal for five new species within the *Burkholderia cepacia* complex. *International Burkholderia cepacia Working Group meeting*. Gent, Belgium, International Burkholderia cepacia Working Group.

Vanlaere, E., A. Baldwin, D. Gevers, D. Henry, E. De Brandt, J. J. LiPuma, E. Mahenthiralingam, D. P. Speert, C. Dowson and P. Vandamme (2009). "Taxon K, a complex within the *Burkholderia cepacia* complex, comprises at least two novel species,

Burkholderia contaminans sp. nov. and *Burkholderia lata* sp. nov." *International Journal of Systematic and Evolutionary Microbiology* 59(1): 102-111.

Vikstrom, E., K.-E. Magnusson and A. Pivoriunas (2005). "The *Pseudomonas aeruginosa* quorum-sensing molecule N-(3-oxododecanoyl)-l-homoserine lactone stimulates phagocytic activity in human macrophages through the p38 MAPK pathway." *Microbes and Infection* 7(15): 1512-1518.

Wopperer, J., S. T. Cardona, B. Huber, C. A. Jacobi, M. A. Valvano and L. Eberl (2006). "A quorum-quenching approach to investigate the conservation of quorum-sensing-regulated functions within the *Burkholderia cepacia* complex." *Applied and Environmental Microbiology* 72(2): 1579-1587.

Zhang, Y., H. Lu and C. I. Bargmann (2005). "Pathogenic bacteria induce aversive olfactory learning in *Caenorhabditis elegans*." *Nature* 438(7065): 179-184.

In: Nematodes
Editors: F. Boeri and J. A. Chung
ISBN: 978-1-61470-784-4

Chapter VI

Management and Control of the Pine Wood Nematode, *Bursaphelenchus Xylophilus*, in Portugal

***E. Sousa*[1], *J. M. Rodrigues*[2], *L. F. Bonifácio*[1], *P. M. Naves*[1] and *A. Rodrigues*[1]**

[1] Instituto Nacional dos Recursos Biológicos (National Institute for Biological Resources), Quinta do Marquês, 2780-159 Oeiras, Portugal.

[2] Prolunp - Autoridade Florestal Nacional (National Forestry Authority), Av. João Crisóstomo, 26-28, 1069-040 Lisboa, Portugal.

Abstract

The Pine Wood Nematode (PWN) *Bursaphelenchus xylophilus* (Steiner and Buhrer) Nickle is the causal agent of the pine wilt disease, and a quarantine organism within the European Union. This pathogenic organism was detected in Portugal in 1999 associated to dead maritime pines (*Pinus pinaster*), with its vector among trees being the native insect *Monochamus galloprovincialis* (Oliv.). Over the years, management and control of both the nematode and the vector has been implemented by the Portuguese National Programme to Control the Pine Wood Nematode (PROLUNP). This programme set the guidelines for phytosanitary measures and procedures incorporating the latest scientific knowledge available. In the period from 1999 to 2008 the nematode was restricted to Setúbal Peninsula. Subsequently new pine wilt focuses were detected in Central Continental Portugal and Madeira Island. The current management strategy is focused on three core points: (i) identification of nematode-affected areas; (ii) prevention of dispersal of wilt disease to new locations, (iii) and reduction of wilt incidence in affected areas. Identification of affected areas (i) is based on periodic surveys in pine forest during the autumn and winter periods (when wilting symptoms are conspicuous) and on the collection of wood samples from symptomatic trees for PWN lab identification. The detection of nematode in new locations implies the delimitation of an Intervention Zone encompassing the infected area and a bordering buffer zone (nematode-free). The spread of wilt disease (ii) occurs by the flight of the insect vector and by transportation of wood infected with the PWN and the vector. Scientific evidence suggests that *M. galloprovincialis* makes short-distance flights of only a few hundred meters if adequate hosts are available. Dispersal of wilt disease can be avoided by trunk-injection with insecticides in healthy pines, along with an intensive monitoring system in the buffer zone. To prevent anthropogenic dispersion, transport of wood between infected and non-infected areas is officially regulated, and technological procedures such as heat-treatment of wood are also applied. Reduction of pine wilt incidence in affected areas (iii) is achieved by the harvesting of wilted trees and biomass (including branches of all dimensions) from November to April, when the insect vector is inactive. In the period May-October an active complementary management approach against other declining agents such as bark beetles is implemented, consisting on the immediate cut of affected trees and the installation of a trap network baited with pheromones and host kairomones to monitor and diminishing both the bark beetle and the *Monochamus* populations. Our case-study in Tróia consisting in a 10-year management program in a 400ha maritime pine forest has successfully diminished wilt incidence and *M. galloprovincialis* populations to residual levels. This is clearly demonstrative that the aforementioned control and management strategies are successful, when implemented with efficiency and within the correct time schedules.

1. Introduction

Maritime pine (*Pinus pinaster* Aiton) is a native tree from the Iberian Peninsula and a major forest species in Portugal, covering over 890.000 ha (27% of the global forest area) (AFN, 2010a), being the most relevant pine species in economic terms. Its main products are manufactured wood, pellets, boards, pallets and resin, representing about 306 million euros to the total exports (eg. Mendes *et al.,* 2004).

Portuguese maritime pine forests were initially restricted to the coastline, north of Tagus River (Fabião, 1987) but as early as XIII century Portuguese kings ordered the forestation of wide areas of center Portugal. The pines continued to be planted in low productivity areas, mountain regions of Northeast and Center Portugal in the 40s of the twentieth century, reaching around 1.3 million ha (Costa, 1995). However, the plantation of large continuous pine stands made them very susceptible to forest fires and bark beetles attacks, as reported by Neves (1964).

By that time are the first studies of pine pests, namely processionary moth, (*Thaumetopoea pytiocampa* Schiff.) (Lepidoptera: Thaumetopoeidae) (Cabral *et al.*, 1965; Cabral and Heitor, 1966) and the bark beetles *Ips sexdentatus* (Boern.), *Orthotomicus erosus* (Woll.), *Tomicus piniperda* (L.) and *Tomicus destruens* (Boern.), but only in the eighties was the sanitary condition of pine ecosystems conveniently assessed (Ferreira and Ferreira, 1990; Ferreira and Cabral, 1999).

In 1999 a new biotic agent of pines' decline was detected at Marateca, Setúbal, and identified as the Pine Wood Nematode (PWN) *Bursaphelenchus xylophilus* (Steiner et Buhrer, 1934) Nickle 1970 (Nematoda: Aphelenchoididae), the casual agent of pine wilt disease.

PWN is endemic in North America where's found in equilibrium with native pines. However, its accidental spread to Japan, in the early twentieth century, caused an abnormal mortality on endemic pine species with heavy losses that reached 2.4 million m^3 of wood, representing 1% of the country's pine forests (Mamiya, 1984). Much later the causal agent of the tree mortality was identified and all process was called Pine Wilt Disease (PWD) (Mamiya and Kiyohara, 1972). Afterwards it as also introduced in China, Korea and Taiwan, through the international trade of infected wood (Dwinell, 1997). In fact, molecular studies (RFLP) carried out on PWN Portuguese populations showed genetic similarities with the Asian Far East populations (Metge and Burgermeister, 2005).

Nowadays, PWD is known as complex interaction between three very distinct organisms: the pine wood nematode, an insect vector and a tree host. Worldwide, the PWN is the only common element of this tri-parted interaction, while the *Monochamus* vectors and the *Pinus* hosts may vary from one location to another, and can even differ in the same geographic area.

After the detection of *B. xylophilus* in Portugal an intensive survey to detect the local insect vectors was initiated and more than 1,400 insects from 21 species and six families were collected from wilt diseased pines and individually analysed. In the end the PWN was only found associated with the native pine sawyer *Monochamus galloprovincialis* (Olivier, 1795) (Sousa *et al.*, 2001, 2002) and the sole host of *B. xylophilus* is the maritime pine *P. pinaster* (Sousa *et al.*, 2002). (Figure 1).

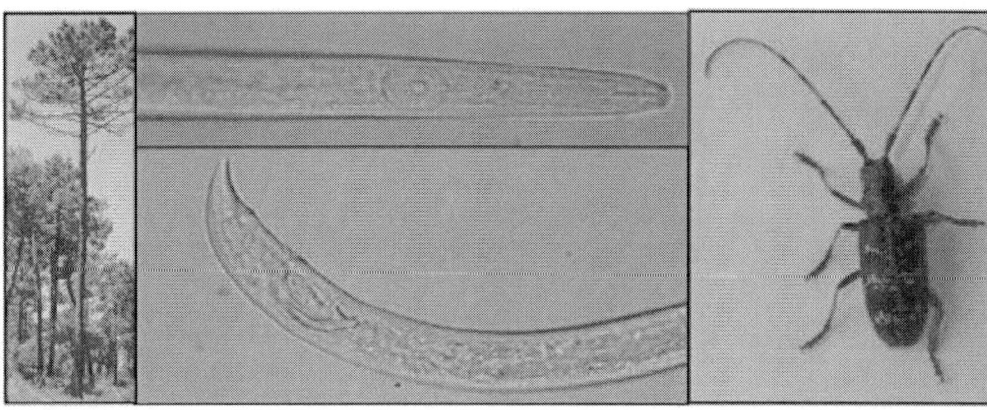

Figure 1. Maritime pine *Pinus* pinaster (a); Pine Wood Nematode, *Bursaphelenchus xylophilus*, detail of head (b); detail of tail of male (c). Insect vector in Portugal *Monochamus galloprovincialis* (x 0,7)(d).

Although the introduction pathway of PWN to Portugal is unknown, most probably some Asian infected *Monochamus* emerged from contaminated wood at Setúbal harbour and were able to infect (through maturation feeding) nearby healthy maritime pine. No exotic *Monochamus* was ever caught in the area and the infection cycle was completed when native *M. galloprovincialis* laid eggs on these newly infected pines, starting a novel *B. xylophilus* - *M. galloprovincialis* interaction.

For wilt disease to express the simultaneous presence of the pinewood nematode, the *Monochamus* vector and the pine host is required. In Portugal, the introduction of *B. xylophilus,* disrupted a long-term equilibrium between the native insect *M. galloprovincialis* and its local host *P. pinaster* (Figure 2).

The PWN adult is a microscopic organism, threadlike and the adult form measures less than 1,5 mm in length (Braasch, 1991) and is transmited to a healthy host tree occurs by the feeding activity of the insect vector (Naves *et al.*, 2007a).

When the nematode is inside the tree host it has very short generations cycles and high reproduction rates, so that 20 to 30 days after infection, leaf transpiration ceases and partial interruption in the sapwood obstructs sap ascent, causing discoloration and wilting of leaves in all or part of crowns. In the process, needles turn yellow and brown starting with the younger ones, remaining attached to the tree for a long period of time (Mamiya, 1984).

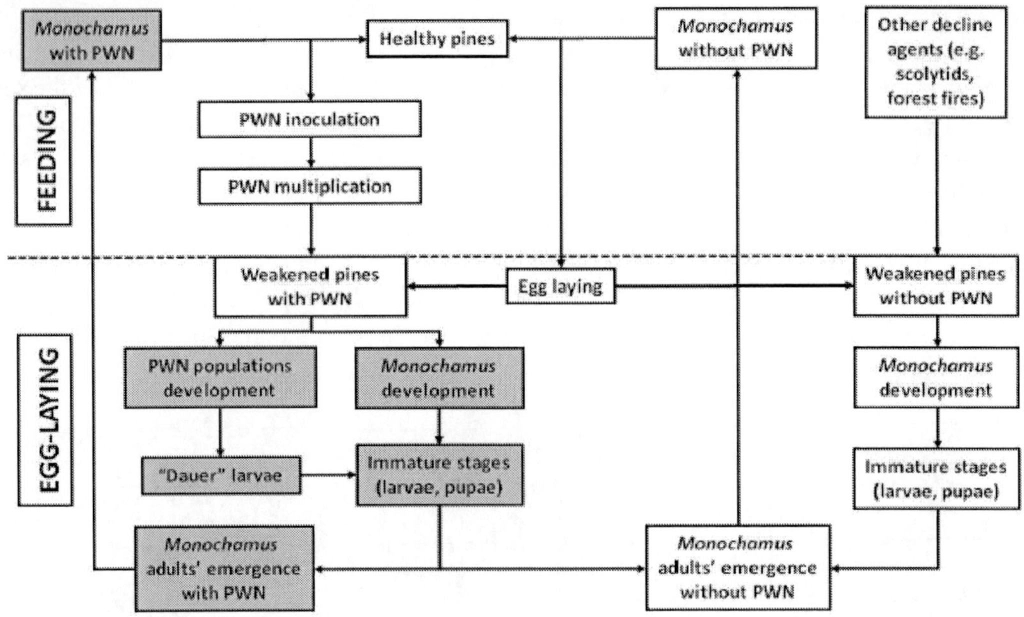

Figure 2. Interactions established between *Monochamus* beetles, the host- trees and *Bursaphelenchus xylophilus*.

Field trials in Southern Portugal showed that the induced production of resin is an effective qualitative methodology to early detect the infection by *B. xylophilus*. Infected pines stopped producing resin about one and a half months before significant yellowing of the needles was visible (>25% canopy), only observed after July (Bonifácio and Sousa, 2011). Tree death due to lack of water occurs in two to three months (Kuroda, 1991; Evans, 1992).

It is however important to emphasize that the discoloration of the crown is not specific for infection with PWN, because identical symptoms occur also in trees attacked by other biotic agents, such as bark beetles. Laboratory analysis of the wood either by morphological identification (Penas *et al.*, 2002) and/or molecular techniques is the only way to confirm the infection by the nematode (Hu *et al.*, 2011).

The effect of nematode infection on wood physical properties was unknown until recently. Research conducted on symptomatic maritime pine trees intended to quantify a possible deterioration of mechanical properties in thermally treated pinewood, reported decrease in basic density (from 548 Kgm^{-3} to 537 Kgm^{-3}), in axial compression stress (from 58Mpa to 55Mpa) and in module of rupture of bending strength (from 131Mpa to 125Mpa). However, these changes were not relevant to downgrade the technological classification of wood specimens. A multiple logistic approach, considering symptomatology as the binary dependent event showed the independent variables: number or rings per unit radial length and axial compression stress, as the main contributes to the highest expected values π_i. Thresholds of logistical responses of symptomatology for new observations in specimens collected in trees located in nematode prone forest trees were found (Rodrigues *et al.*, 2010).

The pine sawyer populations profited from the availability of large number of PWN infected dying trees and forest's bad sanitary conditions to increase significantly, becoming a primary (although indirect) agent of mortality of adult healthy pine trees, with significant economic and ecological impacts.

In Portugal, the *M. galloprovincialis* beetles were considered secondary insects that feed on the bark of healthy pines, not causing any damage to the tree. They do not breed on healthy trees and are attracted only to stressed, dying or recently killed trees in result of the attack of other biotic agents, such as the bark beetles, or from the effect of abiotic factors, such as drought-related stress and forest fires, important occurrences in Mediterranean forest ecosystems (Cabral, 1995; Sousa et al. 2001). Trees that are dead for longer time and already have desiccated phloem and sapwood are considered unsuitable hosts (Linit, 1987; Hanks, 1999).

Eggs are laid in the scars the females chew on the bark of the tree hosts (Evans, et al., 1996) and during this process secondary transmission of PWN can occur, however to a lesser extent than through feeding (Naves *et al.*, 2007b).

The *M. galloprovincialis* larvae hatch after six to nine days depending on temperature, and go through four larval instars in about three months, reaching about 4cm in length (Koutroumpa *et al.*, 2008). The last instar larva digs a gallery deep into the wood, with an oval entrance hole, ending on the pupal chamber where it overwinters.

Nematode and vector coexist in the tree hosts for several months without interaction. As the wood decays the nematodes shift to "dauer-larvae", a resistant stage of development. In late spring, nematodes aggregate around the insect-vector pupal chambers.

As soon as the callow adult is formed, the nematodes transfer to the insect's tracheas or in fewer quantities under the elytra, the antennae and legs of insects (Naves *et al.*, 2006b). Both the number of infested beetles and the mean number of nematodes per insect varies largely, due to intrinsic population factors or to environmental conditions influencing the moisture and fungi complex in the wood. Insect infection rate of *Monochamus* is extremely variable and most of the beetles were found to carry less than 5.000 nematodes. The emergence of new adults occurs digging a new round hole, with about four to eight millimeters in diameter, at the surface of the wood (Sousa *et al.*, 2002).

2. Management and Control of Pine Wilt Disease

The pine wilt disease is difficult to control because the pine wood nematode has a short life cycle under favourable conditions, high fecundity and fertility, alternate phytophagous and micophagous phases and resilient larval stages.

Any management or control strategy against the pine wilt disease has to focus on preventing the spread and dissemination of the nematode, which in short-distances (at a local scale) is done by the flight of infected insect vectors. Before the introduction of the PWN, the vector *M. galloprovincialis* was considered a secondary forest insect and had never been studied for the most important aspects of its biology and ecology, in Portugal or elsewhere. These knowledge gaps became a significant impediment for the development of management strategies ever since the detection of the nematode in 1999, and delayed the development and implementation of successful control measures against the vector.

Therefore, and simultaneously with the implementation of management and control operations in the terrain, research teams conducted scientific studies both in the laboratory and in the terrain on multiple aspects of the vector′s biology and life cycle, the nematode′s impact and behaviour in the field, and the nematode-vector-host interaction, which have been an essential component in the development, optimisation and consolidation of the control procedures in the affected zones.

Over the last eleven years the strategies to manage and control the pine wilt disease have changed and evolved following the progressive accumulation of knowledge about the disease and the detection of new wilt focus.

After the detection of the PWN in 1999, Portuguese authorities following Council Directive 2000/29/EC, implemented a phytosanitary strategy aiming to control and eradicate the pest under a framework denominated as National Eradication Programme for Pinewood Nematode (PROLUNP), based on three core points which over the years have benefited from newly-acquired scientific knowledge:

1) identification of nematode-affected areas;
2) prevention of wilt dispersal to new locations;
3) reduction of wilt incidence in affected areas.

These points are interconnected and subjected to periodic critical evaluation and refinement, with the objectives of preventing further dissemination of the PWN, promoting the control of the insect vector and disrupting the interaction between the three organisms.

2.1. Identification of Nematode-Affected Areas

The correct and rigorous identification and delimitation of forests affected by the PWN was essential since the beginning, for preventing the dissemination of wilt disease and to control the insect vector. A specific survey of PWN was implemented on an annual basis, through all the country, mainly in potential risk areas, such as, places with high concentration of wood (surroundings of entry, wood yards and transformation facilities); pine stands with symptomatic trees and borders of forest fires (DGRF, 2003).

Nematode survey is based on field samplings during the autumn and winter (when the wilting symptoms on the pines are most conspicuous), with the collection of wood samples from the trees (each with around 100g) with an autonomous drill. In the laboratory the nematodes are extracted by keeping the small pieces of wood in a tray with water for 48 hours, at ambient temperature. The saturation of the wood makes the nematodes to leave its host into the water, where they can be sieved with a 400 mesh (38μm), extracted and identified by experienced nematologists and/or molecular techniques (Penas, *et al.*, 2002).

Conifers other than maritime pine, such as *P. sylvestris*, *P. halepensis*, and *P. pinea*, though to be attractive to the vector and/or susceptible to the nematode, are also sampled because they have been found to be suitable for the maturation feeding by the adult beetles, in laboratory conditions (naves *et al.*, 2006c).

Studies also showed that insect vectors laid a very low number of eggs on *P. pinea* and could not complete their development on such logs, suggesting that this pine specie is not an adequate host for the insect′s development. Inversely, species such as *P. halepensis, P. pinaster, P. radiata,* and *P. sylvestris* were suitable hosts for development (Naves *et al.*, 2006c), and forests with these pines should be monitored to evaluate the population levels of the vector and detect new wilt focus. In spite of this, the PWN has never been found in other pine species than maritime pine.

Until 2008, the PWN was confined to the Setúbal Península at Southern Portugal (where a Demarked Area was delimitated) and rest of the country was considered a Free Zone (Rodrigues, 2008). The Demarked area included an Affected Zone surrounded by a 20km radius nematode free Buffer Zone (to prevent natural spread of the insect vector outside the affected area). The limits and area of both zones were annually assessed, and the disease was found to spread mainly towards the southern limits and along the coastal pine forests (Figure 3).

From 2000 the Demarcated Area increased from 309.000ha to over one million hectares in 2007, with a 510.000ha Affected Zone and 500.000ha of Buffer Zone. During this seven-year period the affected zone increased from 48.000ha to 80.000ha, and the number of felled pine trees with wilting symptoms (and not necessarily with the PWN) increased from 53.487 to over 200.000 (Table I).

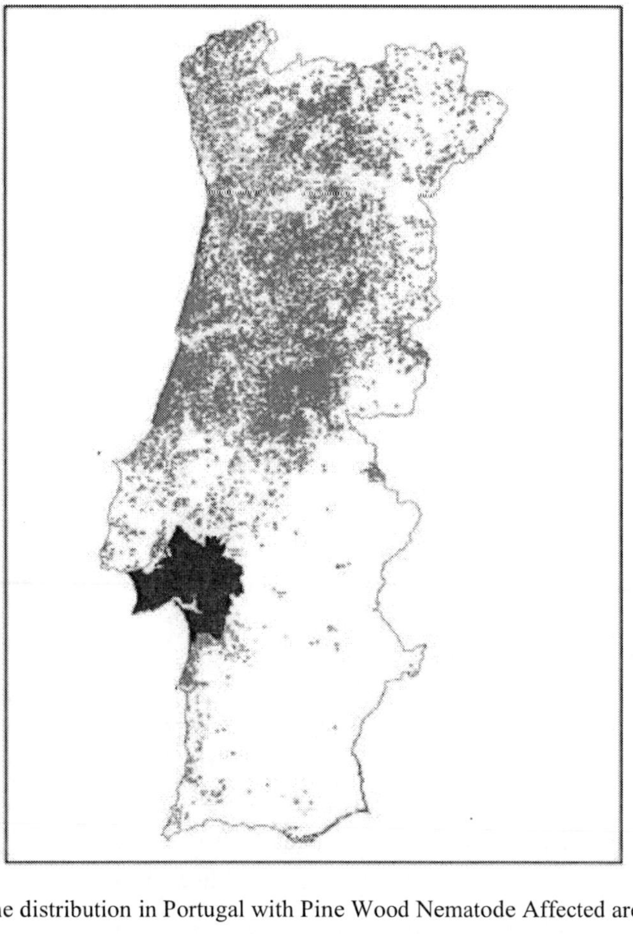

Figure 3. Maritime pine distribution in Portugal with Pine Wood Nematode Affected area delimited in 1999 (black area).

Table I. Evolution of area (hectares) and number of felled pines during eradication of symptomatic pines against pine wilt disease, in Portugal DGRF, 2007b)

	2000	2001	2002	2003	2004	2005	2006	2007
Total area	309.000	564.000	564.000	564.000	617.000	617.000	641.000	1.010.000
Total area of pine stands	48.000	60.000	60.000	60.000	69.000	69.000	69.500	80.000
Area of stands with symptoms	12.392	16.793	11.623	14.076	16.391	19.519	30.247	28.667
Nº felled pines	53.487	54.314	43.014	53.338	74.889	93.308	268.211	196.530
Affected zone Buffer zone	(1)	4.720	7.528	8.120	10.197	9.509	20.774	17.770

(1) – In accordance with European Commission Decision nº 2000/58/EC.

As result of the continuous expansion of the demarcated area, in 2007, the Portuguese and European authorities tried to prevent the natural dissemination of wilt disease with a shift in their strategy (Figure 4). Nearly five million healthy conifer trees were cut to establish a Clear Cut Belt of containment in the periphery of the original Demarcated Area of Setúbal peninsula. This was a 3km-wide corridor free of the vector´s host trees (*Picea orientalis*, *Pinus halepensis*, *Pinus nigra*, *Pinus nigra laricio*, *Pinus pinaster*, *Pinus radiata* and *Pinus sylvestris*).

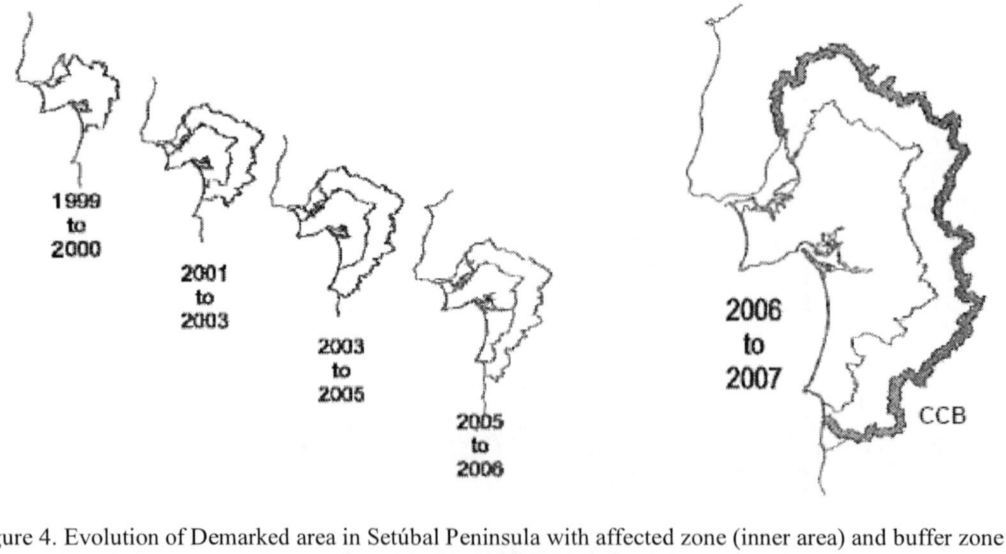

Figure 4. Evolution of Demarked area in Setúbal Peninsula with affected zone (inner area) and buffer zone (outer area). Clear cut belt (CCB), implemented in the limit of buffer zone in 2007.

Until this moment all PWD control strategy was based on avoiding the flight of nematode infected insects to new areas and this massive enterprise was successful for several years, but new focus eventually appeared in late 2007, more than two hundred kilometres north of Setúbal peninsula. They were clearly linked to illegal or accidental transport of wood infected with both the nematode and its vector.

More intensive survey was implemented based on a 2x2 Km grid originally developed for the National Forest Inventory (NFI). From the 12.000 NFI field points, 2.170 points (with an area of 500 m2 each) were classified as "coniferous forest cover", being mostly maritime pine. All the symptomatic conifer trees located in these plots were screened for the PWN presence according to EU established methodology.

The initial systematic grid was shifted annually, with reference to one of the four cardinal points (North, South, East and West). Consequently, after five years the midpoint and the cardinal points will be prospected (accounting for 10.850 observations), thus maximizing the detection of the PWN.

As not all points of the grid can be simultaneously sampled for the presence of the nematode, priority is given to forests with higher risk of suffering wilt disease, namely points located at:

- Proximity to PWN-affected forests;
- Proximity to pine forests recently affected by forests fires (which usually have a high abundance of insect vectors);
- Other pine forests with wilt symptoms, being necessary to validate in the terrain observations from satellite images.

A new restriction zone was then delimited and lead to the classification of the entire Continental Portugal as an Affected Area, with a Buffer Zone of 20km surrounding the Portuguese-Spanish border. Nevertheless there are still numerous maritime pine forests where *B. xylophilus* is absent.

A more detailed characterization of the PWN distribution suggests a concentration in the Central Region with a South to Southeast extension and a focus at north of the 'Setubal Demarcated Area'. With regard to the North region, no additional PWN positive cases were detected in the 2009 and 2010 campaigns. To keep that area free of PWN it is important to stabilize the focus in central Portugal, which requires a continuous, proactive and responsible approach.

In summary, and considering the scattered distribution of the wilt focus, there was a need for a differential action at the regional level, given the distribution of the PWN, the nature of the conifers stands, the landscape features and the logistic problems. Following the positive locations in 2008, 2009 and 2010, Intervention Zones (IZ) were created in geographic areas with minimal reference to the parish level (AFN, 2010b) (Fig. 5).

Each IZ is subject to strict phytosanitary requirements imposed by Portuguese and EU laws, and to national and international standards for the management, harvesting, transport and processing of timber (logs, and round wood) resulting from the cut of symptomatic trees, with the primary objective of preventing further dissemination of wilt disease.

Whenever isolated positive cases of PWN are detected the correspondent parish is also classified as IZ, but the procedure of Local Eradication is applied, encompassing the focus zone and safety zone (nematode-free), all subjected to an intensive sampling to delimitate and eradicate the new focus.

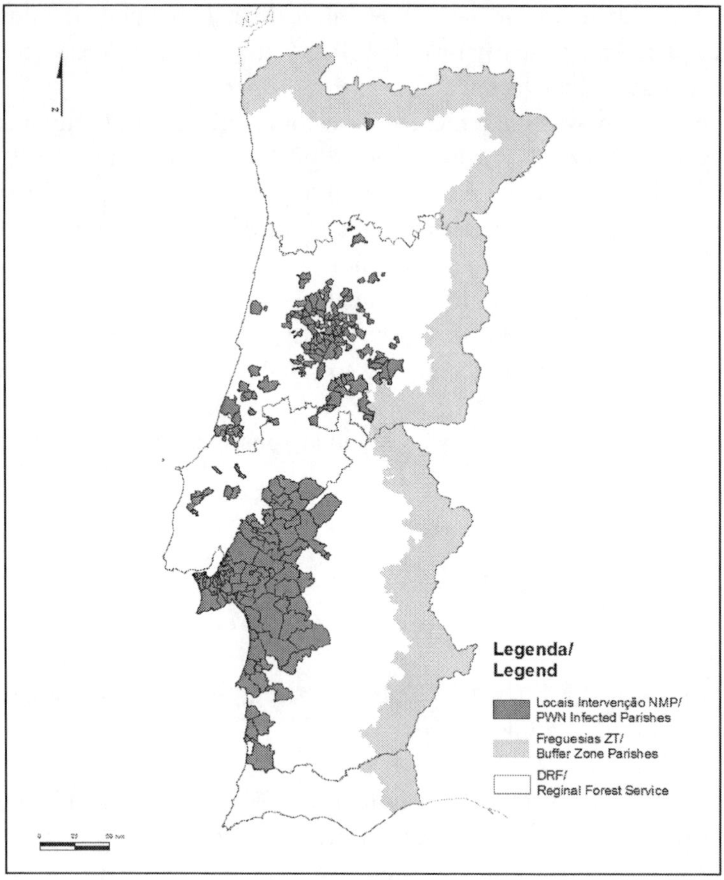

Figure 5. Distribution of parishes with Pine Wood Nematode positive samples, in 2010.

Meanwhile and despite the buffer zone established along the Portuguese-Spanish border, in two occasions nematode-infected trees were reported from Spain, in Extremadura (late 2008) and Galicia (late 2010) (Abelleira *et al.*, 2011). Additionally, the PWN was also detected in the Portuguese Island of Madeira in late 2009.

2.2. Prevention of Wilt Dispersal to New Locations

After 2008, the prevention of PWD dispersal to new locations was focused on the two most important dispersal pathways: the flight of infected insects and the human transport of wood infected with both the PWN and the vector. These dispersal pathways must be minimised and avoided as much as possible.

2.2.1. Dispersal by the Insect Vector

To avoid the dispersal of infected *M. galloprovincialis* adults out of the affected areas it's necessary to develop strategies, based on its flight behaviour and creation of effective barriers. These barriers can be the implemented with the use traps, removal of suitable host trees or with preventive chemical treatment. However the size of these barriers is dependent on the distance of the flight that insect carrying nematodes can achieve.

It was based on these criteria that 20 km buffer zones were established (considered to be an unlikely distance for an insect to fly), and also lead to the implementation of the 2008 clear cut belt (4km wide).

It's still unknown the maximum flight distance for an infected *M. galloprovincialis* adult. Field assays were carried out in Portugal with the release and recapture of marked adult insects, and the low recapture rates (less than 1%) of newly emerged and sexually immature insects suggest the occurrence of an obligatory migratory flight in search of new habitats/hosts just after emergence and possibly influenced by genetic factors (similarly to other insect species – e.g., Shivankar *et al.*, 2006), and which was previously unknown for beetles of this genus.

Preliminary studies with older, marked insects (9-12 days) suggests that *M. galloprovincialis* makes short-distance flights for feeding of only a few hundred meters if adequate hosts are available. The average flight distances were 4,97m and 4,30m for males and females (Bonifácio, 2009), half the distance reported for *Monochamus alternatus* Hope in Japan (10,6m and 12,3m, respectively) (Togashi, 1990), although the maximum flights were identical for both species (around 14m).

The length of the flight depended on the dominant wind direction (*downwind*), increasing with the insect's weight and higher physiological capacities. In open pine stands the insects tended to fly more, preferably landing in a pine with large canopy (older trees), while in dense pine stands with uniform trees the insects fly less or remained longer time the same tree (Bonifácio, 2009).

The efficacy of emamectin benzoate (a semi synthetic and second generation avermectin derivate insecticide) for preventing the spread of wilt disease to new areas was recently tested in Portugal, after being employed with success in Japan (Kazuya *et al.*, 1999; Takai *et al.*, 2003) and the United States (James *et al.*, 2006).

Trunk-injection trials were made in Portugal in 2009 and included three dose-rates (0.032 g a.i./cm DBH, 0.064 g a.i./cm DBH and 0.128 g a.i./cm DBH), along with an untreated control plot, with the sanitary condition of the inoculated pines being periodically assessed for two years. In time none of the inoculated trees dyed, while in the control plot 28% of the pines died due to PWN and bark beetle (*O. erosus*) attacks. Analysis of branch tissues from the treated pines showed that emamectin benzoate had been successfully injected and translocated at an effective concentration, with the quantity increasing accordingly to the injection dose rate.

Additional studies in the laboratory found that the tested dose rates had a clear effect on the longevity and feeding activity of *M. galloprovincialis* exposed to branches collected from the treated trees, with a clear dose rate response and a repellent effect for the med dose and above. Overall, the experiments found a preventive effect of inoculating emamectin benzoate by trunk-injection into adult maritime pines, and the dose rate of 0.064 g a.i./cm DBH is currently regulated and recommended for preventing wilt dissemination and protecting individual trees.

The strategy involving this preventive treatment can only be applied to healthy pines, as declining and symptomatic trees do not effectively translocate the inoculated product through the entire trunk and canopy. Additionally, only pines with a breast height diameter higher than 20cm should be inoculated, as these are the trees preferentially selected by *M. galloprovincialis* for its feeding and breeding activity.

2.2.2. Dispersal by Human Transport of Wood

The human transport of wood infected with both the nematode and the vector is the most important long-distance dispersal pathway for pine wilt disease worldwide (e.g., Kawai *et al.*, 2006).

The risk of PWD dissemination by human activity is related to the transport of wood materials containing both the nematode and the vector, which must not occur during the flight season of the *Monochamus* vectors.

To avoid this, all intervention zones are subjected to strict phytosanitary requirements and standards for the management, harvesting, transport and processing of timber (logs, and round wood) from symptomatic trees felled either in the affected area or in neighbouring buffer zones. Transport of contaminated wood to outside these zones can only be done outside the flight period of the vector insect (from May to October) (Naves *et al.* 2008) and to destinations guarantying treatment or processing of the wood before the emergence of the beetles begin. During the spring and summer the transport of infected wood (with diameters above 20cm) can only be done in close containers or roofed trucks, after the application of an insecticide and to destinations assuring immediate processing of the wood. High-risk materials, such as wood and branches with a diameter below 20cm (which is where most of the insect population completes its larval development) (Bonifácio, 2009), can only be transported in close containers or roofed trucks from one affected region to another and only outside the flight season of the beetle, to avoid creating new wilt focus.

Wood destined for export and packaging materials such as pallets are regarded to have high risk for disseminating the PWN, although it can be mitigated by application of ISPM 15 treatments to reduce introductions and/or spread of quarantine pests associated with wood packaging material (ICPM, 2002). Among the treatments recommended, the correct use of heat treatment (HT) kills both vector and nematodes as the wood reaches a core temperature of 56°C for a minimum of 30 minutes (Dwinell, 1990; 1997), and is obligatory applied by Portuguese exporting companies. Inversely, and to avoid new introduction of the PWN, sampling of imported wood packaging material is also systematically made by the official authorities, involving a minimum of 600 samples per year.

Despite the existing legislation and the collaboration of various official and private entities, the human dissemination of infected wood is extremely difficult to avoid, and the illegal or accidental transport of small quantities of wood with both the nematode and the vector can easily compromise all the efforts and successfully disseminate wilt disease to new locations several kilometres. This risk can be minimised if in the affected areas the nematode incidence and vector populations are kept at low levels, which can be achieved through a multi-integrated and complementary approach of management and control strategies.

2.3. Reduction of Wilt Incidence in Affected Areas

To control PWD efficient methods must be implemented to rapidly decrease the population levels of *M. galloprovincialis*. However, the different methods to control this insect usually have success in localized, small-dimension areas, but are difficult to implement at low cost and have reduced efficiency over large forested areas.

In Portugal the most important management and control strategy consists in the elimination of symptomatic trees during late autumn, winter and early spring, while the insect vector is inside the host (as late-instar larvae or pupae) and there are no adults flying in the forests (Naves *et al.*, 2008). Trees with decline and wilt symptoms are visually identified and marked with ink late in the year, when the vast majority of the nematode-infected pines are already dead or shown conspicuous wilting symptoms through the canopy. Wilted pines are subsequently felled and high-risk materials, such as wood and branches with a diameter below 20cm must be locally destroyed or remove from the stand.

Since the *M. galloprovincialis* larvae can be found inside very small branches and in the thinner sections of the trunk (Sousa *et al.*, 2001), this otherwise neglected material must be thoroughly collected and locally destroyed (by shredding to wood sections with less than 3cm, as this is the minimum size to low survival of *M. galloprovincialis* larvae or pupae in the wood) or transported for processing into materials such as pellets and others. This attention to the branches and the thinner trunk is not required on normal silvicultural operations outside the nematode affected zones, and consequently delays and increase the costs of the control procedures in the areas affected by the PWN.

As the pine sawyer is never found in the stumps or roots of *P. pinaster*, there is no need to remove, destroy or fumigate the stumps of the infected trees in Portugal.

Complementary to these silvicultural practices, the vector's populations can also be diminished during the beetle´s flight season, from late spring to early autumn. This can be achieved with the use of traps baited with attractive lures placed at variable densities in affected pine forests. During the last decade studies have improved the efficiency of trapping against *M. galloprovincialis*, with the comparative testing of various models of traps, resulting in the transparent cross vane trap having the highest captures, followed by the 12-element multi-funnel trap (Bonifácio, 2009) (Figure 6).

Figure 6. Traps more efficient to catch *Monochamus galloprovincialis*. Transparent cross-vane (a); Multifunnel trap (b).

The most effective chemical lures used to attract the flying beetles are a combination of pine odours (α-pinene and ethanol) and bark beetles pheromones (Ipsdienol, Ipsenol and Methyl-butenol) (Pajares *et al.*, 2004; Bonifácio *et al.*, 2011). This attractive mixture reveals an interesting co-evolution between the *M. galloprovincialis* and the scolitid bark beetles, as the pine sawyer responds to the sexual and aggregation pheromones released by the bark beetles when attacking and killing pines, which would be suitable hosts to the pine sawyer´s oviposition. Recently a specific *M. galloprovincialis* pheromone was described and employed to increase captures in traps (Ibeas *et al.*, 2008), although its effect still needs to be further evaluated in traps placed in PWN affected areas, with high densities of *M. galloprovincialis* insects.

Despite having been used to control *Monochamus* vectors in Asia (e.g. Mamiya, 1984; Kobayashi, 1988), large-scale aerial spraying with insecticides was never a viable option in Portugal, because *M.* galloprovincialis has a long emergence period (from May to September) (Naves *et al.*, 2008) and pine forests are scattered through an extremely humanized territory and are located in close proximity to human settlements, agricultural fields, aquatic ecosystems and nature reserves, which are incompatible with repeated applications of generalist chemical insecticides.

The use of specific and efficient natural enemies (bio-control) would also be an interesting and environmental-friendly option to control the insect vector in Portugal, but until now there are no adequate species for such control program. Field surveys already conducted in Portugal did not detect any egg parasitoids (Naves *et al.,* 2005). Although some larval parasitism was found, the low mortality rates for both the phloem and xylem larvae (with a maximum of 10%) and the generalist character of the parasitoids detected (*Coeloides sordidator* Ratzeburg, *Cyanopterus flavator* F. and *Iphiaulax impostor* Scopoli) are not encouraging to promoting bio-control strategies based on these species (Naves *et al.,* 2005). Similar results have been described elsewhere in Europe (Campadelli and Dindo, 1994).

The wood sampling of symptomatic trees in 2010 revealed that only a small part of those trees were infected by the Pine Wood Nematode (9,7%). Taking advantage of the generalised declining sanity of pine forests in the nematode-affected areas, bark beetle outbreaks begun to be noticed from 2006 onwards, with heavy pine mortality due to *O. erosus* and *I. sexdentatus* massive attacks. Having multiple generations within a year and being generally ignored by the control measures (which were focused on the pine wood nematode and *M. galloprovincialis*), these bark beetles were responsible for much higher mortality than the PWN in some areas. Nevertheless, the outbreaks also contributed to increase the incidence of wilt disease, as higher numbers of dead pines suitable for breeding contributed to the build up of the vector populations.

Considering these developments, an active complementary management approach against these new declining agents is now implemented from May to October, when the bark beetles are most active and breeding. This consists on the immediate felling of any wilted pine in the field (regardless of being or not infected with the PWN), and the placement of traps baited with specific sexual and aggregation pheromones, to capture flying beetles and directly diminish the insect´s populations.

Overall, the management and control strategies against both the PWD and the associated mortality agents has already proved that can be successful when implemented with rigor, efficiency and within the correct timing, as the evolution of the sanitary condition of Tróia peninsula pine forests demonstrate.

3. The Tróia Peninsula Case-Study

Tróia peninsula is a private touristic settlement in the south bank of the Sado River, near the city of Setúbal, and is located in the first area affected by the PWN. The peninsula is characterised by its sandy and very poor soils, and the forest cover is mainly composed of maritime and stone pines, with some marginal areas of eucalyptus and acacias. The maritime pine stands cover an area of less than 400ha, having an approximate age of 30-40 years. Suffering from *B. xylophilus* induced mortality since the original detection of the nematode in 1999, the company responsible became increasingly concerned with the generalised decaying sanity and the long-term sustainability of the maritime pine forests in the peninsula.

In the first years the strategy established to improve the sanity of Troía´s forests consisted on the timely identification, marking, felling and removal of all wilted pines in the field during the autumn and winter months, alongside with the placement of traps baited with attractants to capture flying *M. galloprovincialis* during the spring and summer months. To guarantee the rigorous removal from the field of the smaller branches and other woody material which could support larvae of the insect vector, this procedure was mandatory in the formal contracts for the cut and removal of the trees, and only contractors' which compromised with this techniques would be selected for the sanitary operations. To ensure this was executed, a member of the INRB team or the local touristic company would accompany the feeling campaign each year.

In the first year (2000) a total of 4.226 dead pines were felled, although this figure also included the accumulated mortality during the previous years. The oldest and biggest pines were more affected by the mortality, and samplings of the felled trees revealed that the pines were mostly infected with *B. xylophilus* (85%). Differential mortality affected more some zones of the peninsula than others, and the golf course, with its massive pines benefiting from regular fertilizations and constant watering, was one of the areas most affected by wilt disease.

With the careful cut and destruction of infected pines and the systematic placement of traps for the capture of the insect vector, the yearly number of dead maritime pine trees decreased by two-thirds in the second year (1.364), and again by half in the third year (644), representing only 15% of the initial values and a mean of less than two dead pines per hectare (Figure 7).

During 2005-2007, a severe drought affected the region, and along with an outbreak of the pine processionary and bark beetles (*O. erosus* and *I. sexdentatus*) moth it was responsible for a significant increase in the number of dead pines, totalling 1.139 trees. The increasing mortality affected mainly the younger pines, which were more susceptible to the hidric stress and the heavy defoliation caused by the pine processionary.

In the subsequent years tree mortality remained high with around one thousand pines/year, and wood samplings made after 2006 revealed an unexpected pattern: the PWN was now responsible for only 13% of the mortality (Figure 8).

Observations in the field soon detected large bark beetle populations, which were associated with most dead trees. The sanitary strategy of Tróia had to be adapted and adjusted, and a more intense trapping effort with sex pheromones as attractants was used to diminish the bark beetle and pine processionary moth populations during early spring and late summer, eventually leading to a decrease in the annual mortality below 500 pines (Bonifácio, 2009), a value which still stands today.

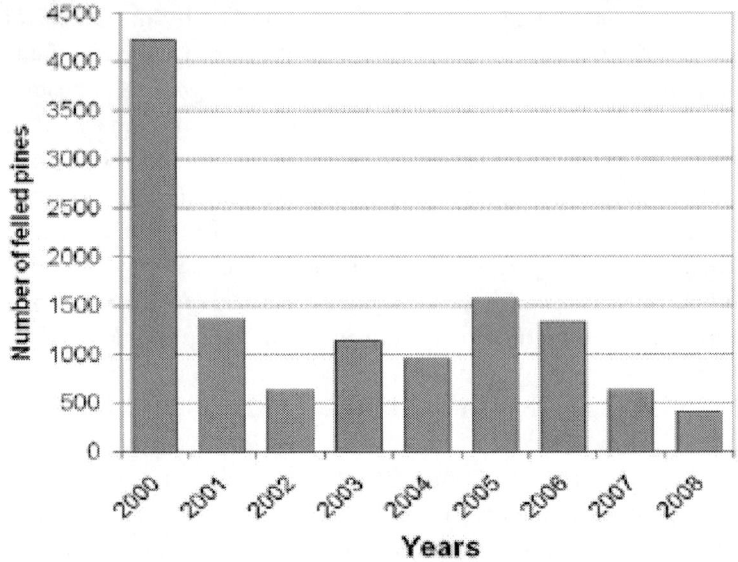

Figure 7. Number of dead maritime pines felled each year at Tróia peninsula.

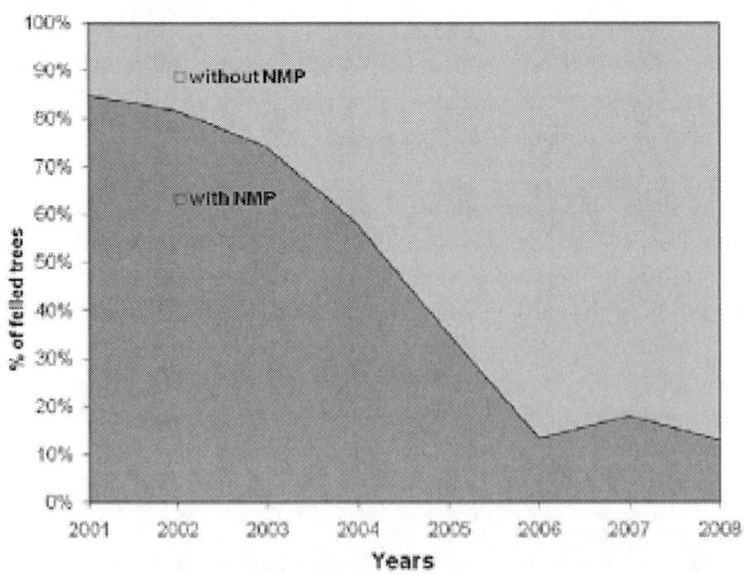

Figure 8. Evolution of Pine Wood Nematode infections in overall pine mortality observed at Tróia peninsula.

Although in residual numbers the PWN is still present in Tróia, along with the bark beetles and pine processionary moth. Nevertheless, the complementary approach of sanitary feelings with the removal of all wood materials, and the timing placement of traps ensures these biotic agents of mortality keep their populations levels low, and give an optimistic perspective for the maritime pine forest of Tróia peninsula, in a zone where the PWN has been present for over a decade.

Conclusion

The experience obtained from more than a decade of managing and controlling the PWN in Portugal, and the scientific knowledge gathered during this period emphasize the necessity to control the immature stages of the insect vector before the emergence of the adults, and the importance of developing efficient methods of capturing/eliminating the recently-emerged beetles before they begin to disperse *B. xylophilus* into new hosts. In fact, the combination of the insect's three-month emergence period, its mean adult life span of about two months and the efficient and prolonged period of nematode transmission help to explain how such a large number of trees are each year infected with *B. xylophilus* and how wilt disease is such an important mortality factor for susceptible pine species.

Additionally, control options are limited, costly and need to be executed with rigor and efficiency in order to be effective. There is still margin for developments and improvements on some techniques, namely on increasing the efficiency of the traps and lures to capture adult *M. galloprovincialis*, and on the development and manipulation of nematode resistant (or less susceptible) varieties or clones of maritime pine.

Despite all the effort, there are still important scientific questions which remain unanswered, such as the flight behavior of the vector insect and aspects of the nematode-vector-host interaction, which clarified would open new possibilities of limiting the disease. As an example, the comprehension of the process that triggers the nematode entry and departure from the vector could allow for the development of methods to disrupt this essential interaction, breaking the infection cycle.

Overall, and despite the joint efforts of various public and private entities, more than ten years after its detection the PWN has disseminated and new focus have appeared in central and northern Portugal, in Madeira Island and even in neighboring Spain. These isolated and scattered focuses were caused by human activity, and emphasize the importance of illegal or accidental transport of wood infected with both the nematode and its vector on the large-range dissemination of the disease, in contrast with the limited and short-range dissemination by the vector′s activity and flight. Controlling and impeding this human-induced dispersal is essential in order to prevent the appearance of new wilting focus in the future.

The presence of the PWN in Portugal for more than a decade has caused considerable ecological, social and economical impact. The pallet industry, for instance, has been severely affected by the obligation to heat-treat all wood destined for exportation, and the many small-dimension industries closed due to financial problems originated by the presence of *B. xylophilus* in Portugal.

The treatment activities of wood and wood-based products carried out by industrial units are being followed and controlled, in a continuous way, by phytosanitary inspectors with training of the Ministry of Agriculture, Rural Development and Fisheries. Meanwhile, governmental measures to support industry were addressed in order to guarantee the fulfillment of the International Standards for Phytosanitary Measures, applicable to international commerce and thus adequate coniferous wood products market flow, such as:

modernization of sawmills and increase of the capacity to perform heat treatment; promotion (and adoption) of innovative forms of coniferous wood products economic valuation.

Despite all the problems, studies in some selected areas such as Tróia peninsula have found that when management and control measures are applied properly and timely, it is possible to maintain a healthy pine forest with the presence of *B. xylophilus*, and that eventually this pathogenic nematode will one day be just one of the many organisms associated with the pine forests in Portugal.

Acknowledgments

The authors would like to manifest their gratitude for colleagues and institutions which, over the years, have assisted and contributed for the development of the scientific studies and the implementation of management operations in the field. Thanks to Drª. Lurdes Inácio, Drª Joana Henriques, Drª Margarida Vieira, Drª Maria Antónia Lampreia, Adérito Matos, Margarida Fontes and Vitor Gonçalves. The INRB, I.P. wish to acknowledge the constant support and prolonged cooperation extended by the directors of TróiaResort (Sonae group), in the person of Drª Célia Ferreira, and Herdade da Comporta in the person of Eng. Pedro Serrasqueiro, Eng. António Moreira and Eng. Francisco Goulão.

Research lines have been funded by: the Foundation for Science and Technology (Ministry of Science and Technology-Portugal) through the projects Parle D and REEQ/1289/AGR/2005, the National Forestry Authority with the project ''O nemátode-da-madeira-do-pinheiro (NMP), Bursaphelenchus xylophilus - Área de trabalho 4: Controlo do insect vector" (funding by the Fundo Florestal Permanente) and the European PHRAME project QLK5-CT-2002-00672.

References

A.F.N. Inventario Florestal Nacional. Portugal Continental; Direcção Nacional de Gestão Florestal: Lisboa, 2010a; 5 ed, pp 209.

A.F.N. National action plan for Pinewood nematode control; Autoridade Florestal Nacional: Lisboa, 2010b; pp 27.

Abelleira, A.; Picoaga, A.; Mansilla, J.P. and Aguin, O. Detection of *Bursaphelenchus xylophilus*, causal agent of pine wilt disease on *Pinus pinaster* in northwestern Spain. Plant Disease. 2011, 95, 776.

Aguiar, F.; Costa, J.C.; Lousã, M.; Moreira, I. Vegetação aquática e ribeirinha da bacia do Sado. In Gestão ambiental dos sistemas fluviais. Aplicação à bacia hidrográfica do rio Sado. Moreira, I.; Saraiva, M.G.; Correia, F.N.; Ed.; ISA Press: Lisboa, 2004, pp 227-246.

Bonifácio, L. Impacte e evolução da doença da murchidão dos pinheiros (pine Wilt Disease) na zona afectada a sul do rio Tejo; PhD thesis presented at Science Faculty of Lisbon University: Lisboa; 2009, pp 209.

Bonifácio, L.; Sousa, E. Early detection methods for the detection of Pine Wood Nematode infections in standing maritime pines, in Portugal. Silva Lusitana. 2011, 19, in press.

Bonifácio, L.; Praias, F.; Sousa, E. Trapping *Monochamus galloprovincialis* (Coleoptera: Cerambycidae), vector of the pine Wood Nematode, with pine allelochemicals, in Portugal. Silva Lusitana, 2011, 19, in press.

Braasch, H. (2001) – *Bursaphelenchus* species in conifers in Europe: distribution and morphological relations. *EPPO Bulletin*. 31, 127-142.

Cabral, M.T.; Heitor, F.P.C. Meios actuação do *Bacillus thuringiensis* sobre a processionária do pinheiro (*Thaumetopoea pytiocampa* Schiff.). Considerações sobre o pH do intestino e da hemolínfa. *Estudos e Informação*. 1966, 236, pp 13.

Cabral, M.T.E.C.; Heitor, F.P.C.; Figo, M.L.N.C.B.P.M. Meios de luta contra a processionária. *Thaumetopoea pytiocampa* Schiff.; Secretaria de Estado da Agricultura. Direcção Geral dos Serviços Florestais e Aquícolas: Lisboa, 1965, pp 52.

Campadelli, G.; Dindo, M.L. *Monochamus galloprovincialis* (Oliv.) (Coleoptera Cerambycidae) nella pineta di S. Vitale. Informatore Fitopatologico. 1994, 1, 31-34.

Costa, M.A.S. Pinheiro bravo e pinheiro manso, exploração e tratamentos; Litexa Ed: Lisboa, 1995, 160pp.

D.G.R.F. Prolunp - Plano Estratégico para a Erradicação do Nemátodo da Madeira do Pinheiro; MADRP: Lisboa, 2003, pp 22.

Dwinell, L.D. Heat-treating and drying southern pine lumber infested with pinewood nematodes. Forest Prod. J. 1990, 40, 53-56.

Dwinell, L.D. Pinewood nematode: Regulation and mitigation. *Annu. Rev. Phytopathol.* 1997, 35, 153-166.

Evans, H.F. Pinewood nematode: biology and international plant health implications. J. Inst. Wood Sci. 1992, 12, 317-320.

Fabião, A.M.D. Árvores e florestas; Europa-América Publ.: Lisboa, 1987, pp 228.

Ferreira, M.C.; Ferreira, G.W.S. Pragas das resinosas. Guia de campo; DGPA/ Ministério Agricultura, Pescas e Alimentação: Lisboa, 1990, pp 108.

Ferreira, M.C.; Cabral M.T. Pragas do Pinhal; Estação Florestal Nacional: Lisboa, 1999, pp 159.

Hanks, L.M. Influence of the larval host plant on reproductive strategies of cerambycid beetles. *Annu. Rev. Entomol.* 1999, 44, 483-505.

Hu, Q.; Kong, X.C.; Wang, X.R.; Zhong, T.K.; Zhu, X.W.; Mota, M.M.; Ren, L.L.; Liu, S.; Ma, C. Direct PCR-based method for detecting *Bursaphelenchus xylophilus*, the pine wood nematode in wood tissue of *Pinus massoniana* Forest Path. 2011, 41, 165-168.

I.C.P.M. International Standards for Phytosanitary Measures. Guidelines for regulating wood for packaging material in international trade; Secretariat of the International Plant Protection Convention, Food and Agriculture Organization of the United Nations: Rome, 2002; 15, pp 15.

I.M.A.R. Flora e comunidades vegetais. Sector norte da península de Tróia; Relat. Síntese. Sonae Turismo/ Imoareia/ IMAR: Lisboa, 2001, pp 72.

Ibeas, F.; Diez, J.J. and Pajares, J. Olfactory Sex Attraction and Mating Behaviour in the Pine Sawyer *Monochamus galloprovincialis* (Coleoptera: Cerambycidae). *J. Insect Behavior.* 2008, 21, 101-110.

James, R.; Tisserat, N.; Todd, T. Prevention of pine wilt of scots pine (*Pinus sylvestris*) with systemic Abamectin injections. *Arboriculture Urban Forestry*. 2006, 32, 195-201.

Kawai, M.; Shoda-Kagaya, E.; Maehara, T.; Zhou. Z.; Lian, C.; Iwata, R.; Yamane, A.; Hogetsu, T. Genetic structure of pine sawyer *Monochamus alternatus* (Coleoptera:

Cerambycidae) populations in northeast Asia: consequences of spread of pine wilt disease. *Environ. Entomol.* 2006, 35, 569-579.

Kazuya, T.; Soejima, T.; Suzuki, T.; Kawazu, K. Studies on development of a novel trunk-injection agent against the pine wilt disease. *Pest Management Science.* 1999, 56, 10.

Kobayashi, F. The Japanese pine sawyer. In Dynamics of forest insect populations; Berryman A.A.; ed.; 1988, 431-454.

Koutroumpa, F.A.; Vincent, B.; Roux-Morabito, G.; Martin, C.; Lieutier, F. Fecundity and larval development of *Monochamus galloprovincialis* (Coleoptera Cerambycidae) in experimental breeding. *Ann. For. Science.* 2008, 65, 707.

Kuroda, K. Mechanisms of cavitation development in the pine wilt disease. *Eur. J. For. Path.* 1991, 21, 82-89.

Linit, M.J. The insect component of the pine wilt disease in the United States; In Pathogenicity of the Pine Wood Nematode; Wingfield, M.J.; ed.; 1987, 26-39.

Mamiya, Y. The pine wood nematode; In Plant and Insect Nematoda; Nickle, W.R.; ed.; 1984, 589-626.

Mendes, A.M.S.C.; Feliciano, D.; Tavares, M. and Dias, R. The Portuguese forests; Country report delivered to the EFFE Project "Evaluating financing of forestry in Europe"; Catholic Univ.: Porto, Porto, 2004, pp 226.

Metge, K.; Burgermeister, W. Molecular identification and pathway analysis of the introduced pinewood nematode Bursaphelenchus xylophilus; In Proc. Symp. Plant Protection and Plant Health in Europe. *Introduction and Spread of Invasive Species,* 2005, pp 2.

Mota, M.; Braasch, H.; Bravo, M.A.; Penas, A.C.; Burgermeister, W.; Metge, K.; Sousa, E. First report of *Bursaphelenchus xylophilus* in Portugal and in Europe. *Nematology.* 1999, 1, 727-734.

Naves, P.; Kenis, M.; Sousa, E. Parasitoids associated with *Monochamus galloprovincialis* (Oliv.) (Coleoptera: Cerambycidae) within the pinewood nematode affected zone in Portugal. *J. Pest Science.* 2005, 78, 57–62.

Naves, P.M.; Sousa, E.; Quartau, J.A. Reproductive traits of *Monochamus galloprovincialis* (Coleoptera: Cerambycidae) under laboratory conditions. *Bull. Ent. Res.* 2006a, 96, 289-294.

Naves, P.; Camacho, S.; Sousa, E.; Quartau, J. Entrance and distribution of the pinewood nematode, *Bursaphelenchus xylophilus*, on the body of its vector in Portugal, *Monochamus galloprovincialis* (Coleoptera: Cerambycidae). *Entomol. Gen.* 2006b, 29, 71-80.

Naves, P.M.; Sousa, E.; Quartau, J.A. Feeding and oviposition preferences of *Monochamus galloprovincialis* for some conifers under laboratory conditions. *Entomol. Exp. Appl.* 2006c, 120, 99-104.

Naves, P.; Camacho, S.; Sousa, E.; Quartau, J. Transmission of the pine wood nematode *Bursaphelenchus xylophilus* through feeding activity of *Monochamus galloprovincialis* (Coleoptera; Cerambycidae). *J. Appl. Entomol.* 2007a, 131, 21-25.

Naves, P.M.; Camacho, S.; Sousa, E.; Quartau, J.A. Transmission of the pine wood nematode *Bursaphelenchus xylophylus* through oviposition activity of *Monochamus galloprovincialis*. *Entomol. Fenn.* 2007b, 18, 193-198.

Naves, P.M.; Sousa, E.; Rodrigues, J.M. Biology of *Monochamus galloprovincialis* (Coleóptera: Cerambycidae) in the pine wilt disease affected zone, Southern Portugal. *Silva Lusitana*. 2008, 16, 133-148.

Neves, C.M.B. Sobre a representação da família Scolytidae (Coleoptera) na entomofauna florestal de Portugal metropolitano, continental. *Revista Agronómica*. 1964. 47, pp 8.

Pajares, J.A.; Ibeas, F.; Diez, J.J. and Gallego, D. Attractive responses by *Monochamus galloprovinciallis* (Col., Cerambycidae) to host and bark beetle semiochemicals. *J. Appl. Entomol.* 2004, 128, 633-638.

Penas, A.C.; Dias, L.S.; Mota, M.M. Precision and selection of extraction methods of Aphelenchid nematodes from maritime pine wood, *Pinus pinaster* L. *J. Nematology*. 2002, 34, 62-65.

Rodrigues, J.M. National eradication program for the Pinewood Nematode. In Pine Wilt Disease: a Worldwide Threat to Forest Ecosystems. Mota, M.M.; Vieira, P. Ed. Springer *Science*, 2008; pp 5-14.

Rodrigues, A.; Casquilho, M.; Oliveira, H.; Bordado, J. A statistical analysis of the impact of nematode attack symptomatology on the mechanical behaviour of *Pinus Pinaster* Ait. wood. *European Journal of Forest Research*, 2010, 129, 145-153.

Shivankar, V.J.; Rao, C.N. and Singh, S. Insect pest dispersal. in Encyclopedia Pest Manag. 2006, 2, 255-257.

Sousa, E.; Bravo, M.A.; Pires, J.; Naves, P.M.; Penas, A.C.; Bonifácio, L.; Mota, M. *Bursaphelenchus xylophilus* (Nematoda; Aphelenchoididae) associated with *Monochamus galloprovincialis* (Coleoptera; Cerambycidae) in Portugal. *Nematology*. 2001, 3, 89-91

Sousa, E.; Naves, P.; Bonifácio, L.; Bravo, M.; Penas, A.; Pires, J.; Serrão, M. Preliminary survey for insects associated with the pine wood nematode *Bursaphelenchus xylophilus* in Portugal. *EPPO Bull.* 2002, 32, 499-502.

Takai, K.; Suzuki, T.; Kawazu, K. Development and preventative effect against pine wilt disease of a novel liquid formulation of emamectin benzoate. *Pest Management Science*. 2003, 59, 365-370.

Togashi, K. A field experiment on dispersal of newly emerged adults of *Monochamus alternatus* (Coleoptera: Cerambycidae). *Res. Popul. Ecol.* 1990, 32, 1-13.

In: Nematodes
Editors: F. Boeri and J. A. Chung
ISBN: 978-1-61470-784-4

Chapter VII

Management of Nematode Infection and Resistance against Anthelmintics: Towards a Sustainable Control in Herbivores

Jacques Cabaret and Anne Silvestre
INRA, UR1282, Infectiologie Animale et Santé Publique,
Nouzilly, F-37380, France.

Abstract

Gastro-intestinal nematode infections are highly prevalent in herbivores worldwide and cause serious economical losses in ruminant breeding. Up to 11 classes of anthelmintic drugs exist to control gastro-intestinal nematode infections. Each class of anthelmintic drugs is characterized by a specific mode of action although multidrug resistance mechanisms may be involved. Two main groups can be defined according to their action on membrane ion channels, or on their biochemical action on enzyme inhibition. Anthelmintic resistance has been recorded extensively for only 3 or 4 classes of anthelmintic drugs, namely, benzimidazoles, macrocyclic lactones and tetrahydropyrimidines or levamisole. These anthelmintic drugs classes are efficient against gastro-intestinal and pulmonary nematodes and represent the main portion of anthelmintic drugs used in the world. Newly developed drugs include monepantel and derquantel Despite a remarkably diverse choice of anthelmintic drugs, gastro-intestinal nematode infections are poorly controlled in the field. Anthelmintic resistance has become so frequent worldwide that even if anthelmintic drug classes act in a different manner on nematode biology, it is unlikely they can reverse the increasing reports of multi-resistant populations of nematodes. Mathematical models have been elaborated to address the importance of genetic mechanisms (monogenic or multigenic) and to evaluate the sustainability of breeding management systems (number of treatment, management of pastures...). This theoretical approach allows for the simulation, *in silico*, of very low initial resistance allelic frequencies and over successive years which is not tractable via experimentations. In this chapter, we propose an exhaustive presentation of the available classes of anthelmintic drugs, their mode of action, and if known, the identified drug resistance mechanisms (benzimidazoles), or putative ones (macrocyclic lactones, tetrahydropyrimidines, levamisole). After a presentation of factors associated with the increase in anthelmintic resistance, the available strategies to slow down the development of anthelmintic drugs resistance such as: Targeted Selective Treatment (TST), Targeted-Treatment (TT), and the use of drug combinations (combos) will be evaluated. We also discuss the agricultural practices implemented to control the dynamics of gastro-intestinal nematode infections in field conditions.

Introduction

Anthelmintic resistance is not a recent phenomenon. It first appeared in 1961, just 5 years after benzimidazole was commercialised. The problem likely appeared more than once and was then regularly selected by anthelmintic treatments. Today, the situation is dramatic (James et al., 2009): behind an apparently large selection of drug classes available to control parasitic gastro-intestinal nematode infections in herbivores, a very small number of drug classes are actually used. Due to the constraints of milk withdrawing, meat commercialisation and the narrow spectrum of activity, (see Nielsen, 2009 for review), the reality of anthelmintic usage in herbivores is limited to 2-3 classes of drugs. Such a situation reduces the possibility of alternating drug classes and increases the selection pressure by only one anthelmintic drug class. A recent report indicated that the global commercialisation of anthelmintic drugs was increasing year by year. Breeders have attempted to compensate for the reduction in treatment efficacy with an increase in treatment frequency, thus illustrating the evolutionary arms race with the development of parasite resistance This is the vicious cycle that induced the appearance of multi drug class resistance, and in some cases, the end of entire sheep farms in some parts of the world such as South Africa (Wyk et al., 1997; Wyk et al., 1999). Today, the mechanisms of resistance against the benzimidazoles are mostly deciphered which has allowed for the development of highly sensitive diagnostic tools to detect low level of anthelmintic resistance (Silvestre and Humbert, 2000). However, such tools are mainly restricted for use within the field of scientific research. There is still a lot of work to be done to determine the mechanisms of resistance against levamisole and avermectins Anthelmintic resistance is tricky to quantify in the reality of a farm, as based on phenotypic tests. As a consequence, only high levels of anthelmintic resistance are reliably detected and alternative solutions become more difficult to set up. In some cases, it has been documented that anthelmintic resistant nematodes are as fit as their susceptible counterparts (Elard et al., 1998; Leignel et al., 2010). Accordingly, the anthelmintic resistance level reached in a farm could not be decreased, even if the use of the specific anthelmintic was stopped and replaced with a new drug class (Devos and Cabaret, 2009). Mathematical models have been developed to explore conditions that were not tractable through experiments: evolution of very low resistant gene frequency, long term evolution, etc. In this chapter, we present the available classes of anthelmintics, the factors associated with the increase in anthelmintic resistance , and the available strategies to slow down the development of anthelmintic resistance. We compare the predictions of mathematical models and the effects of agricultural practices implemented to control the dynamics of nematode populations in field conditions.

Anthelmintic Drugs, Their Mode of Action and Identified Resistance Mechanisms

Up to 11 classes of anthelmintic drugs exist to control gastro-intestinal nematode infections. Each class of anthelmintic drugs is characterized by a specific mode of action, although multidrug resistance mechanisms may be involved. Two main groups can be defined according to their action on membrane ion channels, or on their biochemical action on enzyme inhibition (see Table 1).

Table 1. Some anthelmintics efficient against nematode infection in herbivores, their mode of action, the year of first of commercialisation and first published report of resistance for molecules in bold

Classes	Molecule	Mode of action	Year commercialised / First report of resistance
Cholinergic agonist	Levamisole Pyrantel, Morantel Bephenium, tribendimidine Methyridine Monepantel	Open acetylcholine-gated ion channels (AChR) by binding to AChR subunit	Levamisole1970 / 1979 Pyrantel 1974 / 1996 2010
Cholinergic antagonist	Derquantel Phenotiazine	Bind nicotinic receptor	2010; Derquantel is associated with abamectin 1930 / 1945
GluCl and GABA allosteric modulators	Ivermectin, doramectin, eprinomectin, abamectin	Open glutamate-gated ion channels	1981 / 1988
GABA agonist	Piperazine	Open chloride channel	1949 / 1981
Calcium activated K channel activator	Emodepsine	Bind calcium activated K channel	2003
Tubulin ligands	Thiabendazole, mebendazole, fenbendazole, flubendazole, oxibendazole, albendazole, pro-benzimidazoles	Bind colchicine-binding sites of beta-tubulin and of alpha-beta-tubulin dimers	Thiabendazole 1961 / 1964
Salicylanilides	Closantel Rafoxanide	Chitinase inhibitor / ionophore; inhibition of oxidative phosphorylation Uncoupler of oxidative phosphorilation	1977/ 1995
Phenolderivative	Nitroxinil	Uncoupler of oxidative phosphorilation	1960
Innate immune response inhibitor	Diethylcarbamazine	Inhibits both lipoxygenase and cyclooxygenase pathway	1950

Table 2. Constraints of usage for the three main broad spectrum anthelmintic families in 2006 (Cabaret and Silvestre, 2006) and in 2010

Classes of drug	Molecule	Commercialised product	Days before meat commercialisation, milking withdrawal (2006)	Days before meat commercialisation, number of milking withdrawal (2011)
Benzimidazoles	Albendazole	Albendoral®, Bilutac®	10 days, not allowed in dairy	Same
		Disthelm®, Valbazen®	0 day, 0 milking	10 days, not allowed in dairy
		Proftril®,	10 days, 0 milking	Not for sale
		Rumifuge®	10 days, not allowed in dairy	Same
	Thiabendazole	Nemapan®, Stronglozole®	14 days, 6 milkings	Not for sale
	Oxibendazole	Oxibenmix®	14 days, 4 milkings	Not for sale
	Fenbendazole	Panacur®	8 days, 0 milking	Same
	Oxfendazole	Oxfenil®, Synanthic®	10 (cattle) à 14 days (sheep and goat), 0 milking	Same
		Repidose Farmintic®	6 monthes, not allowed in dairy	Same
Pro-benzimidazoles	Febantel	Drontal®, Rintal®	7 (cattle) à 10 days (sheep and goat), 0 milking	
Imidazothiazoles / Tetrahydropyrimidines	Levamisole	Anthelminticide®, Biaminthic®, Levanol®, Lobiavers®	3 days, not allowed in dairy	3 days, not allowed in dairy
		Némisole®	28 days, 10 milkings	3 days, not allowed in dairy
		Capizol®, Lévisole®, Niratil®	3 days, 1 milking	3 days, not allowed in dairy
		Thelmizole®	14 days, 2 milkings	Not for sale
		Chronomintic®	3 days, 4 milkings	16 weeks, not allowed in dairy
		Imena® L, Spectril®	14 days (cattle) 28 days (sheep), not allowed in dairy	14 days, not allowed in dairy
		Ivecide®, Polystrongle®	14 days (cattle) 28 days (sheep), not allowed in dairy	3 days, not allowed in dairy
		Parsifal®	14 days (cattle) 28 days (sheep), not allowed in dairy	28 days, not allowed in dairy
Macrocyclic Lactones	Abamectine	Enzec®	35 days, not allowed in dairy	Not for sale
	Doramectine	Dectomax®	35 (intramuscular) to 56 days (subcutaneous), not allowed in dairy	35 days (pour-on) to 70 days (intramuscular), not allowed in dairy
	Eprinomectine	Eprinex®	28 days, 0 milking	15 days, 0 milking
	Ivermectine	Ivomec®	28 days, not allowed in dairy	28 days, not allowed in dairy
		Oramec®	3 days, not allowed in dairy	6 days, not allowed in dairy
	Moxidectine	Cydectine®	14 (*per os*) to 35 days (injectable), not allowed in dairy	14 days (pour-on) to 108 days (intramuscular), not allowed in dairy

Among these 11 classes, the three main broad spectrum classes are benzimidazoles (beta-tubulin ligands), macrocyclic lactones (GlucCl and GABA allosteric modulators) and tetrahydropyrimidines or levamisole (cholinergic agonists). These anthelmintics are efficient against gastro-intestinal and pulmonary nematodes and represent the main portion of anthelmintic drugs used in the world. Newly developed drugs include monepantel and derquantel (Table 1), but constraints on the use of anthelmintic classes limit the apparently large selection of anthelmintics available (see Table 2). Surprisingly, different delays may be observed for one active product, commercialised under different names. Further constraints are evolving over time ; for instance, levamisole authorization was not renewed in 2006 in France for lactating ruminants, preventing its usage. Data from Table 2 was obtained in France however recommendations may vary between countries within Europe, or the USA.. For example, Nitroxinil (Dovenix©) is an uncoupler of oxidative phosphorilation; its mode of action on haematophagous nematode and flukes has been attributed to this effect. It is sold within certain countries in Europe such as Portugal, France and Luxembourg but is not available for purchase in the USA .

Identifying the physiological targets of anthelmintics facilitated the discovery of the molecular mechanisms responsible for anthelmintic resistance, including those for benzimidazoles and levamisole. (Lubega and Prichard, 1990; Moreno-Guzman et al., 1998) (Table 3). However, even when a drug target was not clearly established, some non specific molecular mechanisms of resistance were identified such as the P-glycoprotein involved in macrocyclic lactone resistance (McCavera et al., 2007).

Table 3. Molecular mechanism of anthelmintic resistance for the three main broad spectrum family drugs. SNP: single nucleotide polymorphism. Adapted from (Beech and Silvestre, 2010)

Classes of drug	Target site	Specific resistance mechanism
Benzimidazoles	Beta-tubulin	Beta-tubulin SNP
Imidazothiazoles / Tetrahydropyrimidines	Nicotinic acetylcholine receptor	Loss of specific acetylcholine gated ion-channels
Macrocyclic Lactones	Ligand gated chloride channel	Ligand gated chloride-channel subunit SNP

Prevalence of Anthelmintic Resistance in Herbivores

Reviewing scientific literature may be a useful tool in estimating the prevalence of anthelmintic resistance in herbivores. Based on a meta-analysis of published data from 1973 to 1999 (Cabaret, 2000) and from 2000 to 2011 recorded in the Commonwealth Agricultural Bureau (CAB), we estimated the reported progression of anthelmintic resistance prevalence according to the host species between both periods (Table 4). The helminths represent one of the most predominant problem's in herbivore health (between 4.9 to 11.1 %) (Table 4), reports of anthelmitnic resistance remain very low (from 0.8 % to 2.8 %) (Table 4). It is commonly acknowledged that anthelmintic resistance has an extremely high prevalence in herbivores worldwide. A comparison of anthelmintic resistance studies from 1973 to 1999 (Cabaret, 2000) and from 2000 to 2011 reveals a 10 fold increase in the reported prevalence of anthelmintic resistance in the latter period. More specifically, reports of anthelmitic resistance rose from 0.06% to 0.8% in goats; 0.07% to 1% in sheep; 0.01% to 0.1% in cattle; and 0.03% to 0.4% in horses. An index of the development of anthelmintic resistance in different herbivore species was determined for both periods (Table 5). Goats ranked the highest (43%) supporting the fact that goats harbour significantly more resistant nematodes populations than all other herbivores. Surprisingly, this index increased notably for sheep and horses reaching 35 %, whereas anthelmintic resistance is still very low in cattle (15 %). The fastest evolution was observed in horses with anthelmintic resistance increasing from 6 % during the first period to 34 % in the second period. A large body of evidence indicates that reducing the intensity of anthelmintic treatment is not usual in the equine industry (Nielsen, 2009) which may explain the increase of observed anthelmintic resistance cases (see below). The main factors explaining the few cases of anthelmintic resistance observed in cattle, at least in Europe (Demeler et al., 2009), are related to the large percentage of parasites in grazing beef cattle that are maintained in refugia - life cycle stages that are not exposed to anthelmintic treatments, which is further compounded by the few anthelminitc treatments actually in use (Coles, 2002).

Table 4. Number of reported studies on herbivores, helminth infections, use of anthelmintics and anthelmintic resistance, between 1973 and 2011 in the Web of Science (CAB database). * Percentages of references

Host species	Total number of references	References on helminths	References on anthelmintics	References on anthelmintic resistance
Goat	56,503	5,028 (8.9 %)*	997 (1.8 %)	427 (0.8 %)
Sheep	140,669	15,527 (11.1 %)	3,973 (2.8 %)	1,380 (1 %)
Cattle	373,116	18,370 (4.9 %)	3,002 (0.8 %)	455 (0.1 %)
Horse	105,133	5,435 (5.2 %)	1,219 (1.2 %)	417 (0.4 %)

Nematodes have genetic features that favour the development of anthelmintic resistance such as the rapid rate of nucleotide sequence evolution and extremely large population sizes that give the parasites an exceptionally high level of genetic diversity. Hence, all nematode species can acquire anthelmintic resistance and have been reported, but some species are over represented in literature.

For instance, closantel is effective only against hematophagous parasites, such as *Haemonchus contortus*. In the same way, particular phenotypic tests to detect anaemia, once again, are only efficient on hematophagous species.

Resistance is not always easy to define, since phenotypic efficacy assessments may fluctuate largely. In some cases, anthelmintic resistance may be overestimated. Main factors associated with "false resistance" (Cabaret, 2010) are: low quality drugs (some generic drugs) and poor absorption or metabolisation of the drug, which is an important issue for goat treatment (see below).

Conversely, anthelmintic resistance may be underestimated. In field studies, anthelmintic resistance diagnosis is largely based on phenotypic evaluation (Eysker and Ploeger, 2000). Those tests, even if they are reliable at the species level, are poorly sensitive at the community level. Species abundance varies greatly throughout the grazing season, altering the sensitivity of the test for rare species or less fertile species.

Management of Anthelmintic Resistance

Management Factors Associated with Anthelmintic Resistance

What are the main factors associated with anthelmintic resistance increase? From field studies conducted on a large geographical scale, it was demonstrated that host movements are responsible for the introduction of anthelmintic resistant strains of parasites of cattle in the USA (Blouin et al., 1995) and goats from Spain (Requejo-Fernandez et al., 1997), Slovakia (Varady et al., 1993) or Italy. This introduction of anthelmintic resistant parasites was also evidenced at a lower geographical scale: between farms from the same region (less than 50 km away) (Silvestre et al., 2009). This factor underlies the importance of classical practices, such as quarantine, in controlling the spread of anthelmintic resistance nematodes.

Once the anthelmintic resistance allele is first introduced in the farm, which are the breeding management factors that act (directly or not) on its frequency? It is largely acknowledged that the development of anthelmintic resistance depends on an *efficient selective pressure* (Silvestre et al., 2002). The contribution of resistant parasites (i.e. those which survived an anthelmintic treatment) to the subsequent generation, is a key factor that controls the spread of resistance. This covers the following factors: 1) under-dosing may favour the establishment of BZ heterozygous parasites that should have been eliminated by a therapeutic dosage of anthelmintic (favouring the selection of the very first anthelmintic resistance alleles in the parasite population), 2) repeated use of one anthelmintic family drug, possibly at a sub-optimal dosage, and 3) the size of the population in refugia at the time of the treatment (i.e. infective larvae on pastures and parasitic stages in untreated hosts).

Under-dosing has been cited as one of the main contributing factors in the spread of resistance. One experiment was designed to determine the effects of under-dosing on the frequency of benzimidazole resistant allele in the nematode *Teladorsagia circumcincta* (Silvestre et al., 2001). The benzimidazole was tested at 1/32, 1/16, 1/8 and 1/4 of the recommended dose for sheep. The fraction of the susceptible homozygote (SS), susceptible heterozygote (RS), and resistant homozygote (RR) genotypes were compared among dose groups to evaluate differences between SS and RS genotype selective advantage. Almost all SS genotype worms were eliminated by 1/4 of the benzimidazole recommended dose whereas a significant fraction of the RS genotype worms survived treatment. The selective advantage was 4±5 times higher for the RS genotype at 1/4 of the manufacturer's recommended dose. A computer simulation was used to study the putative spread of anthelmintic resistance over a range of RS selective advantages (2, 4±5 and 10-fold), with two average sizes of individual host worm population (20 or 2000 worms/host) and two initial low R allele frequencies (equal or below 1%). In all situations, the lowest selective advantage of the RS genotype over the SS genotype was sufficient to promote the spread of resistance in susceptible populations. These results have significance in a very limited situation: a very low presence of alleles for resistance and important under-dosing (1/4 of the dose). Thus, they cannot be extended to all situations of benzimidazole resistance. They also cannot be extended to resistance of other drugs, for which the mechanisms are not well defined (levamisole or macrocyclic lactones). Tetramisole (an imidazothiazole compound being a mixture of dextrogyre and levogyre-levamisole drug) has been used for decades in Morocco and in particular in the Middle-Atlas. Moreover, the recommended dose was half the dose as proposed in other countries (Berrag and Cabaret, 2002). The reason was probably that good results on *Haemonchus contortus* were obtained in Tchad and Senegal at half the dose. The main strongyle species in Middle-Atlas are *Teladorsagia circumcincta* and *Trichostrongylus sp.*, and the strongyles in the area received a half dose for more than 10 years. We noticed that when used at the normal dose, the drug was still efficient and that its efficacy was not eroded by years of under-dosage. Cattle in France has been submitted to normal dose of avermectine (200µg/kg bodyweight, the same dose as the one for strongyles) in young cattle and to micro-doses (2µg/kg bodyweight) in other groups of cattle for durably controlling bovine hypodermosis (Boulard, 2002). Thus, a part of the cattle population has received a clear under-dosage of avermectine over several years. There is still an absence of resistance to ivermectines' reports in French cattle and this does not militate for under-dosage as a selective factor for resistance to macrocyclic lactones. Conversely, several laboratory selections for resistance to ivermectine were obtained by means of under-dosage on putative susceptible strains. The role of underdosage in the building of resistance to anthelmintics has probably been overestimated but cannot be completely ruled out.

The repeated use of the same anthelmintics is also a frequently cited source of resistance to anthelmintics. This is one explanation for the high prevalence of resistance in French dairy-goats (Chartier et al., 2001): benzimidazoles were the only agreed drugs in lactating goats without the withdrawal of milk production, which explained its intensive use for decades.

Usefullness of Mathematical Models

The onset and the selection of anthelmintic resistance have threatened the sustainability of control schemes of nematode infections of herbivores (Table 4 and Table 5). To face this issue, mathematical models have been developed to decipher mechanisms selecting for resistance in gastro-intestinal nematode / ruminant models. For benzimidazoles (BZ), it was observed that fitness of BZ-resistant parasites was as good as the one of BZ-susceptible parasites (Elard et al., 1998; Leignel et al., 2010); indicating that reversion to susceptibility to BZ was not probable and that the anthelmintic resistance level reached in a farm may not be easily counter-selected (Devos and Cabaret, 2009). The alarming challenge was then to limit the appearance and the selection of anthelmintic resistant parasites, and to study the forces that increase anthelmintic resistance allele frequency when the very first resistant parasites appear (i.e. when anthelmintic resistance allele frequency is very low). Neither parasite populations harbouring very low resistant gene frequency are tractable by experimentation, nor following evolution of resistant gene frequency during a long period, from 5 to 20 years (Barnes et al., 1995; Gaba et al., 2006). Moreover, a high number of parameter associations make biological experiment unsuitable.

Mathematical models developed to study the evolution of anthelmintic resistance can be arranged into two classes: 1) model based on population genetics only (genetic models) and 2) models based both on population genetics and epidemiology. Most models rely on a monogenic determinism of anthelmintic resistance; some rare cases investigated multigenic resistance (Barnes et al., 1995). In fact, those models were proposed before genetic determinism of anthelmintic resistance to BZ, levamisole or avermectins was known. As a consequence, different genetic determinisms were investigated (dominance, co-dominance and recessivity of resistance alleles).

Table 5. Index of anthelmintic resistance in herbivores between 1973 and 1999 and between 1973 and 2011. Index of anthelmintic resistance corresponds to the number of references on anthelmintic resistance *100 / number of references on anthelmintics

Host species	Index of helminth resistance (1973-1999)	Index of helminth resistance (1973-2011)
Goat	30 %	43 %
Sheep	22 %	35 %
Cattle	17 %	15 %
Horse	6 %	34 %

Genetic models study the evolution of anthelmintic resistance allele frequency and rely on the Hardy Weinberg equilibrium (Dobson et al., 1987; Gettinby, 1989; Gettinby et al., 1989; Leathwick et al., 1995; Saul, 1995). The frequency p_t of resistance allele (r) and the frequency of susceptible allele q_t ($q_t = 1 - p_t$) at time (t) depend on the selection coefficient or selective advantage (w_{rs}, w_{rr} and w_{ss}). At treatment, w_{rr} is higher than w_{rs} and w_{ss}. At each time step of the model, frequency of the resistant allele is calculated, based on the selective advantage of each genotype, according to equation 1 below (Crow and Kimura, 1970).

$$p_{t+1} = (w_{rr}p_t^2 + w_{rs}p_tq_t) / (w_{rr}p^2 + 2w_{rs}p_tq_t + w_{ss}q_t^2)$$

Selective advantage may be replaced by the survival rate of each genotype ((Barnes et al., 1995; Saul, 1995; Silvestre et al., 2001). Equation 1 follows resistance allele frequency according to intensity of selection pressure due to anthelmintic treatment (i.e. the selective advantage). Those genetic models predict a sigmoid evolution of anthelmintic resistance allele frequency which indicates that a fast and high increase of anthelmintic resistance frequency is expected, as it was the case with BZ resistance (5 years after they were first commercialised). Equation 1 also indicates that the maximal speed of resistant allele frequency increase is reached when p = 0.5, which corresponds roughly to the sensitivity of phenotypic tests (Roos et al., 1995). As a consequence, when resistance is detected in farms, speed of selection of anthelmintic resistance is already too high to be slowed down. All genetic models supported a fast and high increase of anthelmintic resistance in parasite populations but several simplifications were made such as the absence of host and parasite dynamics . After an anthelmintic treatment, the contribution of surviving (resistant) parasites to the subsequent generation depends on parasite fertility, lifetime duration of the free living stages on pastures and the transmission rate of infective stages. Epidemiology was then introduced in mathematical models to test efficiency of control schemes of nematode infections (Anderson and May, 1985).

An efficient control scheme of nematode infection needs to strike the balance between a short term necessity for an efficient and easy control (i.e. using the available anthelmintics in all flock treatment) and the long term necessity of delaying the selection of anthelmintic resistance alleles to control nematode parasite populations (Figure 1). Epidemiology was introduced through the dynamics of both host and parasite populations (Barnes and Dobson, 1990; Smith, 1990) estimating the abundance of parasites populations at different life stages. However, deterministic models cannot be used: each life trait (nematode birth, death and transmission) is a chance event generating variability in time and space in the population and in the selection of anthelmintic resistance alleles. Stochastic models take into account this random aspect of demographic and environmental processes (Cornell et al., 2003; Gaba et al., 2006).

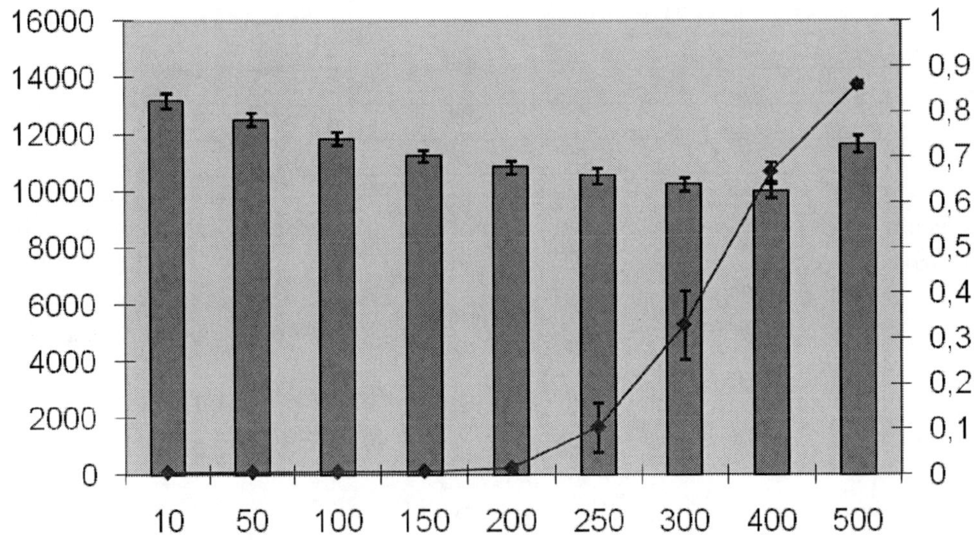

Figure 1. Evolution of gastro-intestinal nematode burden (histogram, up to 13 500 nematodes) and resistance allele frequency (line, up to 0.88), after 5 year simulation, according to the number of treated hosts in a 500 lambs flock.

Moreover, nematode parasites are not evenly distributed in their hosts, and a large number of hosts harbour few parasites whereas a few hosts are heavily infected (Shaw and Dobson, 1995). The heterogeneity of nematode burden in host results from biological processes such as: aggregated spatial distribution of infective stages on pastures (Shaw et al., 1998), differences in host sensitivity to nematode infection and environmental and demographic stochasticity (Anderson and Gordon, 1982). The aggregated distribution of nematodes in hosts modifies the density-dependant regulations in each infra population (parasite population in a host), modulate the individual reproductive success of nematodes (May and Woolhouse, 1993) and the rate of selection of adaptive traits (Criscione and Blouin, 2005). Integrating those aspects of spatial heterogeneity modifies greatly the evolution of rare anthelmintic resistant genotypes (Cornell et al., 2000).

Integrating parasite epidemiology and population dynamics into mathematical models helps to define alternative control strategies to the mass treatment of flocks (Barnes et al., 1995; Gaba et al., 2006). These models explored the effect of different control schemes and allowed the identification of optimal anthelmintic usage to reach best nematode control with the least selection pressure (Figure 1). Maintaining a refugia (i.e. a proportion of the nematode population that is not submitted to the anthelmintic selection pressure) is an alternative to mass treatment (Wyk, 2001). In nematode infections, three refugia are identified: nematodes in untreated hosts, encysted larvae and free-living stages. Encysted larvae may represent a huge population in some cases (Gaba et al., 2010; Leignel et al., 2010) but they cannot be easily manipulated. Conversely, the role of the refugia constituted by free-living stages on pastures was clearly evidenced in anthelmintic resistance evolution (Gaba et al., 2006). The investigation of the refugia constituted by nematodes in untreated hosts resulted in the definition of several selective treatment strategies (Kenyon et al., 2009).

Developing an individual based model (Gaba et al., 2006), indicated that even if frequent anthelmintic treatment facilitates anthelmintic resistance selection, frequent treatments are not sufficient in selecting for resistance. First, the stochastic ingestion of infective stages highlights the impact of genetic drift when the anthelmintic resistance allele is very rare (initial frequency at 0.5%) and anthelmintic selection as low as 2 treatments per year. Second, without a differential survival rate of infective larvae during the summer, the probability that selection for resistance occurs is strongly reduced (less than 10% of model outputs) due to the large size of the population in refugia (i.e. infective larvae on pastures). Another way to maintain refugia is to leave some hosts untreated by Targeted Selective Treatment (TST). The aggregation of nematode parasites in their hosts provides an opportunity to delay anthelmintic resistance, treating only the most heavily infected hosts (i.e. those hosts that make the greatest contribution to pasture infectivity). Individual health indices: FAMACHA©; (Wyk and Bath, 2002), DISCO (Cabaret et al., 2006) or faecal soiling (Brougham and Wall, 2007) are available to target the hosts. This model indicated that treating between 20% and 30% of the flock was efficient both at maintaining resistance at a low level and controlling nematode parasite infection (Gaba et al., 2010). Simulations also show that the optimal percentage of TST was dependant on the size of the flock (the more the flock size, the less the percentage), modifying the number of nematode parasites exposed to treatment. Individual scoring of hosts is time consuming for farmers, and efficiency of random TST was investigated by this same individual based model. Random TST efficiently counter selects for anthelmintic resistance in comparison with all flock treatment. To reduce selection pressure exerted by anthelmintic treatments, it was suggested to alternate anthelmintics from different drug classes. An experimental study was conducted on a BZ-resistant population of sheep nematodes, treated with levamisole. Even if BZ-resistant nematodes are efficiently controlled by levamisole treatments (Overend et al., 1994) the study indicated that for anthelmintic resistance as high as 25%, alternation of anthelmintics was not efficient to extensively delay the evolution of resistance (Leignel et al., 2010). Barnes et al. (1995) evaluated the efficiency of anthelmintic alternation in an initial nematode population, harbouring 0.01% of anthelmintic resistant alleles. Genetic determinism of anthelmintic resistance was 2 independent genes, each anthelmintic killing 99%, 50% and 10% of SS, SR and RR genotypes, respectively (co-dominance of anthelmintic resistance alleles). After 20 year-simulations, the alternation of anthelmintics always results in the selection of resistant nematodes for one or the other anthelmintic used. There was no appreciable development of anthelmintic resistance over 20 years when using combos However, this strategy is challenging: in rare cases, resistance to one anthelmintic may favour the acquisition of resistance to another one (Mottier and Prichard, 2008). Almost all models assume that anthelmintic resistance is determined by one locus, with two alleles, except for levamisole where the genetic determinism is more complex and still needs to be elucidated (Fauvin et al., 2010; Neveu et al., 2010). A polygenic determinism may have consequences on parasite phenotypes: for instance, not only survival or death but survival conditioned by the dosage or possible evolution costs of mutations that may modify life history traits of parasites (Leignel et al., 2002; Melo, 2005). Although modelling is a powerful tool for deciphering rare events (the onset of resistance) or long term results of treatment strategies, there is a real need to also consider field data.

Field Perspectives

Which animals should be treated?

Internal parasites are almost always aggregated across their host population, with a large number of hosts harbouring few parasites and few heavily infected hosts. In sheep nearly 80% of the gastro-intestinal nematodes are found in 20% of the sheep and in 2 to 3 years old horses only 50% of the animals harbour such nematodes (Cabaret et al., 2011). It is then crucial to detect the infected animals in order to selectively treat them if we want to reduce the unworthy use of anthelmintics. The most obvious indicator is the faecal egg counts but, due to the laboratory recourse the delay for transport and response, and finally the cost (often higher than anthelmintic cost), it has never really been adapted to a large scale. A unique situation is found in Denmark: the delivery of anthelmintics for horses to farmer must be comforted by a faecal egg counts evaluation. This legal constraint although probably useful for reducing the use of anthelmintics, has not spread to other countries. There are several indirect pathophysiological indicators, in the hand of the farmers that have been suggested to selectively target anthelmintic treatments. The FAMACHA© system (Wyk and Bath, 2002) in small ruminants can be used to identify sheep suffering from anaemia (likely caused by *Haemonchus contortus*). FAMACHA© is the most used indicator and has been tested in a large variety of conditions (among others: Burke et al., 2007; Koopmann et al., 2006). A diarrhoea score (DISCO) can be a good indicator of actual nematode infection during the end of spring and autumn in a temperate climate (Cabaret et al., 2006) or throughout the year in two regions of Morocco (Ouzir et al., 2011). Body condition scoring (Russel, 1984) is a potential method for identifying lambs requiring anthelmintic treatment (Wyk and Bath, 2002). The latter is difficult to use in young lambs (our unpublished data) and is not useful to detect infected ewes (Ouzir et al., 2011). The main problem to use indirect indicators is that they are fairly labour intensive. They are accepted in countries where labour cost is low as South Africa (Wyk and Bath, 2002) or Morocco (Berrag et al., 2009), but are not accepted in countries with higher costs of labour (Cabaret et al., 2009). The organic farmers are more inclined than conventional farmers to use the TST based on pathophysiological indicators in developed countries (Cabaret et al., 2009). The TST indirect indicators were mostly developed for meat sheep and none are available in cattle or horses. There is also a difficulty for farmers to use these indirect indicators: they are in the best cases statistically efficient to detect animals in need of a treatment, but they may fail to detect a particular animal or conversely wrongly detect an animal to be treated. Regardless of the indicator, these false positives or negatives (Cabaret, 2008), could be a source of mistrust for the farmer in using the TST.

When to treat?

The pathophysiological indicators could also be used differently: sentinel sheep could be checked at regular intervals which are intended to help for the treatment decisions of all the flock at a particular month and against a specified group of internal parasites. This strategy (Targeted Treatment) could be compatible with many farming systems due to low cost and facility of sampling few animals. In fact, this strategy has never been evaluated in the field. Much remains to be determined such as the optimal size of the sentinel group and which indicator should be used.

What anthelmintics should be used?

The diagnosis of anthelmintic resistance should be done before any treatment protocol is undertaken. A single effective drug is the rule in countries or species where gastro-intestinal nematodes' resistance remains low, such as in European cattle (Demeler et al., 2009). Drug alternation may slow the development of resistance but will not solve the problem (Leignel et al., 2010): the increase of resistance to benzimidazole was proportional to the use of benzimidazoles. The reversion of resistance is not expected: during a two-year study, the benzimidazole resistant gastro-intestinal nematodes remained at the same level when treated with another drug or when they remained untreated. Another study on benzimidazole resistance of gastro-intestinal nematodes in sheep and goats showed that the level of resistance to benzimidazole did not change much 10 years after changing the anthelmintic (from benzimidazoles to ivermectin). The change of drug is then intended for a better control of infection but not to reduce the resistance to the former drug. The association of drugs (up to four different classes of anthelmintics) is proposed in the countries with a high level of gastro-intestinal nematode resistance to drugs. These associations (combos) are useful in cases where animals are introduced from another farm, as a quarantine treatment. Their value for all flock treatment may be more questionable. Apart from the ecological (many drugs possibly have residues in meat and milk products) or economical (the cost of treatment will be high) considerations, it is not obvious that it will reduce the spread of resistance in sites where gastro-intestinal nematodes anthelmintic resistance is common. Combos cannot be used as a "blanket" treatment to stop resistance in practice: farmers will use combinations of two or three drugs and they may maintain resistance to a fourth or a fifth drug.

Conclusions: Towards Comprehensive Recommendations for Anthelmintic Use?

There is a need to clarify the recommendations for the use of anthelmintics since they have been largely modified over the last 20 years - mostly in relation to the acquisition of resistance by gastro-intestinal nematodes. The "treat herbivores and move them to a new clean pasture" strategy was recommended when controlling susceptible gastro-intestinal nematodes. It has since changed to "move the herbivores to a previously grazed pasture and then treat them" when considering resistant gastro-intestinal nematode control, in order to maintain refugia and consequently dilute the resistant alleles. Previously, the treatment of the entire flock was the rule. This too has altered to treat only those animals, or groups of animals in need of treatment. The British Veterinary Association (BVA) has produced a poster on the responsible use of anthelmintics in grazing animals (http://www.bva.co.uk/public/documents/BVA_Anthelmintics_poster.pdf, (accessed on the 30th of May 2011). This was probably intended to bring together the requirements of past strategies (for susceptible gastro-intestinal nematodes) and new strategies (for resistant ones). The BVA advised that newly treated animals should not necessarily be moved onto clean pasture; a clear strategy of pasture management should be proposed to the farmer and decision trees should be established in relation with the gastro-intestinal situation in each farm. The BVA also promoted the use of faecal egg counts to adapt the drenching strategy to the actual nematode infection and thus avoid blanket treatments. The use of pathophysiological indicators should be developed as an alternative to costly faecal egg counts, however, further research is needed to advance their efficacy, particularly in temperate regions. The only currently available data are in sheep and the feasibility of the indicators' use in cattle and horses has to be studied. The weight of an animal could be an alternative indicator but its use is dependant on the availability of electronic weighing machines. Understanding the mechanisms of resistance to most drugs (e.g. levamisole, ivermectin) is yet limited and thus modelling the evolution of resistance remains uncertain, except for benzimidazoles. Nevertheless, the bases for the sustainable use of anthelmintics against gastro-intestinal nematodes are in part available to breeders, should they choose to incorporate them. These include;the correct choice of anthelmintic targeted to the parasites present in a farm, at the right dose, preferably to the susceptible young animals and avoidance of clean pastures after treatment when resistance to a drug has been demonstrated.

Acknowledgements

Many data presented here are issued from the European Parasol programme or from discussions with colleagues within this programme (Boumadiane Berrag of Agronomic and Veterinary Institute Hassan II in Morocco, Fiona Kenyon and Frank Jackson of the Moredun Institute, UK). We also thank Vincent Leignel (Université du Maine, France) for useful comments on the structure of this chapter. We gratefully acknowledge the careful reading and comments on the last version of the manuscript by Caroline Chylinski (Marie Curie PhD on Nematode System Health, INRA Nouzilly, France).

References

Anderson, R.M., Gordon, D.M., (1982). Processes influencing the distribution of parasite numbers within host populations with special emphasis on parasite-induced host mortalities. *Parasitology,* 85, 373-398.

Anderson, R.M., May, R.M., (1985). Helminth infections of humans: mathematical models, population dynamics, and control. *Advances in Parasitology*, 24, 1-101.

Barnes, E.H., Dobson, R.J., (1990). Population dynamics of *Trichostrongylus colubriformis* in sheep: computer model to simulate grazing systems and the evolution of anthelmintic resistance. *International Journal for Parasitology*, 20, 823-831.

Barnes, E.H., Dobson, R.J., Barger, I.A., (1995). Worm control and anthelmintic resistance: adventures with a model. *Parasitology Today*, 11, 56-63.

Beech, R.N., Silvestre, A., (2010). Mutations associated with anthelmintic drug resistance. *Anti-Infective Agents in Medicinal Chemistry*, 9, 105-112.

Berrag, B., Cabaret, J., (2002). Efficacité insuffisante d'un anthelminthique à base de Tétramisole : une histoire de dose. *Espace Vétérinaire*, 22, 10-11.

Berrag, B., Ouzir, M., Cabaret, J., (2009). A survey on meat sheep farms in two regions of Morocco on farm structure and the acceptability of the targeted selective treatment approach to worm control. *Veterinary Parasitology*, 164, 30-35.

Blouin, M.S., Yowell, C.A., Courtney, C.H., Dame, J.B., (1995). Host movement and the genetic structure of populations of parasitic nematodes. *Genetics*, 141, 1007-1014.

Boulard, C., (2002). Durably controlling bovine hypodermosis. *Veterinary Research*, 33, 455-464.

Brougham, J.M., Wall, R., (2007). Faecal soiling and gastrointestinal helminth infection in lambs. *International Journal for Parasitology*, 37, 1255-1268.

Burke, J.M., Kaplan, R.M., Miller, J.E., Terrill, T.H., Getz, W.R., Mobini, S., Valencia, E., Williams, M.J., Williamson, L.H., Vatta, A.F., (2007). Accuracy of the FAMACHA system for on-farm use by sheep and goat producers in the southeastern United States. *Veterinary Parasitology*, 147, 89-95.

Cabaret, J., (2000). Anthelmintic resistance in goats: from fiction to facts. In: Proceeding of the 7th international conference on goats, Tours-Poitiers, (France), 15th-21st May 2000, pp. 793-794.

Cabaret, J., (2008). Pro and cons of targeted selective treatment against digestive-tract strongyles of ruminants. *Parasite*, 15, 506-509.

Cabaret, J., (2010). False Resistance to Antiparasitic Drugs: Causes from Shelf Availability to Patient Compliance. Anti-Infective *Agents in Medicinal Chemistry*, 9, 161-167.

Cabaret, J., Benoit, M., Laignel, G., Nicourt, C., (2009). Current management of farms and internal parasites by conventional and organic meat sheep French farmers and acceptance of targeted selective treatments. *Veterinary Parasitology*, 164, 21-29.

Cabaret, J., Gonnord, V., Cortet, J., Sauve, C., Ballet, J., Tournadre, H., Benoit, M., (2006). Indicators for internal parasitic infections in organic flocks: the diarrhoea score DISCO proposal for lambs. *In Organic congress 2006: organic farming and european rural development* (Odense, Denmark), pp. 552-553.

Cabaret, J., Guerreo, M.C., Duchamp, G., Wimel, L., Kornas, S., (2011). Distribution agrégée du parasitisme interne par les nématodes chez les équins: intérêt pour le diagnostic et la gestion antiparasitaire. In 37e journée de la recherche équine (Paris, France), pp. 49-54.

Cabaret, J., Silvestre, A., (2006). La résistance aux anthelminthiques chez les ruminants : diagnostic et gestion en ferme. Groupement Technique Veterinaire, Hors Série, 192-198.

Chartier, C., Soubirac, F., Pors, I., Silvestre, A., Hubert, J., Couquet, C., Cabaret, J., (2001). Prevalence of anthelmintic resistance in gastrointestinal nematodes of dairy goats under extensive management conditions in southwestern France. *Journal of Helminthology*, 75, 325-330.

Coles, G.C., (2002). Cattle nematodes resistant to anthelmintics: why so few cases? *Veterinary Research*, 33, 481-489.

Cornell, S.J., Isham, V.S., Grenfell, B.T., (2000). Drug-resistant parasites and aggregated infection - early-season dynamics. *Journal of Mathematical Biology*, 41, 341-360.

Cornell, S.J., Isham, V.S., Smith, G., Grenfell, B.T., (2003). Spatial parasite transmission, drug resistance, and the spread of rare genes. *Proceedings of the National Academy of Sciences of the United States of America*, 100, 7401-7405.

Criscione, C.D., Blouin, M.S., (2005). Effective sizes of macroparasite populations: a conceptual model. *Trends in Parasitology*, 21, 212-217.

Crow, J., Kimura, M., (1970). An introduction to the theory of population genetics. New York.

Demeler, J., Zeveren, A.M.J.v., Kleinschmidt, N., Vercruysse, J., Hoglund, J., Koopmann, R., Cabaret, J., Claerebout, E., Areskog, M., Samson-Himmelstjerna, G.v., (2009). Monitoring the efficacy of ivermectin and albendazole against gastro intestinal nematodes of cattle in Northern Europe. *Veterinary Parasitology*, 160, 109-115.

Devos, J., Cabaret, J., (2009). Résistance des strongles aux benzimidazoles chez les ovins et les caprins : présentation d'une essai terrain sur la réversibilité. In Journées nationales des GTV. Le nouveau-né : abord individuel et gestion sanitaire du troupeau. (Nantes, France), pp. 861-866.

Dobson, R.J., Griffiths, D.A., Donald, A.D., Waller, P.J., (1987). A genetic model describing the evolution of levamisole resistance in Trichostrongylus colubriformis, a nematode parasite of sheep. *IMA Journal of Mathematics Applied in Medicine and Biology*, 4, 279-293.

Elard, L., Sauve, C., Humbert, J.F., (1998). Fitness of benzimidazole-resistant and - susceptible worms of *Teladorsagia circumcincta*, a nematode parasite of small ruminants. *Parasitology,* 117, 571-578.

Eysker, M., Ploeger, H.W., (2000). Value of present diagnostic methods for gastrointestinal nematode infections in ruminants. *Parasitology,* 120, 109-119.

Fauvin, A., Charvet, C., Issouf, M., Cortet, J., Cabaret, J., Neveu, C., (2010). cDNA-AFLP analysis in levamisole-resistant Haemonchus contortus reveals alternative splicing in a nicotinic acetylcholine receptor subunit. *Molecular and Biochemical Parasitology*, 170, 105-107.

Gaba, S., Cabaret, J., Ginot, V., Silvestre, A., (2006). The early drug selection of nematodes to anthelmintics: stochastic transmission and population in refuge. *Parasitology*, 133, 345-356.

Gaba, S., Cabaret, J., Sauve, C., Cortet, J., Silvestre, A., (2010). Experimental and modeling approaches to evaluate different aspects of the efficacy of Targeted Selective Treatment of anthelmintics against sheep parasite nematodes. *Veterinary Parasitology*, 171, 254-262.

Gettinby, G., (1989). Computational veterinary parasitology with an application to chemical resistance. *Veterinary Parasitology*, 32, 57-72.

Gettinby, G., Soutar, A., Armour, J., Evans, P., (1989). Anthelmintic resistance and the control of ovine ostertagiasis: a drug action model for genetic selection. *International Journal for Parasitology*, 19, 369-376.

James, C.E., Hudson, A.L., Davey, M.W., (2009). Drug resistance mechanisms in helminths: is it survival of the fittest? *Trends in Parasitology*, 25, 328-335.

Kenyon, F., Greer, A.W., Coles, G.C., Cringoli, G., Papadopoulos, E., Cabaret, J., Berrag, B., Varady, M., Wyk, J.A.v., Thomas, E., Vercruysse, J., Jackson, F., (2009). The role of targeted selective treatments in the development of refugia-based approaches to the control of gastrointestinal nematodes of small ruminants. *Veterinary Parasitology*, 164, 3-11.

Koopmann, R., Holst, C., Epe, C., (2006). Experiences with the FAMACHA-Eye-Colour-Chart for identifying sheep and goats targeted for anthelmintic treatment. *Berliner und Munchener Tierarztliche Wochenschrift,* 119, 436-442.

Leathwick, D.M., Vlassoff, A., Barlow, N.D., (1995). A model for nematodiasis in New Zealand lambs: the effect of drenching regime and grazing management on the development of anthelmintic resistance. *International Journal for Parasitology*, 25, 1479-1490.

Leignel, V., Cabaret, J., Humbert, J.F., (2002). New molecular evidence that *Teladorsagia circumcincta* (Nematoda: Trichostrongylidea) is a species complex. *Journal of Parasitology*, 88, 135-140.

Leignel, V., Silvestre, A., Humbert, J.F., Cabaret, J., (2010). Alternation of anthelmintic treatments: a molecular evaluation for benzimidazole resistance in nematodes. Veterinary Parasitology, in press.

Lubega, G.W., Prichard, R.K., (1990). Specific interaction of benzimidazole anthelmintics with tubulin: high-affinity binding and benzimidazole resistance in *Haemonchus contortus*. *Molecular and Biochemical Parasitology*, 38, 221-232.

May, R.M., Woolhouse, M.E.J., (1993). Biased sex ratios and parasite mating probabilities. *Parasitology,* 107, 287-295.

McCavera, S., Walsh, T.K., Wolstenholme, A.J., (2007). Nematode ligand-gated chloride channels: an appraisal of their involvement in macrocyclic lactone resistance and prospects for developing molecular markers. *Parasitology,* 134, 1111-1121.

Melo, A.C., (2005). Resistance to benzimidazoles in nematoda *Haemonchus contortus* in Ceara state. Fortaleza university, Ceara, Brazil,

Moreno-Guzman, M.J., Coles, G.C., Jimenez-Gonzalez, A., Criado-Fornelio, A., Ros-Moreno, R.M., Rodriguez-Caabeiro, F., (1998). Levamisole binding sites in *Haemonchus contortus*. International *Journal for Parasitology*, 28, 413-418.

Mottier, M.D., Prichard, R.K., (2008). Genetic analysis of a relationship between macrocyclic lactone and benzimidazole anthelmintic selection on *Haemonchus contortus*. *Pharmacogenetics and Genomics*, 18, 129-140.

Neveu, C., Charvet, C.L., Fauvin, A., Cortet, J., Beech, R.N., Cabaret, J., (2010). Genetic diversity of levamisole receptor subunits in parasitic nematode species and abbreviated transcripts associated with resistance. *Pharmacogenetics and Genomics*, 20, 414-425.

Nielsen, M.K., (2009). Restrictions of anthelmintic usage: perspectives and potential consequences. *Parasites and Vectors*, 2, S1-S7.

Ouzir, M., Berrag, B., Benjouad, A., Cabaret, J., (2011). Use of pathophysiological indicators for individual decision of anthelmintic treatment of ewes against gastrointestinal nematodes in Morocco. *Veterinary Parasitology*, in press.

Overend, D.J., Phillips, M.L., Poulton, A.L., Foster, C.E.D., (1994). Anthelmintic resistance in Australian sheep nematode populations. *Australian Veterinary Journal*, 71, 117-121.

Requejo-Fernandez, J.A., Martinez, A., Meana, A., Rojo-Vazquez, F.A., Osoro, K., Ortega-Mora, L.M., (1997). Anthelmintic resistance in nematode parasites from goats in Spain. *Veterinary Parasitology*, 73, 83-88.

Roos, M.H., Kwa, M.S.G., Grant, W.N., (1995). New genetic and practical implications of selection for anthelmintic resistance in parasitic nematodes. *Parasitology Today*, 11, 148-150.

Russel, A.J.F., (1984). Body condition scoring of sheep. In Practice, 6, 91-93.

Saul, A., (1995). Computer model of the maintenance and selection of genetic heterogeneity in polygamous helminths. *Parasitology*, 111, 531-536.

Shaw, D.J., Dobson, A.P., (1995). Patterns of macroparasite abundance and aggregation in wildlife populations: a quantitative review. *Parasitology*, 111, 111-133.

Shaw, D.J., Grenfell, B.T., Dobson, A.P., (1998). Patterns of macroparasite aggregation in wildlife host populations. *Parasitology,* 117, 597-610.

Silvestre, A., Cabaret, J., Humbert, J.F., (2001). Effect of benzimidazole under-dosing on the resistant allele frequency in *Teladorsagia circumcincta* (Nematoda). *Parasitology*, 123, 103-111.

Silvestre, A., Humbert, J.F., (2000). A molecular tool for species identification and benzimidazole resistance diagnosis in larval communities of small ruminant parasites. *Experimental Parasitology*, 95, 271-276.

Silvestre, A., Leignel, V., Berrag, B., Gasnier, N., Humbert, J.F., Chartier, C., Cabaret, J., (2002). Sheep and goat nematode resistance to anthelmintics: pro and cons among breeding management factors. *Veterinary Research*, 33, 465-480.

Silvestre, A., Sauve, C., Cortet, J., Cabaret, J., (2009). Contrasting genetic structures of two parasitic nematodes, determined on the basis of neutral microsatellite markers and selected anthelmintic resistance markers. *Molecular Ecology*, 18, 5086-5100.

Smith, G., (1990). A mathematical model for the evolution of anthelmintic resistance in a direct life cycle nematode parasite. *International Journal for Parasitology*, 20, 913-921.

Varady, M., Praslicka, J., Corba, J., Vesely, L., (1993). Multiple anthelmintic resistance of nematodes in imported goats. *Veterinary Record*, 132, 387-388.

Wyk, J.A.v., (2001). Refugia - overlooked as perhaps the most potent factor concerning the development of anthelmintic resistance. *Onderstepoort Journal of Veterinary Research*, 68, 55-67.

Wyk, J.A.v., Bath, G.F., (2002). The FAMACHA(C) system for managing haemonchosis in sheep and goats by clinically identifying individual animals for treatment. *Veterinary Research,* 33, 509-529.

Wyk, J.A.v., Malan, F.S., Randles, J.L., (1997). How long before resistance makes it impossible to control some field strains of *Haemonchus contortus* in South Africa with any of the modern anthelmintics. *Veterinary Parasitology*, 70, 111-122.

Wyk, J.A.v., Stenson, M.O., Van der Merwe, J.S., Vorster, R.J., Viljoen, P.G., (1999). Anthelmintic resistance in South Africa: surveys indicate an extremely serious situation in sheep and goat farming. *Onderstepoort Journal of Veterinary Research*, 66, 273-284.

In: Nematodes
Editors: F. Boeri and J. A. Chung
ISBN: 978-1-61470-784-4

Chapter VIII

Slug Pest Management through Nematodes in Agricultural and Horticultural Crops

Tarique Hassan Askary*[*], *Akhtar Ali Khan*[**], *Mohammed Islam Shah Waliullah*, *Shafat Ahmad Banday**, *Umar Iqbal** and *Mohammed Maqbool Mir**

Division of Entomology, Sher-e-Kashmir University of Agricultural Sciences and Technology of Kashmir, Shalimar, Srinagar, Jammu and Kashmir, India.
*Division of Pomology, Sher-e-Kashmir University of Agricultural Sciences and Technology of Kashmir, Shalimar, Srinagar, Jammu and Kashmir, India.

Abstract

Slugs are considered a persistent problem in a variety of agricultural and horticultural crops. They are widely dispersed, prolific and damage the crops reducing their yield and quality, thus making them unmarketable. One individual slug can kill up to a 50 wheat seeds in the first week of sowing. An yield loss of 50-90% in corn and soybean has been reported due to slugs. The high cost of chemical pesticides and their hazardous effect on the environment demand an alternative approach for slug pest management that should be eco-friendly and cost effective. The nematode belonging to the species *Phasmarhabditis* serve as a biological control agent of slugs. The infective juveniles of *Phasmarhabditis hermaphrodita* move through the soil, locate and infect the slug by penetrating through a natural opening at the posterior end of the mantle. Once inside the host the nematode releases a bacteria *Moraxella osloensis*, the main killing agents which multiply profusely and produce endotoxin which is tolerant to heat and protease treatments. Thus, the nematode just acts as a vector in transporting the bacterium into the shell cavity of the slug. The nematode-bacterium complex is a lethal combination

[*] E mail: tariq_askary@rediffmail.com
[**]E mail: akhtaralikhan47@rediffmail.com

which leads to the death of slug within 7-21 days. Although this bioagent has been commercialized in some countries under different trade names, however, there is still a need to find out its vital optimum application rates, timing and proper application technology to make it cost effective and include it in the pest management programme.

Keywords: *Phasmarhabditis, Moraxella osloensis,* bioagent, pest management, slug, application technology.

1. Introduction

Slug is a common name normally used for any gastropod mollusc. They differ from snail (large external shell) in lacking a shell or has a reduced or small internal shell [Wilson and Grewal, 2005]. Slugs are considered serious pests of agricultural and horticultural crops [Barker, 2001, 2002], destroying foliage faster than plants can grow, thus killing even fairly large plants. They destroy fruits and vegetables prior to harvest, making small holes in them during feeding, thus left an individual item unsuitable for consume and market. Such affected crop becomes more vulnerable to rot and other diseases. An yield loss of 50-90% in corn and soybeans have been reported in the years having favourable conditions for slug development [Wilson, 1992]. Report reveals that one individual slug can kill up to 50 wheat seeds in the first week of sowing, with smaller slugs killing more seeds than larger slugs. In dry autumn, 35% of early cereal plants were reported to be damaged by slugs and this figure went up to 50% in wet conditions [Members Agri-Bulletin, 2009].

Slugs belong to the family Arionidae, Agriolimacidae, Limacidae, Milacidae, and Helicidae [Glen, 1997; Koch *et al.*, 2000]. Several slug species such as *Deroceras reticulatum*, *D. panormitanum*, *Arion circumscriptus*, *A. silvaticus* and *Lehmannia valentiana* are reported to cause damage to crops [Ester *et al.*, 2004], however among them the most widespread and serious slug pest species is *Deroceras reticulatum* [Ester and Wilson, 2005], commonly known as grey garden slug, a small air breathing terrestrial pulmonate gastropod mollusc belonging to family Agriolimacidae [Table-1]. *D. reticulatum* are omnivorous, feeding mainly on fresh leaves and fruits or seedlings and generally active at night. The native distribution of this slug species is Europe, North Africa, Australasia, North and South America. Keeping in view the importance of this crop pest, there is a demand to search out a safe method for its management that would be eco-friendly and cost-effective.

Table 1. Scientific classification of slug, *Deroceras reticulatum*

Kingdom: Animalia
Phylum: Mollusca
Class: Gastropoda
Super family: Limacoidea
Family: Agriolimacidae
Genus: Deroceras
Species: *D. reticulatum* (O.F. Muller, 1774)

Phasmarhabditis hermaphrodita are microscopic nematodes belonging to family Rhabditidae [Table-2] that are capable of effectively parasitizing and killing several slug species from the families Agriolimacidae, Lamacidae, Milacidae and Arionidae [Wilson *et al.*, 1993a].

Table 2. Scientific classification of slug parasitic nematode *Phasmarhabditis hermaphrodita*

Kingdom: Animalia
Phylum: Nematoda
Class: Secernentea
Order: Rhabditida
Subfamily: Peloderinae
Genus: Phasmarhabditis
Species: *P. hermaphrodita* (A. Schneider, 1859)

They are bacterial feeding nematodes; a facultative parasite, capable of living saprobically on dead slugs, slug faeces [Tan and Grewal, 2001a] and leaf litter [Mac Millen *et al.*, 2009]. Schneider [1859] was the first who described the nematode associated with the slug *Arion ater*. The infective juveniles of *P. hermaphrodita* move through the soil, locate and infect the slug by entering through a natural opening at the posterior end of the mantle. Once inside the host, the nematode releases bacteria, *Moraxella osloensis* that causes septicemia resulting in death of the slug within 7-21 days [Tan and Grewal, 2001a]. The mantle region of slug swells which is the characteristic symptom of nematode infestation. In the recent years *P. hermaphrodita* has been commercialized and exploited as a biological molluscicide in several countries of the world [Askary, 2010].

This chapter summarizes the information regarding *P. hermaphrodita*, important from biological control point of view, limitations in their use and ideas to overcome the problem in the present context.

2. Life Cycle

The life cycle of *Phasmarhabditis* has not been extensively studied till now; however, the little information available on this aspect indicates that the life cycle of this nematode is dependent upon the slug species it encounters. Three distinct life cycle of *Phasmarhabditis* sp. have been described below:

- Saprobolic

Maupas [1900] was the first to establish cultures of *P. hermaphrodita* on rotting flesh and found this nematode to be a protandrous autogamous hermaphrodite with one male and 715 females. Tan and Grewal [2001a] reared this nematode on slug faeces and reported that *P. hermaphrodita* is a facultative parasite of the slug which can complete its life cycle under non-parasitic conditions in association with the host. Wilson *et al.* [1995a] are also of the same opinion that this nematode can reproduce on a wide range of bacteria and can recover and reproduce saprobically on dead invertebrate in the soil. Rae *et al.* [2006] exposed *P. hermaphrodita* to a range of cues associated with highly susceptible host *Deroceras reticulatum*.

Cues included foot and mantle mucus and faeces from live *D. reticulatum* and the attractiveness of infected and non-infected *D. reticulatum*. It was observed that *P. hermaphrodita* was attracted more to dead than live hosts but was equally attracted to infected and non-infected *D. reticulatum*. These report confirm that *P. hermaphrodita* is a facultative parasite that is capable of growing and reproduce on decaying plant and animal material present in the soil.

– Necromenic

Under natural conditions, the infective juveniles of *Phasmarhabditis* enter into a slug and remain there without further development till the death of the host [Mengert, 1953]. As the host dies, the infective juveniles start feeding on the cadaver, develop and multiply there. When the food starts depleting formation of new infective juveniles takes place which can be found in this arrested state of development in the mantle cavity or the digestive tracts of the slugs. Mengert [1953] found necromenic life cycle of *P. hermaphrodita* on larger species of slugs (*Limax cinereoniger*) whereas Maupas [1900] observed it on *Arion ater*, the smaller member of Arionidae family. However, it is still debatable whether the entrance of nematode into slug and completion of life cycle there is parasitic or necromenic.

– Parasitic

The infective juveniles of *P. hermaphrodita* represent a stage in the development of nematode which is specially adopted for survival under unfavourable conditions. In this stage the harboured bacteria *Moraxella osloensis* colonize in the gut of nematode. The infective juveniles actively search for potential slug hosts. They make entry into the slug through the breathing pore (pneumostome) below the mantle and move to the shell cavity [Tan and Grewal, 2001a].

The bacteria are released here which start multiplying and the nematode feeds on them and resume their growth. The bacteria cause septicemia in the slug which lead to swelling of the rear half of the mantle where the nematodes reproduce.

Once the second generation is produced these offspring spread throughout the slug's body and develop. It has been noted that from the very time after infection there is a marked reduction in the feeding activity of the slug (Glen *et al.*, 2000). On an average 250-300 offspring of nematode is produced.

The infected slug eventually dies and third generation nematodes are produced which feeds on slug cadaver. The death of the slug generally occur between 7-21 days (Wilson *et al.*, 1993a). The nematodes take several days to complete their life cycle. It is supposed that the nematode is able to modify the host's behaviour so that it remains below the ground surface before death and thus escape from predators and scavengers [Pechova and Foltan, 2008]. When the conditions are favourable, the infective juveniles move off through the soil in search of new slug hosts [Tan and Grewal, 2001a].

3. Nematode-Bacteria Association

The information on the association of slug parasitic nematode, *Phasmarhabditis hermaphrodita* with bacteria, *Moraxella osloensis* has not been extensively carried out, therefore, a meager information is available on this aspect. The infective juveniles of *P. hermaphrodita* invade slug through the dorsal integumental pouch just posterior to the mantle, enter the shell cavity via a short canal, releases bacteria and develop into self-fertilizing hermaphrodites that produce progeny resulting in the death of the host [Tan and Grewal, 2001a]. It is also realized that important organs such as kidney, lung and heart are located in the mantle region, therefore, the bacteria alone may kill the slug without entering the hemocoel. The nematode may take more than seven days to kill slugs, however, feeding by slugs is strongly inhibited within a few days of infection, providing rapid crop protection [Georgis *et al.*, 2006]. The association of *P. hermaphrodita* with the bacteria *M. osloensis* is mutualistic [Tan and Grewal, 2003].

Moraxella osloensis is a gram negative aerobic bacterium that is coccal or rod shaped but tends to be pleomorphic. The bacteria which are now considered to be *M. osloensis* would previously have been considered to be *Moraxella nonliquefaciens* or *Mima polymorpha* (var). oxidans [Sugarmann and Clarridge, 1982]. The scientific classification of *M. osloensis* is presented in Table-3. This bacterium is occasionally isolated from the upper respiratory tract, genitourethral specimens, blood, cerebrospinal fluid and pyogenic manifestations in joints, bursae and other sites in humans [Bovre, 1984]. *M. osloensis* is considered an opportunistic human pathogen and has been found to cause diseases, such as endocarditis [Stryker *et al.*, 1982], osteomyelites [Sugarmann and Clarridge, 1982], meningitis [Fijen *et al.*, 1994] and central venous catheter infection [Buchmann *et al.*, 1993].

Table 3. Scientific classification of *Moraxella* sp

Kingdom: Bacteria
Phylum: Proteobacteria
Class: Gammaproteobacteria
Order: Pseudomonadales
Family: Moraxellaceae
Species: *M. osloensis* (Bovre and Henriksen, 1967)

No highly specific mutualistic association of the nematode, *P. hermaphrodita* with bacteria has been reported so far. Lack of bacterial specificity as food source as well as lack of rigid cuticle in slugs indicates that more or less there is general association of bacteria with *P. hermaphrodita* [Wilson and Grewal, 2005]. Researchers have described different species of bacteria associated with *P. hermaphrodita*. Wilson *et al.* [1995a] recovered more than 150 bacterial isolates from inside the juveniles of *P. hermaphrodita* found from living and dead *Deroceras reticulatum* and from xenic foam chip cultures of the nematodes. However, the nematodes grown with *M. osloensis* or *Pseudomonas fluorescens* were found pathogenic to the slug only. This indicates that differences in pathogenicity between nematode and bacterium combinations did not result from differences in bacterial pathogenicity alone. The bacterium *M. osloensis* produces an endotoxin, which is tolerant to heat and proteinase treatments. Therefore, it is suggested that there may be a proteinaceous substance, structurally associated with the endotoxin that masks its toxity in the young bacterial cells [Tan and Grewal, 2002]. Later on Tan and Grewal [2003] discovered that purified *M. osloensis* lipopolysaccharide is an endotoxin which possesses a lethal toxity to *D. reticulatum*. *M. osloensis* lipopolysaccharide possesses a lethal toxicity to *D. reticulatum* when administered by injection but no contact or oral toxicity to this slug. Coinjetion of galactosamine with lipopolysaccharide increased the toxicity of lipopolysaccharide to the slug to two to four fold. Tan and Grewal [2001b] reported that pathogenicity of nematode varies with the age. Aged *P. hermaphrodita* are less virulent than young ones because new infective juveniles of *P. hermaphrodita* carry more viable cells of *M. osloensis* than the older one.

4. Behavioural Ecology

Soil depth, temperature and moisture level are some of the factors affecting the survival and pathogenicity of slug parasitic nematode *Phasmarhabditis hermaphrodita*. Glen [1997] found the best performance of nematodes between 2-10 cm soil depth. In an experiment, incorporation of nematodes into the soil by cultivation at a depth of 2-10 cm proved beneficial in the management of slug population in chinese cabbage and winter wheat whereas those left on the surface had not effect [Wilson *et al.*, 1996b]. *P. hermaphrodita* were most efficacious when applied to moist soil followed by up to 5 cm of irrigation [Georgis *et al.*, 2006]. The best optimum temperature for the survival of infective juveniles of *P. hermaphrodita* ranges between 5-15°C. A temperature above 25°C is considered detrimental for nematodes [Grewal and Grewal, 2003].

Phasmarhabditis hermaphrodita use a range of cues released by slugs to locate hosts. Rae *et al.* [2006] exposed infective juveniles of *P. hermaphrodita* to cues which included foot, mantle mucus and faeces from live *Deroceras reticulatum* and also volatile cues released from this slug. The attractiveness of live and dead *D. reticulatum* and the attractiveness of infected and non-infected *D. reticulatum* were also compared in the experiment. It was observed that *P. hermaphrodita* responded poorly to volatile cues but was strongly attracted to live *D. reticulatum*, its faeces and mucus from both the foot and mantle. The attraction of *P. hermaphrodita* was more to dead than live hosts; however, it was equally attracted to infected and non-infected host. Hapca *et al.* [2007] investigated the response of *P. hermaphrodita* to the presence of slug mucus on plates of agar. The nematodes were found significantly attracted by filter paper impregnated with slug mucus as compared to the paper impregnated with water. It was also noted that different mucus treatments induced significantly different effects on the speed of nematode movement and the distribution of the turning angles, as well as the fractal dimension of nematode foraging trail. This laid to the hypothesis that the nematodes exhibit both chemotactic and chemokinetic response to a signal emanating from slug mucus. Recently, behavioural responses of the infective juveniles of *P. hermaphrodita* was analyzed using video-taped sequences. The nematodes during contact with mucus samples, decreased duration and frequency of forward crawling and increased duration and frequency of head thrusting and also duration of head waving. The other host recognition behaviour was body waving and pushing. Thus, it was suggested that *P. hermaphrodita* forages by the 'cruise' method until mucus from potential hosts is encountered, but then concentrates its search behaviour, characterized by the decrease in forward crawling [Small and Bradford, 2008].

5. Host Range and Effects

In the recent years, field and laboratory studies have been carried out by various research workers to know the parasitic behaviour of *Phasmarhabditis hermaphrodita* against different slug species [Wilson *et al.*, 2000a; Grewal *et al.*, 2003]. This nematode is reported to control slug infestations in a wide range of crops such as wheat, lettuce and strawberries (Hapca *et al.*, 2007). The use of *P. hermaphrodita* against the different slug species such as *Deroceras reticulatum, D. panormitanum, Arion circumscriptus, A. silvatics and Lehmannia valentiana* has also been assessed with special attention to its efficacy in sugarbeet, green asparagus, Brussels sprouts, iceberg lettuce, cymbidium and Alastroemeria [Ester *et al.*, 2004]. In an experimental study it was found that *P. hermaphrodita* has a parasitic effect on the slug families belonging to Arionidae, Limicidae and Milacidae [Glen, 1997]. Greater number of nematodes resulted in greater mortality of slugs, however, smaller number of nematodes greatly reduced the feeding of the host. Wilson *et al.* [1993a] reported that a single high dose of nematode applied to the slugs under field conditions caused significant mortality to three different pest families of slugs i.e, *D. reticulatum, D. panormitatum, A. silvaticus, A. distinctus, A. intermedius, A. ater, Tandonia budapestensis and T. sowerbyi.* Speiser and Andermatt [1994] conducted a field experiment to investigate the potential of *P. hermaphrodita* against slugs. The slug species *A. rufus* and *D. reticulatum* were infected by *P. hermaphrodita* and often died. Wilson *et al.* [1995c] prepared culture of *P. hermaphrodita* under *in vitro* conditions and applied it as a drench to soil at four different dose rates (3×10^8, 1×10^9, 3×10^9 and 1×10^{10}/ha) one or four days before planting lettuce seedlings in a polythene tunnel. The effects of the four nematode doses on slug damage during the first three weeks after planting as well as the number of slugs found within and below lettuce plants at harvest were measured. It was observed that the highest two doses of nematodes significantly reduced the number of slugs found within lettuce plants at harvest and on the soil surface below the plants. In another experiment Wilson *et al.* [1994a] applied *P. hermaphrodita* to assess the damage to seeds and seedlings of winter wheat due to slugs. The nematodes were applied immediately after seed sowing and slug damage to wheat seeds and seedlings were assessed six and thirteen weeks after drilling. It was observed that seedling survival increased and slug grazing damage to seedlings decreased linearly with increasing log nematode dose. After six weeks there were significantly more undamaged plant equivalents on plots treated with the two highest nematode doses (3×10^9 and 1×10^{10}/ha) than on untreated plots. Thus it was suggested that the efficiency of *P. hermaphrodita* in protecting wheat from slug damage was through inhibition of feeding by infected slugs. Application of *P. hermaphrodita* in reducing the population of *D. reticulatum* and the damage caused to chinese cabbage seedlings due to this slug has also been reported (Wilson *et al.*, 1996b).

Tan and Grewal [2001a] observed that *P. hermaphrodita* invade *D. reticulatum* within 8-16 hours following external exposure with the posterior mantle region containing the shell cavity serving as the main portal entry. Rae *et al.* [2008] demonstrated a slug defence mechanism to inhibit *P. hermaphrodita.* In this experiment three slug species (*D. reticulatum,*

Milax gagates and *Limax pseudoflavus*) were exposed to the parasitic nematode *P. hermaphrodita.* The nematode was able to cause mortality as well as feeding inhibition to both *D. reticulatum* and *M. gagates* but did not negatively affect *L. pseudoflavus.* When surviving *L. pseudoflavus* was dissected large numbers of *P. hermaphrodita* were found encapsulated in the shell of the slug. Thus, it was concluded that increasing shell size was able to trap invading nematodes, which could be an immune response to *P. hermaphrodita* invasion. Grimm [2002] in a soil bioassay, exposed adult *D. reticulatum* and three different weight classes of young *A. lusitanicus* (0.15, 0.24 and 0.45 g) to a single dose (170 infective juveniles per g of soil) of *P. hermaphrodita*. In all the weight classes of *A. lusitanicus*, food consumption was reduced by at least 50%, however, the greater reduction in feeding (90%) was noted in the smallest *A. lusitanicus*. The nematodes successfully killed *D. reticulatum* but were less efficient in killing young *A. lusitanicus*. At the final stage of experiment highest mortality was recorded in *D. reticulatum* (98%) and the smallest weight class of *A. lusitanicus* (47%). There was almost no mortality in the largest weight of *A. lusitanicus*.

6. Effects on Other Organisms

Phasmarhabditis hermaphrodita is a parasitic nematode capable of killing several slug species; however, its affect on non-target organisms has not been extensively studied. Whatever, the little work done on this aspect makes the evidence clear that this nematode is safe for non-target snails, beneficial predators and earthworms. Koch *et al.* [2000] carried out an experiment to test the effect of *P. hermaphrodita* on two snails, *Helix pomatia* and *Capaea nemoralis*. Both the snails survived the nematode attack without any sign of infection. In another experiment susceptibility of seven common hedgerow species (*Monancha cantiana, Capaea hortensis, C. nemoralis, Pomatias elegans, Oxychilus helveticus, Clausilia bidentata and Discus rotundatus*) were tested against *P. hermaphrodita* under laboratory condition. Soil treatment with recommended application rate of nematodes caused significant mortality only for the snail *M. cantiana* but with five times the recommended rate. A significant mortality was also found in *C. hortensis*. However, no mortality under field condition was observed in any of the snail species used in the experiment [Wilson *et al.*, 2000a]. Morley and Morrit [2006] investigated the effects of *P. hermaphrodita* on two common fresh water snails, *Lymnaea stagnalis* and *Physa fontinalis* at 'spray tank' concentration and a 50% diluted 'spray tank concentration' over a period of 14 days. Mortality of *L. stagnalis* was observed at both application levels but *P. fontinalis* suffered no mortalities over the experimental period. Wilson *et al.* [1994b] conducted a bioassay of *P. hermaphrodita* against tenebrionid beetles, *Zophoba morio* and *Tenebrio molitor*. The results indicated no infection of nematodes in either of the two organisms. Under laboratory condition adults of *Pterostichus melanarius*, the beneficial predatory carabid beetle were unaffected when exposed even a high dose of *P. hermaphrodita* [Wilson *et al.*, 1993c].

De Nardo *et al.* [2004] evaluated the effect of commercial formulation of *P. hermaphrodita* on the earthworm *Eisenia fetida*. Adults of *E. fetida* were exposed in one-litre glass beaker to *P. hermaphrodita* at three concentrations (1x, 10x and 50x of the field recommended rate of 3×10^9 billion nematodes/ha) during a 14 days period in an artificial soil substrate. The injured earthworms with posterior ends removed were exposed to the 10x field at recommended rate of the commercial formulation. Neither intact nor injured *E. fetidae* showed susceptibility to *P. hermaphrodita* during the 14 days of exposure even at concentrations 10 and 50 times higher than the label dose. Iglesias *et al.* [2003] also found earthworm, *Lumbricus* spp. unaffected by *P. hermaphrodita* under field conditions. Rae *et al.* [2005] exposed five earthworm species (*L. terrestris, E. hortensis, E. andrei, E. fetida* and *Dendrodrilus rubidus*), one flatworm spcies (*Arthurdendyus triangulatus*) and the susceptible slug *Deroceras reticulatum* to the recommended field application rate (3×10^9 billon /ha) and five times this rate of *P. hermaphrodita* . A significant mortality was observed only in *D. reticulatum* whereas the earthworms and flatworm were unaffected.

7. Production and Formulation

The slug parasitic nematodes *Phasmarhabditis hermaphrodita* are generally multiplied on a semi-synthetic diet (*in vitro*) which have their own advantages and limitations. Wilson *et al.* [1995b] used five species of bacteria which supported vigorous growth of *P. hermaphrodita* on kidney agar in liquid cultures; however, there were significant differences in final number of nematodes and number of infective juveniles formed. *Providencia rettgeri* was found to give the best yields of infective juveniles in liquid culture. In both liquid and solid phase, monoxenic growth of *P. hermaphrodita* was better on media based on pig kidney than on wholly soluble media. *P. hermaphrodita* has also been grown successfully in xenic culture using solid foam chip culture, based on the methods of Bedding [1984] and also in deep liquid culture on a flask shaker suitable for scale up to fermenter production [Wilson *et al.*, 1993b]. In monoxenic culture an yield of one lakh infective juveniles per ml has been achieved by Wilson *et at.* [1995a]. When maximum yields of infective juveniles are obtained they are concentrated by centrifugation before formulation [Young *et al.,* 2002]

Formulation refers to the preparation of a product from an active ingredient by the addition of certain active (functional) and non active (inert) substances [Grewal, 2002]. It provide a means to improve the activity, storage stability, delivery, ease to use and field efficacy of the nematodes [Askary, 2010]. After a five years research programme funded by Micro Bio limited, the use of *Phasmarhabditis* as a biological molluscicide was patented by Wilson *et al.* [1993d]. Later on in1994 MicroBio limited (now Becker Underwood), launched a product which are being sold commercially under the trade name Nemaslug® (Glen *et al.*,1994,1996) in several European countries like France, Ireland, Italy, Netherlands, Switzerland and Germany. *P. hermaphrodita* can be formulated by using any of Becker-Underwood's vermiculite, polymer or clay formulation, however, the shelf life of this product is very less as compared to other entomopathogenic nematodes [Ester and Wilson, 2005].

8. Application Technology

Application technology is one of the most important aspect which aims at minimum loss during transfer of nematodes to the target slug pest. Nematodes being a live material have to be handled very carefully and with caution so that the deleterious effect of surrounding is reduced to the minimum and they work efficiently. Several factors affect the delivery of infective juveniles in close proximity to the target slug pest for achieving optimal results at minimal possible cost. Nematode culture is mixed with water and once the suspension is prepared, it is advised to apply within thirty minutes as a drench to the surface of soil. Application of nematodes to moist soil followed by up to 5 cm of irrigation will provide an effective result [Georgis *et al.*, 2006]. The best optimum soil temperature for the application of nematodes is 15°C with a maximum of 20°C. Soil temperatures above 25°C is considered detrimental to the nematode [Georgis *et al.*, 2006]. Researchers suggest that application of nematodes in the early evening will avoid the ill effects of ultra violet rays and a light irrigation in the soil immediately after application [Wilson *et al.*, 1996a; Hass *et al.*, 1999] will save the nematode from desiccation. Application of nematodes one week in advance of sowing or planting can protect the young seedlings and cuttings from slugs for at least six weeks.

In the recent years *Phasmarhabditis hermaphrodita* has been field tested for the management of slugs in different crops such as oilseed rape [Wilson *et al.*,1994a], wheat [Wilson *et al.*, 1994c, 1996b], lettuce [Wilson *et al.*,1995a], strawberry and sugar beet [Wilson and Gaugler, 2000]. Speiser and Andermatt [1994] suggested an effective dose of *P. hermaphrodita* against slugs to be $1\times10^6/m^2$ whereas Ester *et al.* [2003] recommended one dose of $3\times10^5/m^2$ or three times at a dose of $50{,}000/m^2$ as row application for the control of slugs in green asparagus. However, a dose of 3×10^9 infective juveniles/ha is required for effective management of slugs in the field [Wilson *et al.*, 1996b; Iglesias *et al.*, 2001].

Infective juveniles of *Phasmarhabditis hermaphrodita* can be successfully applied with back pack sprayers and conventional hydraulic spray equipment with reduced pressure (100 kPa) and a larger filter mesh as well as through irrigation lines [Wilson and Gaugler, 2000]. Other equipments used for application are watering can but for larger areas sprayer is used with maximum spraying pressure maintained up to 5 bar. The sprayer nozzle and filters should have an opening of at least 1mm.

Compatibility of *Phasmarhabditis hermaphrodita* has been reported with chemical molluscicide such as metaldehyde bait pellets, therefore, the nematodes can also be applied in combination with these pellets [Wilson *et al.*, 2000b]. Grewal *et al.* [2001] suggested the application of *P. hermaphrodita* to slug shelter as a novel approach in increasing the efficiency and reducing the dose of nematodes. Spot treatment of nematodes around individual plant and in narrow bands centered on the crop rows in row crops has also been reported [Wilson *et al.*, 1999].

Discussion and Conclusion

In the present context it is very clear that slug parasite nematode *Phasmarhabditis hermaphrodita* is a promising biocontrol agent which is being commercialized in several countries of the world. This bioagent has been patented since 1994 but some points are still debatable which provide hindrance in its utilization on a large scale. High product cost, limited market share, limited product availability, lack of suitable production technology, poor shelf life, lack of proper technique among the users, low pest application survival and low efficiency under unfavourable conditions are some of the reasons which have still kept the slug parasitic nematode under-utilized in pest management programmes. For commercial purposes nematodes are mass produced but under such cases pathogenicity varies among different batches. It has also been proved that aged *P. hermaphrodita* nematodes are less virulent than the young ones which restrict the development of the product [Tan and Grewal, 2001b]. Also, these beneficial nematodes always require a low temperature for its survival and application, thereby adding an additional expense for producers, thus making the final product costly. Therefore, post-application survival, longevity of the molluscicidal effect and price of the nematodes are the important aspects which need to be brought under consideration before going for commercialization of the product.

In the past few years some progress has been made in formulation and application technologies, however, further improvements are still required to make the programme effective. Experiments have proved that *P. hermaphrodita* can interact synergistically with some chemical molluscicide, therefore a combination of multiple tactics should be prepared to achieve a satisfactory result. Since, commercial product of nematodes is expensive, it is vital to optimize its application dose rates and timing to save the excessive use of the bioagent. To obtain maximum benefits, the shelf life of nematodes needs to be increased and proper formulation and delivery system needs to be worked out. In addition to this, an improvement in transport logistic and marketing is also required.

Acknowledgements

The authors are highly indebted to all the researchers who have generated the valuable information presented in this chapter.

References

Askary, T.H. (2010). Nematodes as biocontrol agents. In: E. Lichtfouse (Ed.), *Sociology, Organic Farming, Climate Change and Soil Science*, pp. 347-378. Heidelberg, Germany, Springer.

Barker, G.M. (2001). *The Biology of Terrestrial Gastropods.* pp. 468. Wallingford, UK, CAB International.

Barker, G.M. (2002). *Molluscs as Crop Pests.* pp. 468. Wallingford, UK, CAB International.

Bedding, R.A. (1984). Large-scale production, storage and transport of the insect-parasitic nematodes *Neoaplectana* spp. and *Heterorhabditis. Annals of Applied Biology*, 101: 117-120.

Bovre, K. and Henriksen, S.D. (1967). A new *Moraxella* species, *Moraxella osloensis,* and a revised description of *Moraxella nonliquefaciens. International Journal of Systematic Bacteriology*, 17: 127-135.

Bovre, K. (1984). Genus II. *Moraxella* Lwoff 1939, 173 emend, Henriksen and Bovre 1986, 391. In: N.R. Krieg and J.G. Holt (Ed.), *Bergey's Manual of Systematic Bacteriology*, pp.296-303. Baltimore, Md., Williams and Wilkins.

Buchmann, A.L., Pickett, M.J. and Ament, M.E. (1993). Central venous catheter infection caused by *Moraxella osloensis* in a patient receiving home parenteral-nutrition. *Diagnostic Microbiology and Infectious Disease*, 17: 163-166.

De Nardo, E.A.B, Sinderman, A., Grewal, S.K. and Grewal, P.S. (2004). Non-susceptibility of earthworm *Eisenia fetida* to the rhabditid nematode *Phasmarhabditis hermaphrodita,* a biocontrol agent of slugs. *Biocontrol Science and Technology*, 14: 93-98.

Ester, A. and Wilson, M.J. (2005). Application of slug-parasitic nematodes. In: P.S. Grewal, R-U Ehlers and D.I. Shapiro-Ilan (Eds.), *Nematodes as Biocontrol Agents*, pp. 431-444, Wallingford, U.K., CAB International.

Ester, A., Arkema, M., Gruppen, R., Hazendonk, A., Huisman., M., Huiting, H.F., Rozen, k.van, Vlaswinkel, M.E.T., van Rozen, K. (2004). Control of Slugs with molluscophage nematodes. *Gewasbescherming*, 35 : 1-5.

Ester, A., Rozen, van K. and Molendijk, L.P.G. (2003). Field experiments using the rhabditid nematode *Phasmarhabditis hermaphrodita* or salt as control measures against slugs in green asparagus. *Crop Protection*, 22: 689-695.

Fijen, C.A.P., Kuijper, E.J., Tjia, H.G., Daha, M.R. and Dankert, J. (1994). Complement deficiency predisposes for meningitis due to nongroupable meningococci and neisseria-related bacteria. *Clinical Infectious Diseases*, 18:780-784.

Georgis, R., Koppenhofer, A.M., Lacey, L.A., Belair, G., Duncan, L.W., Grewal, P.S., Samish, M., Tan, L., Torr, P. and van Tol, R.W.H.M. (2006). Successes and failures in the use of parasitic nematode for pest control. *Biological Control*, 38: 103-121.

Glen, D. (1997). Nematodes make war with slugs. *Bifur*, 170: 27-29.

Glen, D.M., Wilson, M.J., Brain, P. and Stroud, G. (2000). Feeding activity and survival of slugs, *Deroceras reticulatum*, exposed to the rhabditid nematode, *Phasmarhabditis hermaphrodita*: a model of dose response. *Biological Control*, 17: 73-81.

Glen, D.M., Wilson, M.J., Hughes, L., Cargeeg, P. and Hajjar, A. (1996). Exploring and exploiting the potential of the rhabditid nematode *Phasmarhabditis hermaphrodita* as a biocontrol agent for slugs. In: I.F. Henderson (Ed.), *Slug and Snail Pests in Agriculture*, Monograph No. 66, pp. 271-280. Thornton Health, UK., British Crop Protection Council.

Glen, D.M., Wilson, M.J., Pearce, J.D. and Rodgers, P.B. (1994). Discovery and investigation of a novel nematode parasite for biological control of slugs. In: *Proceedings of the Brighton Crop Protection Conference, Pests and Diseases,* pp. 617-624. Brighton, UK.

Grewal, P.S. (2002). Formulation and application technology. In: R. Gaugler (Ed.). *Entomopathogenic Nematology,* pp. 265-287. Wallingford, UK., CAB International.

Grewal, P.S., Grewal, S.K., Taylor, R.A.J., Hammond, R.B. (2001). Application of molluscicidal nematodes to slug shelters: a novel approach to economic biological control of slugs. *Biological Control*, 22: 72-80.

Grewal, S.K. and Grewal, P.S. (2003). Effect of osmotic desiccation on longevity and temperature tolerance of *Phasmarhabditis hermaphrodita* (Nematoda: Rhabditidae). *Journal of Parasitology*, 89: 434-438.

Grewal, S.K., Grewal, P.S., Hammond, R.B. (2003). Susceptibility of North American native and non-native slugs (Mollusca: Gastropoda) to *Phasmarhabditis hermaphrodita* (Nematoda: Rhabditidae). *Biocontrol Science and Technology*, 13: 119-125.

Grimm, B. (2002). Effect of the nematode *Phasmarhabditis hermaphrodita* on young stages of the pest slug *Arion lusitanicus*. *Journal of Molluscan Studies*, 68: 25-28.

Hapca, S., Crawford, J., Rae, R., Wilson, M. and Young, I. (2007). Movement of the parasitic nematode *Phasmarhabditis hermaphrodita* in the presence of mucus from the host slug *Deroceras reticulatum*. *Biological Control*, 41: 223-229.

Hass, B., Hughes, L.A. and Glen, D.M. (1999). Overall versus band application of the nematode *Phasmarhabditis hermaphrodita* with and without incorporation into soil for biological control of slugs in winter wheat. *Biological Control Science and Technology*, 9: 579-586.

Iglesias, J., Castillego, J., Castor, R. (2003). The effects of repeated applications of the molluscicide metaldehyde and the biocontrol nematode *Phasmarhabditis hermaphrodita* on molluscs, earthworms, nematodes, acarids and collembolans: a two year study in north west Spain. *Pest Management Science*, 59: 1217-1224.

Iglesias, J., Castillejo, J. and Castor, R. (2001). Mini-plot field experiments on slug control using biological and chemical control agents. *Annals of Applied Biology*, 139: 285-292.

Koch, R., Jackel, B. and Plate, H.P. (2000). Controlling pest slugs: new methods and the verification of their effects. *Gesunde Pflanzen*, 52: 1-10.

Mac Millen, K., Haukeland, S., Rae, R.G., Young, I.M., Crawford, J.W., Hapca, S. and Wilson, M.J. (2009). Disposal patterns and behaviour of the nematode *Phasmarhabditis hermaphrodita* in mineral soils and organic media. *Soil Biology and Biochemistry*, 41 :1483-1490.

Maupas, E. (1900). Modes et formes de reproduction des nematodes. *Archives de Zoologie Experimentale et Generale*, 7: 563-628.

Members' Agri-Bulletin (2009). Metaldehyde slug pellet stewardship. *Royal Agricultural Society of England*, pp. 276.

Mengert, H. (1953). Nematoden und Schnecken. *Zeitschrift fur Morphologie und Okoligie der Tiere,* 4: 311-349.

Morley, N.J. and Morritt, D. (2006). The effects of the slug biological control agent, *Phasmarhabditis hermaphrodita* (Nematoda), on non-target aquatic molluscs. *Journal of Invertebrate Pathology,* 92: 112-114.

Muller, O.F. (1774). Vermivm terrestrium et fluviatillium, seu animalium infusoriorum, helminthicorum, et testaceorum, non marinorum, succincta historia. *Volumen alterum*, pp. I-XXXVI (=1-36), (1-214, (1-10). Havniae and Lipsiae. (Heineck and Faber).

Pechova, H. and Foltan, P. (2008). The parasitic nematode *Phasmarhabditis hermaphrodita* defends its slug host from being predated or scavenged by manipulating host spatial behaviour. *Behavioural* Processes, 78: 416-420.

Rae, R.G., Robertson, J.F, Wilson, M.J. (2005). Susceptibility of indigenous UK earthworms and an invasive pest flatworm to the slug parasitic nematode *Phasmarhabditis hermaphrodita. Biocontrol Science and Technology*, 15: 623-626.

Rae, R.G., Robertson, J.F., Wilson, M.J. (2008). Susceptibility and immune response of *Deroceras reticulatum*, *Milax gagates* and *Limax pseudoflavus* exposed to the slug parasitic nematode *Phasmarhabditis hermaphrodita*. *Journal of Invertebrate Pathology*, 97: 61-69.

Rae, R.G., Robertson, J.F. and Wilson, M.J. (2006). The chemotactic response of *Phasmarhabditis hermaphrodita* (Nematoda: Rhabditida) to cues of *Deroceras reticulatum* (Mollusca: Gastropoda). *Nematology*, 8: 197-200.

Schneider, A.F. (1859). Uber eine nematodenlarve und gewisse verscheidenheiten in dn geschlechsorganen der nematoden. *Zeitchrifts fur Wissenschaftliche Zoologie*, 10: 176-178.

Small, R.W., Bradford, C. (2008). Behavioural responses of *Phasmarhabditis hermaphrodita* (Nematoda: Rhabditida) to mucus from potential hosts. *Nematology*, 10: 591-598.

Speiser, B. and Andermatt, M. (1994). Biocontrol of slugs with nematodes. *Agrarforschung*, 1: 115-118.

Stryker, T.D., Stone, W.J. and Savage, A.M. (1982). Renal-failure secondary to *Moraxella osloensis* endocarditis. *Johns Hopkins Medical Journal*, 151: 217-219.

Sugarmann, B. and Clarridge, J. (1982). Osteomyelitis caused by *Moraxella osloensis*. *Journal of Clinical Microbiology*, 15: 1148-1149.

Tan, L. and Grewal, P.S. (2001a). Infection behaviour of the rhabditid nematode *Phasmarhabditis hermaphrodita* to the grey garden slug *Deroceras reticulatum*. *Journal of Parasitology*. 87: 1349-1354.

Tan, L. and Grewal, P.S. (2001b). Pathogenicity of *Moraxella osloensis*, a bacterium associated with the nematode *Phasmarhabditis hermaphrodita*, to the slug *Deroceras reticulatum*. *Applied and Environmental Microbiology*, 67: 5010-5016.

Tan, L. and Grewal, P.S. (2002). Endotoxin activity of *Moraxella osloensis* against the grey garden slug, *Deroceras reticulatum*. *Applied and Environmental Microbiology*, 68: 3943-3947.

Tan, L., Grewal, P.S.(2003). Characterization of the first molluscicidal lipopolysaccharide from *Moraxella osloensis*. *Applied and Environmental Microbiology*, 69: 3646-3649.

Wilson, H.R. (1992). Slugs on ohio field crops. Ohio Pest Management and Survey Program. *Field Crops Pest Management Circular*, 20.

Wilson, M., Hughes, L., Hamacher, M., Barahona, D., Glen, D., Gwynn, R.L., Smits, P.H., Griffin, C., Ehlers, R.U., Boemare, N. and Masson, J.P. (1996). Post-application incorporation of *Phasmarhabditis hermaphrodita* into soil for slug control. In: *Proceedings on Cost 819- Entomopathogenic Nematodes : Application and Persistence of Entomopathogenic Nematodes*, May 16-20, pp. 29-36. Todi, Perugia, Italy.

Wilson, M.J. and Gaugler, R. (2000). Terrestrial mollusc pests. In: L.A. Lacey and H.K. Kaya (Eds.), *Field Manual of Techniques in Invertebrate Pathology*, pp. 787-804. Dordrecht, The Netherlands. Kulwer Academic Publishers.

Wilson, M.J. and Grewal, P.S. (2005). Biology, production and formulation of slug-parasitic nematodes. In: P.S. Grewal, R-U. Ehlers and D.I. Shapiro-Ilan (Eds.), *Nematodes as Biocontrol Agents*, pp. 421-429. Wallingford, UK., CAB International.

Wilson, M.J., George, S.K., Glen, D.M., Pearce, J.D. and Rodgers, P.B. (1993c). Biological control of slug and snail pests with a novel parasitic nematode. In: *ANPP Third International Conference on Pests in Agriculture*, pp. 425-432, Montpellier, France.

Wilson, M.J., Glen, D.M. and George, S.K (1993a). The rhabditid nematode *Phasmarhabditis hermaphrodita* as a potential biological control agent for slugs. *Biocontrol Science and Technology*, 3: 503-511.

Wilson, M.J., Glen, D.M. and George, S.K. and Butler, R.C. (1993b). Mass cultivation and storage of the rhabditid nematode *Phasmarhabditis hermaphrodita*, a biocontrol agent for slugs. *Biocontrol Science and Technology*, 3: 513-521.

Wilson, M.J., Glen, D.M., George, S.K. and Pearce, J.D. (1995a). Selection of a bacterium for the mass production of *Phasmarhabditis hermaphrodita* (Nematoda, Rhabditidae) as a biocontrol agent for slugs. *Fundamental and Applied Nematology*, 18: 419-425.

Wilson, M.J., Glen, D.M., and George, S.K. and Hughes, L.A. (1995c). Biocontrol of slugs in protected lettuce using the rhabditis nematode *Phasmarhabditis hermaphrodita*. *Biocontrol Science of Technology*, 5: 233-242.

Wilson, M.J., Glen, D.M. and Pearce, J.D. (1993d). Biological control of mollusks. World Intellectual Property Organization, Patent No. WO 93/00816, pp. 38.

Wilson, M.J., Glen, D.M., and George, S.K., Pearce, J.D and Rodgers, P.B. (1994a). The potential of the rhabditid nematode *Phasmarhabditis hermaphrodita* for biological control of the slugs. In: *Proceedings of a Symposium and Workshop on Genetics of Entomopathogenic Nematode-Bacterium Complexes*, October 23-27, pp. 230. St. Patrick's college, Maynoot, Co. Kildare, Ireland.

Wilson, M.J., Glen, D.M., George, S.K. and Pearce, J.D.(1995). Selection of a bacterium for the for the mass production of *Phasmarhabditis hermaphrodita* (Nematode : Rhabditidae) as a biocontrol agent for slugs. *Fundamental and Applied Nematology*, 8: 419-425.

Wilson, M.J., Glen, D.M., Hughes, L.A., Pearce, J.D. and Rodger, P.B. (1994b). Laboratory tests of the potential of entomopathogenic nematodes for the control of field slugs (*Deroceras reticulatum*). *Journal of invertebrate Pathology*, 64: 182-187.

Wilson, M.J., Glen, D.M., George, S.K., Pearce, J.D. and Wiltshire, C.W. (1994c). Biological control of slugs in winter wheat using the rhabditid nematode *Phasmarhabditis hermaphrodita. Annals of Applied Biology*. 125: 377-390.

Wilson, M.J., Glen, D.M., Pearce, J.D. and Rodgers, P.B. (1995b). Monoxenic culture of the slug parasite *Phasmarhabditis hermaphrodita* (Nematoda, Rhabditidae) with different bacteria in liquid and solid phase. *Fundamental and Applied Nematology*, 18: 159-166.

Wilson, M.J., Hughes, L.A., Hamacher, G.M. and Glen, D.M. (2000a). Effects of *Phasmarhabditis hermaphrodita* on non-target molluscs. *Pest Management Science*, 56: 711-716.

Wilson, M.J., Hughes, L.A., Hamacher, G.M., Barahona, L.D. and Glen, D.M. (1996b). Effects of soil incorporation on the efficacy of the rhabditid nematode, *Phasmarhabditis hermaphrodita,* as a biological control agent for slugs. *Annals of Applied Biology*, 128: 117-126.

Wilson, M.J., Hughes, L.A., Jefferies, D. and Glen, D.M. (1999). Slugs (*Deroceras reticulatum*) and *Arion ater* agg.) avoid soil treated with the rhabditid nematode *Phasmarhabditis hermaphrodita. Biological Control*, 16: 170-176.

Wilson, M.J., Major, B., Hughes, L.A. and Glen, D.M. (2000b). Integrating biological and chemical control of slugs. In: *Proceedings of an International Conference on Pests and Diseases*, November 13-26, pp. 1007-1012. Brighton, UK.

Young, J.M., Dunnill, P. and Pearce, J.D. (2002). Separation characteristics of liquid nematode cultures and the design of recovery operations. *Biotechnology Progress*, 18: 29-35.

In: Nematodes
Editors: F. Boeri and J. A. Chung
ISBN: 978-1-61470-784-4

Chapter IX

Diversity and Distribution of Nematode Communities in Grassland in Relation to Its Establishment Age and Utilization

*A. Čerevková**
Parasitological Institute of the Slovak Academy of Sciences,
Hlinková 3, 040 01 Košice, Slovak Republic.

Abstract

The communities of nematodes in six orographic units of the Slovak Republic and in three different types of grassland were investigated. The types of grassland included new meadow established on arable soil, permanent meadow and permanent pasture. Out of a total nematode specimens isolated from investigated grassland, 90 species and 66 genera were identified. Majority of identified species belonged to the orders Tylenchida, Dorylaimida and Rhabditida. The highest diversity of nematode species and genera was recorded at localities with lover altitude and higher average annual temperature. Cluster analysis showed that the structure of nematode communities and diversity depended more on meadows age and on their utilization than on geographical, climatic and soil conditions of sites. Plant parasitic nematodes were dominant trophic group in all types of grassland. Bacterial feeders were subdominant trophic group only in new meadows in arable soil, where soil was characterised by higher resources of organic matter. On the other hand, omnivores characterised by a long life cycle, and low abundance especially in cultivated soil, were subdominant trophic group in permanent meadows and pastures. The Maturity Index (MI) values were lover in new meadows than in permanent meadows and pastures. Low proportion of Plant Parasitic Index and Maturity Index (PPI/MI) in permanent meadows and pastures is characteristic for natural ecosystems, without environmental destruction. Using the Structure Index and Enrichment Index, all localities

* e-mail: cerev@saske.sk

of new meadows on arable soil were characterised as moderately disturbed with matured food web and N-enriched conditions, and many localities with permanent pastures were characterised as undisturbed with structure food web conditions.

Keywords: Nematoda, community structure, grassland, nematode diversity.

Introduction

Nematode communities in grassland ecosystems in Europe, was recently studied by several authors, e. g. Wasilewska (1994), de Goede and Bongers (1998), Popovici and Ciobanu (2000) and Háněl (2003). Studies dealing primarily on nematode species diversity, the impact of biotic and abiotic factors on nematode communities and investigated the relationships between trophic groups of nematodes in grassland soil. Ekschmitt et al. (1999, 2001) compared diversity of nematodes and representation of trophic groups of nematodes in grasslands of six European countries with different climatic and soil conditions. Studies show the impact of biotic and abiotic environmental factors on the structure of nematodes. The number of nematode genera in all trophic groups' increased from the colder northern regions towards warmer areas. Grassland ecosystems, such as natural and rich representation of grasses and herbs were characterized by high and diverse nematode species composition (Hodda and Wanless, 1994a,b; De Goede and Bongers, 1998; Ettema and Yeates, 2003). Different way of management of meadows and their age had an impact not only on the representation of plants but also on the presence of soil organisms and thus to the nematodes. Effect of grazing pasture on the structure of nematode communities studied Wall-Freckman and Huang (1998) and McSorley and Frederick (2000) in the U.S. and Zolda (2006) in Austria. All these authors found that grazing had no or very little effect on nematode abundance, but they found an effect on nematode abundance and on the composition of nematode fauna throughout the year in connection with the change of the precipitation and the temperature.

The aim of this study was to determine and compare the qualitative and quantitative structure of nematode communities, trophic structure and some ecological parameters in grassland with different establishment of its age and utilization.

Material and Methods

Communities of soil free living and plant parasitic nematodes were studied at six localities in the mountain and submountain areas where grassland is a widespread ecosystem type in the Slovak Republic. A characteristic of the localities is in Table 1.

Characteristic of meadow types:

New meadow - newly established on arable soil during last 3 years, fertilized and regularly mown usually two to three times per year, with dominant grass species *Dactylis glomerata* L., *Festuca rubra* L. and *Lolium perene* L. but also herbs, especially *Capsella bursa-pastoris* (L.) Medik., *Taraxacum officinale* auct. non Weber., very often, with a strong presence of *Elytrigia repens* (L.) DESV. New meadow was sowing of grasses on arable soil or natural grassing of arable land.

Table 1. Characteristics of investigated localities

Localities	Coordinates	Altitude (m a.s.l.)	Average annual rainfall (mm)	Average annual temperature (°C)	pH (H_2O)	Soil
Stropkov	49°12´N	220-250	600-700	8	5.4	loamy soil
	21°38´E					
Vrbovce	48°47´N	360-400	600-700	8	5.8	loamy, gravelled soil
	17°28´E					
Ľubietová	48°45´N	560-650	700-800	8	5.6	loamy, gravelled soil
	19°22´E					
Folkmár	48°51´N	640-680	700-800	9	5.6	clay, loamy soil
	21°10´E					
Hybe	49°02´N	740-900	700-800	6	5.9	clay, loamy soil
	19°49´E					
Telgárt	48°51´N	1000 -1100	800-1000	5	6.5	loamy soil
	20°11´E					

Permanent meadow - covered with indigenous multispecies vegetation. Usual plant species were *Alchemilla* sp., *Alopecurus pratensis* L., *Bromus arvensis* L., *Dactylorhiza majalis* (Rchb.) P. F. Hunt et Summerh., *Dactylis glomerata* L., *Festuca pratensis* Huds., *Festuca rubra* L., *Plantago lanceolata* L., *Polygonatum odoratum* (Mill.) Druce, *Salvia pratensis* L., *Trifolium pratense* L., *Trifolium repens* L., and *Ranunculus polyanthemos* L.. Permanent meadow was irregularly mown.

Permanent grassland - regularly grazed by cattle or sheep. Pastures were in higher altitude, usually on marginal soils, with shallow humus layer, mostly skeletal, with slopes up to 25°. Representation of grasses and herbs was poorer and it was influenced by the soil type. Usually occurred *Eupatoria agrimonia* L., *Anthyllis vulneraria* L., *Bromus arvensis* L., *Bromus inermis* Leyss., *Carlina acaulis* L., *Chrysanthemum leucanthemum* L., *Dactylis glomerata* L., *Nardus stricta* L., *Poa pratensis* L., *Trifolium pratense* L., *Trifolium repens* L. and various species of trees and bushes.

Soil samples were collected in each site from all three types of grassland. Soil samples were taken two times: on spring (May) and autumn (October). From each site and sampling date, 5 replicates of soil samples were taken using hand spade down to a depth of 10 cm and mixed into a bulk sample. Nematodes were isolated from 500 g of the mixed soil using the Coob´s flotation-sieving method, fixed in FAA and studied in permanent glycerine slides.

Parameters of nematodes communities were studied as follows:

The Shannon index of diversity was calculated from species and genera (H'spp and H'gen). Nematode species were divided into seven trophic groups according to data on feeding habits compiled by Yeates et al. (1993). The Maturity Index (MI) and Plant Parasitic Index (PPI) were calculated according to Bongers (1990), PPI/MI ratio introduced by Bongers and Korthals (1995), B/F ratio proposed by Wasilewska (1997), Enrichment Index

(EI), Structure Index (SI) and Chanel Index (CI) according to Ferris et al. (2001) with weightings of nematode taxa as suggested by the authors. Statistical calculations were performed by the STATISTICA, ANOVA post-hoc Tukey HSD test and Cluster analysis Ward´s method (StatSoft, 2001).

Results

Species Composition of Nematode Fauna

Out of a total of 11 775 nematodes isolated from investigated grassland, 90 nematode species and 66 genera were identified. From those, 32 species belonged to the order Tylenchida, 25 to the order Dorylaimida, 11 to Rhabditida, 8 to Mononchida, 5 to Aphelenchida, 5 to Araeolaimida and 4 to Enoplida. Number of species and genera varied only a little in all investigated types of grassland. Number of species (73) and genera (56) in permanent meadows was higher than in new meadows in arable soil (68 and 56, respective) or in permanent pastures (66 and 52, respective). Abundance of nematodes was different in all types of grassland, localities and sampling date, but significant differences were not found (Table 3). Higher abundance of nematodes in permanent meadows and pastures were caused by a high abundance of *Helicotylenchus digonicus* in a locality of Folkmár.

Obtained values for Shannon-Weaver Index species and genera diversity indicated different species structure of nematode communities in different climatic, topographic and soil conditions. Index of species diversity H'spp in new meadows in different sites varied from 1.89 to 2.92 and index of genera diversity H'gen from 1.89 to 2.89. In the permanent meadows there were the differences in diversity indexes in the sites situated in higher altitude, H'spp here was from 1.42 to 2.91 and H'gen from 1.40 to 2.85. In the sites with pasture, H'spp was from 1.33 to 3.23, and H'gen from 1.18 to 3.05. A comparison of species and genera diversity indexes for all three types of grasslands at different sites showed that the highest nematode species and genera diversity were recorded in grassland at localities Stropkov, Ľubietová and Vrbovce, regardless of types of grassland, localities with lover altitude, better climatic and soil conditions. On the other hand, the lowest species and genera diversity index was determined at the localities with a higher altitude and colder climatic conditions – Folkmár, Hybe and Telgárt.

Cluster analyses (Figure 1) compared species composition and nematode population densities in six different sites and three types of grasslands. They showed greater impact of type of grassland and utilization compared to the impact of geographical, climatic and soil conditions of sites. This is evidenced by the fact that samples from all sites with grassland on arable soil made a separate cluster. Second major cluster was created by sites of permanent grassland and pasture.

Table 2. Nematode species composition, their total abundance - A, dominance - D (%) and frequency of occurrence - F (%) in three different types of meadows (n=12)

		New meadows			Permanent meadows			Pastures		
		A	D%	F%	A	D%	F%	A	D%	F%
	Bacterial feeders	659			455			480		
1	*Acrobeloides-nanus*	49	1.51	83.33	72	1.77	83.33	23	0.51	58.33
2	*Alaimus-primitivus*	8	0.25	25	54	1.33	58.33	45	1.01	50
3	*Anaplectus-granulosus*	16	0.49	25	5	0.12	16.67	17	0.38	25
4	*Aulolaimus-oxycephalus*	2	0.06	8.33	2	0.05	8.33	1	0.02	8.33
5	*Cephalobus-persegnis-*	81	2.5	100	73	1.8	83.33	53	1.19	83.33
6	*Eucephalobus-mucronatus-*	4	0.12	16.67	8	0.2	8.33	6	0.13	16.67
7	*Eucephalobus-oxyuroides*	-	-	-	4	0.1	8.33	7	0.16	25
8	*Eucephalobus-striatus-*	27	0.83	50	25	0.62	66.67	35	0.78	83.33
9	*Chiloplacus-propinquus-*	4	0.12	16.67	7	0.17	16.67	11	0.25	25
10	*Chiloplacus-symmetricus*	26	0.8	33.33	19	0.47	16.67	3	0.07	8.33
11	*Monhystera*-sp.-juv.	17	0.52	25	-	-	-	-	-	-
12	*Panagrolaimus-rigidus*	176	5.42	91.67	5	0.12	25	10	0.22	50
13	*Paramphidelus-dolichurus-*	9	0.28	25	37	0.91	66.67	59	1.32	83.33
14	*Plectus-opisthocirculus-*	-	-	-	8	0.2	16.67	18	0.4	25
15	*Plectus-parietinus-*	3	0.09	8.33	18	0.44	41.67	53	1.19	66.67
16	*Plectus-parvus*	11	0.34	25	3	0.07	8.33	-	-	-
17	*Prismatolaimus-intermedius*	84	2.59	83.33	6	0.15	16.67	-	-	-
18	*Protorhabditis-postneri*	4	0.12	16.67	-	-	-	-	-	-
19	*Rhabditis*-spp.-juvs.	129	3.97	66.67	101	2.49	100	126	2.82	83.33
20	*Teratocephalus-terrestris*	3	0.09	8.33	5	0.12	16.67	7	0.16	16.67
21	*Tylocephalus-auriculatus*	6	0.18	16.67	3	0.07	8.33	6	0.13	16.67
	Fungal-feeders	189			202			160		
22	*Aphelenchoides-composticola-*	17	0.52	33.33	20	0.49	25	17	0.38	25
23	*Aphelenchoides-parietinus*	-	-	-	-	-	-	6	0.13	25
24	*Aphelenchoides-saprophilus*	-	-	-	4	0.1	16.67	-	-	-
25	*Aphelenchus-avenae-*	112	3.45	100	111	2.73	83.33	86	1.92	91.67
26	*Ditylenchus-intermedius-*	1	0.03	8.33	1	0.02	8.33	1	0.02	8.33
27	*Nothotylenchus-acris-*	2	0.06	8.33	4	0.1	8.33	3	0.07	8.33

Table 2. (Continued).

		New-meadows			Permanent-meadows			Pastures		
		A	D%	F%	A	D%	F%	A	D%	F%
28	*Paraphelenchus-pseudoparietinus-*	40	1.23	66.67	22	0.54	8.33	-	-	-
29	*Tylencholaimus-mirabilis-*	17	0.52	41.67	18	0.44	16.67	15	0.34	8.33
30	*Tylencholaimus-zeelandicus*	-	-	-	22	0.54	25	32	0.72	58.33
	Omnivores	431			790			885		
31	*Allodorylaimus-granuliferus-*	-	-	-	-	-	-	7	0.16	8.33
32	*Aporcelaimellus-obtusicaudatus-*	73	2.25	66.67	73	1.8	91.67	74	1.66	75
33	*Axonchium-coronatum*	4	0.12	16.67	13	0.32	33.33	22	0.49	16.67
34	*Axonchium-propinquum-*	-	-	-	2	0.05	8.33	-	-	-
35	*Diphtherophora-communis-*	25	0.77	50	43	1.06	50	58	1.3	66.67
36	*Discolaimus-texanus-*	-	-	-	11	0.27	8.33	1	0.02	8.33
37	*Enchodelus-macrodorus-*	63	1.94	75	147	3.62	75	209	4.68	100
38	*Eudorylaimus-carteri*	60	1.85	58.33	50	1.23	50	101	2.26	75
39	*Eudorylaimus-iners-*	-	-	-	-	-	-	2	0.04	8.33
40	*Eudorylaimus-opistohystera-*	-	-	-	6	0.15	8.33	-	-	-
41	*Eudorylaimus-* spp.-juvs.	138	4.25	100	179	4.41	91.67	204	4.56	100
42	*Mesodorylaimus-bastiani-*	28	0.86	8.33	34	0.84	66.67	48	1.07	66.67
43	*Mesodorylaimus-bastianoides-*	25	0.77	41.67	3	0.07	8.33	2	0.04	8.33
44	*Mesodorylaimus-centrocercus-*	5	0.15	8.33	11	0.27	33.33	9	0.2	25
45	*Microdorylaimus-parvus-*	3	0.09	8.33	9	0.22	16.67	15	0.34	16.67
46	*Oxydirus-oxycephalus-*	3	0.09	8.33	165	4.06	91.67	103	2.3	83.33
47	*Prodorylaimus-brigdammensis*	2	0.06	8.33	4	0.1	8.33	5	0.11	16.67
48	*Pungentus-silvestris*	-	-	-	15	0.37	8.33	22	0.49	8.33
49	*Thonus-ettersbergensis*	2	0.06	16.67	25	0.62	16.67	3	0.07	8.33
	Insect-parasites	5			6			2		
50	*Heterorhabditis-* sp.-juvs.	5	0.15	8.33	-	-	-	-	-	-
- 51	*Steinernema-affine-*	-	-	-	6	0.15	16.67	2	0.04	16.67
	Predators	90			88			138		
52	*Anatonchus-tridentatus-*	11	0.34	41.67	19	0.47	41.67	12	0.27	41.67
53	*Clarkus-papillatus-*	-	-	-	-	-	-	1	0.02	8.33
54	*Coomansus-parvus-*	28	0.86	58.33	20	0.49	58.33	60	1.34	100

		New meadows			Permanent meadows			Pastures		
		A	D%	F%	A	D%	F%	A	D%	F%
55	*Coomansus zschokkei*	-	-	-	14	0.34	8.33	-	-	-
56	*Miconchus studeri*	40	1.23	50	-	-	-	2	0.04	8.33
57	*Mylonchulus sigmaturus*	5	1.5	8.33	13	0.32	41.67	30	0.67	50
58	*Prionchulus muscorum*	3	0.09	8.33	-	-	-	-	-	-
59	*Tripyla affinis*	-	-	-	12	0.3	41.67	5	0.11	25
60	*Tripyla filicaudata*	3	0.09	16.67	10	0.25	41.67	28	0.63	66.67
	Fungal feeders	460			469			315		
61	*Aglenchus agricola*	123	3.79	100	166	4.09	83.33	108	2.42	91.67
62	*Basiria gracilis*	2	0.06	8.33	6	0.15	16.67			
63	*Basiria similis*	2	0.06	8.33	-	-	-	-	-	-
64	*Boleodorus acutus*	2	0.06	8.33	3	0.07	16.67	-	-	-
65	*Boleodorus thylactus*	26	0.8	33.33	42	1.03	33.33	21	0.47	41.67
66	*Coslenchus costatus*	39	1.2	50	40	0.99	75	24	0.54	41.67
67	*Filenchus vulgaris*	94	2.9	66.67	137	3.37	83.33	156	3.49	91.67
68	*Malenchus bryophilus*	11	0.34	16.67	10	0.25	16.67	-	-	-
69	*Malenchus exiguus*	3	0.09	8.33	23	0.57	33.33	-	-	-
70	*Neopsilenchus magnidens*	8	0.25	25	23	0.57	33.33	-	-	-
71	*Tylenchus davainei*	18	0.55	41.67	13	0.32	33.33	-	-	-
72	*Bitylenchus dubius*	132	4.07	83.33	6	0.15	25	6	0.13	16.67
	Plant parasites	1412			2050			2489		
73	*Criconema princeps*	-	-	-	2	0.05	8.33	1	0.02	8.33
74	*Criconemoides informis*	-	-	-	8	0.2	33.33	11	0.25	16.67
75	*Helicotylenchus canadensis*	36	1.11	8.33	-	-	-	70	1.57	8.33
76	*Helicotylenchus digonicus*	1076	33.15	100	1781	43.87	100	216 7	48.49	100
77	*Helicotylenchus pseudorobustus*	59	1.82	33.33	86	2.12	16.67	68	1.52	50
78	*Heterodera avenae*	-	-	-	-	-	-	2	0.04	8.33
79	*Longidorus leptocephalus*	6	0.18	25	24	0.59	33.33	5	0.11	33.33
80	*Macroposthonia curvata*	3	0.09	8.33	11	0.27	8.33	3	0.07	8.33
81	*Paratylenchus bukowinensis*	-	-	-	40	0.99	58.33	42	0.94	50
82	*Paratylenchus elachistus*	31	0.96	58.33	16	0.39	33.33	2	0.04	8.33
83	*Paratylenchus goodeyi*	4	0.12	16.67	-	-	-	-	-	-

Table 2. (Continued).

		New-meadows			Permanent-meadows			Pastures		
		A	D%	F%	A	D%	F%	A	D%	F%
84	*Paratylenchus hamatus*	-	-	-	4	0.1	8.33	-	-	-
85	*Paratylenchus-nanus-*	11	0.34	25				11	0.25	16.67
86	*Pratylenchoides-crenicauda*	22	0.68	41.67	5	0.12	8.33			
87	*Pratylenchus-pratensis*	4	0.12	8.33	37	0.91	75	51	1.14	66.67
88	*Rotylenchus-goodeyi-*	14	0.43	16.67	16	0.39	33.33	2	0.04	8.33
89	*Rotylenchus-pumilus*	138	4.25	83.33						
90	*Rotylenchus-robustus*	8	0.25	16.67	20	0.49	41.67	54	1.21	50
Total-numer-of-nematodes		3246			4060			4469		

Table 3. Community indices of soil nematodes in three types of grassland, mean ± S.D. (n=12). For explanation of all indexes, please see chapter Materials and Methods

	New meadows			Permanent meadows			Pastures		
	mean	±	S.D.	mean	±	S.D.	mean	±	S.D.
Abundance	270	±	92.2	338	±	306.8	372	±	301.7
H'spp	2.46	±	0.32	2.46	±	0.48	2.45	±	0.65
H'gen	2.37	±	0.32	2.40	±	0.45	2.29	±	0.65
Bacterial feeders (%)	20.52	±	8.74	12.24	±	5.58	13.81	±	8.41
Fungal feeders (%)	6.68	±	5.67	6.61	±	4.63	6.44	±	4.08
Omnivores (%)	14.81	±	9.97	25.95	±	12.99	27.07	±	13.31
Insec parasites (%)	0.20	±	0.65	0.26	±	0.67	0.02	±	0.05
Predators (%)	2.77	±	1.99	3.49	±	3.66	5.36	±	4.89
Root fungal feedre (%)	10.78	±	5.83	13.37	±	6.35	8.78	±	6.35
Plant parasites (%)	44.24	±	15.46	38.08	±	20.01	38.52	±	26.50
MI	2.84	±	0.50	3.59	±	0.34	3.50	±	0.38
PPI	2.69	±	0.12	2.83	±	0.22	2.67	±	0.21
PPI/MI	0.98	±	0.16	0.80	±	0.12	0.77	±	0.11
B/F	4.86	±	3.15	2.79	±	2.35	2.95	±	2.93
EI	72.97	±	14.96	58.07	±	10.85	58.13	±	15.84
SI	87.87	±	5.37	92.99	±	4.17	93.07	±	5.62
CI	18.31	±	18.06	26.18	±	17.45	32.02	±	34.77

Eudominant species *Bitylenchus dubius, Eudorylaimus carteri, Helicotylenchus digonicus, Panagrolaimus rigidus* and *Rotylenchus pumilus* were found in new meadows in arable soil, the species *Enchodelus macrodorus*, *Helicotylenchus digonicus* and *Oxydirus oxycephalus* in permanent meadows and the species *Eudorylaimus carteri, Enchodelus macrodorus, Filenchus vulgaris* and *Helicotylenchus digonicus* in permanent pastures. Frequency of the occurrence at the level of 100 % in new meadows in arable soil were recorded for the species of *Aglenchus agricola, Aphelenchus avenae, Cephalobus persegnis* and *Helicotylenchus digonicus*, in permanent meadows for *Helicotylenchus digonicus* and juvenile *Rhabditis* spp., and in permanent pastures for *Coomansus parvus, Enchodelus macrodorus* and *Helicotylenchus digonicus* (Table 2).

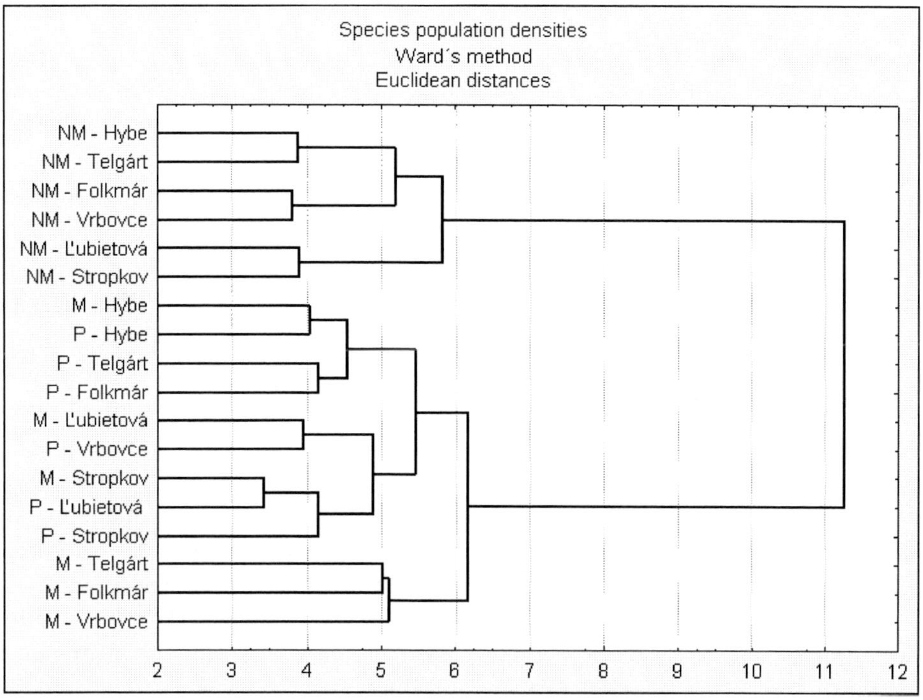

Figure 1. Dendrogram of cluster analysis of nematode species in individual sites and three types of grassland, new meadow NM, permanent meadow – M, pasture – P; data log(x+1), where x = species population density.

Out of a total number of identified nematode species, 31 species were plant parasitic nematodes, 20 belonged to obligate parasites and 11 to facultative parasites. The following obligate parasite species were identified in all investigated types of grassland: *Bitylenchus dubius, Helicotylenchus digonicus, H. pseudorobustus, Longidorus leptocephalus, Mesocriconema curvatum, Pratylenchus pratensis, Rotylenchus goodeyi* and *R. robustus*, facultative plant parasites *Aglenchus agricola, Boleodorus thylactus, Coslenchus costatus* and *Filenchus vulgaris*.

Trophic Groups, Functional Diversity, Food Web Diagnostic

Nematode communities were represented by all trophic groups. Plant parasiting nematodes were dominant trophic group in all types of grassland; in new meadows in arable soil they formed 44%, in permanent meadows 38% and in permanent pastures 39% (Table 3). Bacterial feeders were subdominant trophic group only in new meadows in arable soil where soil was characterised by higher resources of organic mater. In this trophic group it was found significant difference between the monitored types of grassland ($F = 3.58$, $df = 2$, $p < 0.05$), but post-hoc analysis of Tukey's HSD test showed significant difference only between temporary grasses on arable land and grassland ($p < 0.05$). On the other hand, omnivores characterised by a long life cycle, and low abundance especially in cultivated soil were subdominant trophic group in permanent meadows and pastures. In permanent meadows, omnivores were represented by 26% and in the pastures by 27% of the total nematode fauna. In new meadows on arable soil were represented by 15%. Significant difference was found between new meadows and grassland or pasture ($F = 3.40$, $df = 2$, $p < 0.05$). Next trophic groups were root-fungal feeders and fungal feeders. Trophic groups of predators and insect parasites were the least abundant groups in nematode communities in all types of grassland.

The Maturity Index (MI) values were lover in new meadows (2.84) compared with permanent meadows (3.59) and pastures (3.50), which relates to a higher number of omnivores with higher c-p values in permanent meadows and pastures. Using single factor ANOVA, we found significant difference between the monitored types of grassland ($F = 10.59$, $df = 2$, $p < 0.001$), while the post-hoc analysis, Tukey's HSD test, showed significant differrence between new meadows and permanent meadows or pasture ($p < 0.001$,or $p < 0.01$, respectively). In the case of MI it was no found significant difference between permanent meadows and pastures. The Plant Parasitic Index (PPI) value was similar in all investigated types of grassland, and no apparent differences were observed. Low average proportion of PPI/MI in permanent meadows and pastures (less than 0.9) was characteristic for natural ecosystems without environmental destruction. In contrast, for most new meadows this ratio was higher than 0.9 and the highest was found in new meadows in area of higher altitude Telgárt. When comparing different types of grasslands they were confirmed significant differences ($F = 7.49$, $df = 2$, $p < 0.01$) in post-hoc analysis of Tukey's HSD test, between new meadows and permanent meadows ($p < 0.05$) and new meadows and pastures ($p < 0.01$). No significant differences were found between permanent meadows and pastures.

The B/F ratio and Channel Index (CI) did not demonstrate significant differences between investigated types of grassland. The indices EI and SI characterize the food web in the soil. We found significant differences in the index EI ($F = 4.1$, $df = 2$, $p < 0.05$), and pos-hoc analysis of Tukey's HSD test showed significant difference between the new meadows and permanent meadows ($p < 0.05$) as well as new meadows and pasture ($p < 0.05$). In the SI index, we found no significant difference between the monitored types of grassland ($F = 3.76$, $df = 2$, $p < 0.05$) and post-hoc analysis showed significant difference only between new meadows and pasture ($p < 0.05$). Using the Structure Index (SI) and Enrichment Index (EI), all localities of new meadows on arable soil were characterised as moderately disturbed with matured food web and N-enriched conditions, and many localities with permanent pastures were characterised as undisturbed with structure food web conditions.

Discussion

Free-living and plant parasitic nematodes are heterogeneous community in soil, with high species diversity and abundance especially in natural ecosystems. That is probably why most nematode species were identified in permanent meadows, characterized by diverse of plants with rich root system, compared with new meadow established on arable soil or in pasture. Intensively cultivated arable soil is characterized by reduction of nematode species communities and their abundance, especially when compared to natural ecosystems. Several authors, e. g. Parmelee and Alston (1986) and Sohlenius and Sandor (1989) show a significant effect of intensive plowing on nematode communities. Háněl (2003) detected the highest nematode abundance in permanent meadow compared with fallow land and arable soil. Similarly, lower abundance of nematodes in the new meadow on arable soil than in a permanent meadow was found by Valocká and Sabová (1997) and Wasilewska (1979). Vegetation type and age of establishment has a significant impact on the species diversity. Wasilewska (1994) found higher values of index diversity in permanent meadows than in newly established meadow on arable soil. In all investigated meadows, high index of species and genera diversity was found on sites that were located at a lower altitude, with higher average of annual temperatures and light sandy loam soil (e.g. Stropkov and Ľubietová). On the other hand, a low diversity of species was recorded in permanent meadows and pasture in sites with higher altitude (e.g. Hybe, Folkmár). Most balanced value indexes of species and genera diversity were recorded in new meadows and the highest differences were found in pastures, which may be due to different soil conditions of sites. Popovici and Ciobanu (2000) in studying of nematode communities in 36 grassland localities in Romania found H'gen diversity index from 2.38 to 3.47 and Háněl (1996) in the grasslands in southern Bohemia found H'gen from 3.45 to 3.62.

The eudominant (more then 10% from all nematode fauna) nematode species identified in all three types of grasslands and almost in all sites was *Helicotylenchus digonicus*. Brzeski (1998) referred his species as common in different ecosystems in Poland. Several authors (Sumner, 1967, Todd and Tisserat, 1985; Chastagner and McElroy, 1984) considered genus *Helicotylenchus* to be abundant in rhizosphere of grassland. Host plants of *Helicotylenchus* by Towshend (1972) and Towshend and Potter (1973) are especially clover, alfalfa and timothy. Our results suggest certain types of nematodes, which often occur with high abundance in new meadows, and their occurrence in permanent meadows and pastures was no found, or only with very low abundance. In the new meadow it was found a high abundance of species as *Bitylenchus dubius, Panagrolaimus rigidus, Prismatolaimus intermedius* and *Rotylenchus pumilus*. *Panagrolaimus rigidus*, dominant species in new meadows, occurs only sporadically and with very low abundance in permanent meadows and pastures. Háněl (2003) found that *Panagrolaimus* spp. in cultivated arable soil accounted for 23% compared to the permanent meadows accounted only 4% of the total nematode fauna. *Prismatolaimus intermedius* with high abundance in new meadows was found sporadically in permanent meadows and only in very low abundance. In pastures this species was no found at all. On the other hand, some species of nematodes, according to our results, often occurred with high abundance in permanent meadows and pastures and vice versa, did not occur in new meadows. For example, *Oxydirus oxycephalus* was dominated species with high frequency of occurrence in permanent meadow and pastures, but in new meadow we just recorded it with a low abundance and only at one site. In meadows and pastures we recorded high diversity of species from the order Dorylaimida. Many species of this order are sensitive indicators of anthropogenic disturbance (Wasilewska, 1986). Thomas (1978) indicated a lower number of species order Dorylaimida in cultivated soil than in natural ecosystems.

Dominated trophic group in all three types of grassland in six sites in Slovakia were plant parasites represented mostly by genera *Helicotylenchus, Paratylenchus, Pratylenchus* and *Rotylenchus*. Sohlenius and Sandor (1989) referred high occurrence of plant parasiting nematodes, especially genera *Helicotylenchus, Pratylenchus*, *Paratylenchus* and *Trichodorus*, in permanent meadows. High percentage proportion of plant parasites nematodes is also linked to the degradation of soil conditions due to long-term cultivation of monocultures. Plant parasites dominated at 61% of surveyed grasslands in Romania (Popovici and Ciobanu, 2000), with the most represented genera *Paratylenchus*, *Rotylenchus* and *Criconema*. These authors mentioned bacterial feeders as another dominant trophic group at 28% of sites. Similar results were also found by Lišková and Čerevková (2005) in rizosphere of alluvial meadows in Slovakia. Plant parasites were dominating trophic group represented mainly by genera *Helicotylenchus, Tylenchorhynchus, Longidorus* and *Trichodorus*. McSorley and Frederick (2000) found plant parasites as the dominating trophic groups in pastures in the USA, followed by bacterial feeders and fungal feeders. Plant parasites were represented by genus *Tylenchorhynchus, Helicotylenchus* and *Criconemella*.

Bacterial feeders were the second most abundant trophic group in most of the new meadows represented in particular by *Acrobeloides nanus, Cephalobus persegnis, Panagrolaimus rigidus, Prismatolaimus intermedius*. Wasilewska (1995) associated dominance of bacterial feeders with high microbial activity in soil, especially with increased nitrogen content in soil after fertilization. De Goede et al. (1993) considered bacterial feeders to be very sensitive to changes in soil environment. Many of bacterial feeding nematodes are characterized by short-life cycle (several days) and their abundance in the soil can change rapidly especially if there is the change in nitrate content of soil.

According to our results, omnivores were the second dominant trophic group in most of the permanent meadows and pastures represented mostly by high abundance of species *Enchodelus macrodorus*, *Eudorylaimus carteri*, *Oxydirus oxycephalus* and juvenile stages of the genus *Eudorylaimus*. Nematode species from omnivore trophic groups are predominantly characterized by longer life cycle and lower reproductive capacity and, together with the trophic group of predators, are considered as indicators of the natural, undamaged soil environment (Wasilewska, 1997).

Maturity index (MI) is used for characteristics of the soil environment. The lower value of MI is represented in nematode fauna by more adaptable nematode species with high reproductive activity, short-lived cycle and high tolerance to changes in the soil (e.g. Rhabditidae, Panagrolaimidae).

Intensive ploughing and other agro-technical interventions in the soil environment has strong impact on nematode communities. In permanent crops, which include permanent meadows, are minimum interventions into the soil and soil nematode fauna is more represented by nematodes with long-life cycle. Our results confirmed that MI values were significantly lower in new meadows compared with permanent meadows and pastures. Low values of MI in new meadows were caused by high proportion of bacterial feeding nematodes and also low proportion of omnivores and predators. Conversely, a higher proportion of omnivores and predators, and higher MI in permanent meadows and pastures characterize these ecosystems as stable. If comparing permanent meadows and the newly established meadows on arable soil, similar results were recorded by Valocká and Sabová (1997).

The ratio of PPI/MI was examined in different types of grasslands as well as in different locations. The differences were again between new meadows, where the PPI/MI was on average higher than it was in permanent meadows and pastures. According to Bongers et al. (1997) value of the PPI/MI below 0.9 indicates the natural environment, with an adequate supply of nutrients. Decomposition of soil organic matter in soil is carried by bacterial feeders and fungal feeders. According to Wasilewska (1979), McSorley and Frederick (2000), Ruess (2003) and others authors, the decomposition processes in grassland rhizosphere involved particularly bacterial feeders. To determine the way of decomposition of organic matter, they were used the ratio of bacterial feeders and fungal feeders (B/F) and channel index (CI). Low values of CI indicate the way of decomposition of organic matter by nematode families Rhabditidae and Panagrolaimidae. Our results show, that low values of CI were recorded mainly in new meadow e.g. on the site Hybe, with high abundance of species *Panagrolaimus rigidus* and in site Telgárt, with high abundance of bacterial feeders represented by the genus *Rhabditis*. In contrast at pasture it was found higher CI, and decomposition of organic matter was conducted by fungal feeders, mostly by species from families Aphelenchidae and Aphelenchoididae. Similarly, high values of CI was found in permanent grasslands in Austria (Zolda, 2006). Other indices evaluated Enrichment index (EI) and Structure Index (SI) defines the most localities with new meadows, according to Ferris et al. (2001) as disturbed with matured food web condition and most of localities of permanent meadow according EI and SI as undisturbed with structure food web conditions.

References

Bongers, T. (1995). The maturity index: an ecological measure of environmental disturbance based on nematode species composition. *Oecologia*. 83 (1), 14-19.

Bongers, T., and Korthals, G. (1995). The behaviour of MI and PPI under enriched condition. *Nematologica*. 41 (3), 286.

Bongers, T., Van Der Meulen, H. and Korthals, G. (1997). Inverse relationship between the nematode maturity index and plant parasite index under enriched nutrient conditions. *Applied Soil Ecology*. 6 (2), 195-199.

Brzeski, M. W. (1998). *Nematodes of Tylenchina in Poland and temperate Europe*. Warszawa: Muzeum i Instytut Zoologii PAN.

Chastagner, G.A., and McElroy, F.D. (1984). Distribution of plant-parasitic nematodes in putting green turfgrass hosts of some gramineous cyst nematodes. *EPPO. Bulletin*. 12, 399-411.

De Goede, R.G.M., and Bongers, T. (1998). *Nematode communities of northern temperate grassland ecosystems*. Giessen: Focus.

De Goede, R.G.M., Bongers, T. and Ettema, C.H. (1993). Graphical presentation and interpretation of nematode community structure: c-p triangles. *Mededelingen Faculteit Landbouwwetenschappen Universiteit Gent*. 58 (2), 743-747.

Ekschmitt, K., Bakonyi, G., Bongers, M., Bongers, T., Boström, S., Dogan, H., Harrison, A., Kallimanis, A., Nagy, P., O´Donnell, A.G., Sohmenius, B., Stamou, G.P. and Wolters, V. Effects of the nematofauna on microbial energy and matter transformation rates in European grassland soils. *Plant and Soil*. 212 (1), 45-61.

Ekschmitt, K., Bakonyi, G., Bongers, M., Bongers, T., Boström, S., Dogan, H., Harrison, A., Nagy, P., O´Donnell, Papatheodorou, E.M., Sohmenius, B., Stamou, G.P. and Wolters, V. Nematode community structure as indicator of soil functioning in European grassland soils. *European Journal of Soil Biology*. 37 (4), 263-268.

Ettema, C.H., and Yeates, G.W. (2003). Nested spatial biodiverzity patterns of nematode genera in a New Zealand forest and pasture soil. *Soil Biology and Biochemistry*. 35 (2), 339-342.

Ferris, H., Bongers, T., and De Goede, R.G.M. (2001). A framework for soil food web diagnostics: extension of the nematode faunal analysis concept. *Applied Soil Ecology.* 18 (1) 13-29.

Háněl, L. (1996). Composition and seasonal changes of soil nematode community in a South Bohemian meadow. *Acta Societatis Zoologicae Bohemicae.* 60 (2-3), 103-114.

Háněl, L. (2003). Recovery of soil nematode populations from cropping stress by natural secondary succession to meadow land. *Applied Soil Ecology.* 22 (3), 255-270.

Hodda, M., and Wanless, F.R. (1994a) Nematode from an English chalk grassland: species distributions. *Nematologica.* 40 (1), 116-132.

Hodda, M., and Wanless, F.R. (1994b). Nematode from an English chalk grassland: population ecology. *Pedobiologia.* 38 (6), 530-545.

Lišková, M. and Čerevková, A. (2005). Nematode communities of river banks and adjacent grasslands in the Slovak Republic. *Helminthologia.* 42 (4), 223-232.

McSorley, R. and Frederic, J.J. (2000). Short term effects of cattle grazing on nematode communities in Florida pastures. *Nematropica.* 30 (2), 211-221.

Parmelee, R.W. and Alston, D.G. (1986). Nematode trophic structure in conventional and no tillage agroecosystems. *Journal of Nematology.* 18 (3), 403-408.

Popovici, J. and Ciobanu, M. (2000). Diversity and distribution of nematode communities in grassland from Romania in relation to vegetation and soil characteristics. *Applied Soil Ecology.* 14 (1), 27-36.

Ruess, L. (2003). Nematode soil faunal analysis of decomposition pathways in different ecosystems. *Nematology.* 5 (2), 179-181.

Sohlenius, B. and Sandor, A. (1989). Ploughing of a perennial grass ley – effect on the nematode fauna. *Pedobiologia.* 33 (4), 119-210.

STATSOFT, INC. 2001. STATISTICA Cz-Software system for data analysis, Release 6, www.statsoft.cz.

Sumner, D.R. (1967). Nematodes in bluegrass. *Plant disease reporter.* 51, 457-460.

Thomas, S.H. (1978). Population densities of nematode under seven tilliage regimes. *Journal of Nematology.* 10 (1), 24-27.

Todd, T.C. and Tisserat, N.A. (1985). Evaluation of nematicides for the control of plant parasitic nematodes on a bentgrass green. *Fungicide and Nematicide Test.* 41, 73.

Towshend, J.L. (1972). Effect of hay components on the numbers of nematodes. *Nematologica.* 18 (1), p. 149-151.

Towshend, J.L. and Potter, J.W. (1973). Nematode numbers under cultivars forage legumes and grasses. *Canadian Plant Disease Survey.* 53 (1), 194-195.

Valocká, B. and Sabová, M. (1997). Communities of soil and plant nematodes in two types of grassland. *Helminthologia.* 34 (2), 97-103.

Wall Freckman, D. and Huang, S.P. (1998). Response of the soil nematode community in a shortgrass steppe to long-term and short-term grazing. *Applied Soil Ecology.* 9 (1), 39-44.

Wasilewska, L. (1979). The structure and function of soil nematode communities in natural ecosystems and agrocoenoses. *Polish ecological studies.* 5 (2), 97-145.

Wasilewska, L. (1986). Wplyw antropopresji na strukture i funcjonowanie zespolów nicieni glebowych. *Zeszyty problemowe postepów nauk rolniczynych.* 323 (1), 11-31.

Wasilewska, L. (1994). The effect of age of meadows on succession and diversity in soil nematode communities. *Pedobiologia.* 38 (1), 1-11.

Wasilewska, L. (1995). Differences in development of soil nematode communities in single- and multi-species grass experimental treatments. *Applied Soil Ecology*. 2 (1), 53-64.

Wasilewska, L. (1997). Soil invertebrates as bioindicators, with special reference to soil-inhabiting nematodes. *Russian Journal of Nematology*. 5 (2), 113-126.

Yeates, G.W., – Bongers, T., De Goede, R.G.M., Freckam, D.W. and Goergieva, S.S. (1993). Feeding habits in soil nematode families and genera - an outline for soil ecologists. *Journal of Nematology*. 25 (3), 315-331.

Zolda, P. (2006). Nematode communities of grazed and ungrazed semi-natural steppe grasslands in Eastern Austria. *Pedobiologia.* 50 (1), 11-22.

In: Nematodes
Editors: F. Boeri and J. A. Chung
ISBN: 978-1-61470-784-4

Chapter X

Mexican Wild and Domestic Canids: A Potential Risk of Zoonosis? A Review

Norma Hernández-Camacho**, *Robert W. Jones, Raúl F. Pineda-López and Carlos A. López-González
Cuerpo Académico de Ecología y Diversidad Faunística.
Facultad de Ciencias Naturales, Universidad Autónoma de Querétaro, México. Av. De las Ciencias s/n. Delegación Santa Rosa Jáuregui.

Abstract

Anthropogenic habitat loss and global climate change are factors that have been demonstrated to directly affect host parasite systems because they increase the possibility of contact among potential hosts (animals, human). Zoonosis, the transfer of parasites from domestic or wild animals to man, is a problem of medical and economic importance in the Americas. Several studies exist concerning zoonosis in Latin America, principally reported transfer of parasites from domestic animals. However, there are virtually no studies of zoonosis with wild mammals, even those that are relatively tolerant of human presence and frequently found in suburban and urban habitats. This is of special concern in Mexico, which ranks second in total mammal diversity, and the study of parasite communities in Mexican wild mammals remains relatively unexplored. This chapter presents a revision of the information available concerning the parasitic nematodes of wild animals of Mexico with emphasis on the most abundant and ecological generalist wild canids: coyotes (*Canislatrans*) and the gray fox (*Urocyoncinereoargenteus*). These canids are of special interest because of their flexible and generalist feeding behavior and because urban encroachment of their natural habitat has increased the probability of parasite transfer to humans (eg. *Ancylostomacaninum*, *Toxocara* sp., *Strongyloidesstercoralis*) and the encounter chances with domestic and feral dogs. Management strategies, based on basic knowledge of the parasitofauna of these two canid species are needed to monitor, prevent and control of zoonosis including identification of high risk zones, animal treatment and relocation and increase of public awareness. Such

*Corresponding author: minos100@hotmail.com

strategies may help to avoid negative consequences for wild mammals, and parasite monitoring should be part of an integrated conservation and public health program for Mexico and others countries of Latin America.

Introduction: Zoonotic Diseases in American and LatinAmerica

Habitat loss as a result of the anthropogenic activities (social and economic) and global climatic change are confirmed factors that have a direct influence on the composition of communities of parasites in natural ecosystems. This two factors alter and increase the possibility of contact between the original vectors, final and potential hosts resulting in the outbreak of epizootic diseases and zoonosis (Patz et al., 2000; Callum and Dobson, 2002; Patz et al, 2004; Macpherson, 2005, Martínez-Meyer, 2005; Ditchkoff et al., 2006).Zoonosis, or the transmission of wildlife and domestic animal micro and macroparasites to human being, is an important ecological and epidemiological problem with serious consequences and, in many occasions as a result of the response of the human being towards the animals (and their products), is being recognized as a health issue of world-wide importance (Patz et al., 2000; Macpherson, 2005; Reyes-Novelo et al., 2011).

In the Americas, the problem of the zoonotic diseases due to nematodes has been addressed in a series of studies in the last decades, these have focused primarily on the potential impact of parasitism as related to the domestic and animal production, although the potential impact of the wildlife within the cycle of zoonotic transmission is also considered in these studies (Patz et al., 2000; Dobson and Foufopoulos, 2001; Macpherson, 2005; Ditchkoff et al., 2006; Bradley et al. 2006). A formal study has yet to be conducted to consider the effect that land use and global climatic change as key factors in the establishment, modification and dispersal of zoonoses in wild animal communities.

In Mexico, this problem has been little recognized, although these several studies conducted on zoonosis in the last fifty years, many of these have been oriented towards the diseases transmitted through foods and domestic animals, and of these, only a few examine zoonosis involving nematodes. However, one case in particular is of *Gnathostoma* sp. (gnathostomosis), it has received much attention in Mexico due to its high prevalence through the ingestion of badly cooked or raw fish (*ceviche*), in tourist zones on the Pacific coast areas of Guerrero and Colima (Rojas-Molina et al., 1999; Cruz-Aubry, 2002; Martínez-Salazar and León-Règagnon, 2005; Lamothe-Argumedo, 2006; López-Durán, 2006; Reyes-Novelo et al., 2011).Nevertheless, zoonoses involved transmission of nematodes by the terrestrial vertebrates, specifically wild canids, is an incipient field of study although, new descriptions and registries of parasitic nematodes in species of carnivores have recently been published for Mexico (León-Règagnon et al., 2005; Bertoni-Ruiz et al., 2005; Díaz-Camacho et al., 2009, Hernández-Camacho et al., 2010). These extend the basic knowledge on this group of organisms in the country, but the risk of zoonosis between the terrestrial wildlife and human have not been seriously considered.

Mexican Canids: A Risk of Potential Zoonosis?

The wild canids with the greatest distribution and abundance in Mexico are the coyote (*Canislatrans*) and the gray fox (*Urocyoncinereoargenteus*). Both are considered to be ecological generalist carnivores that tolerate human presence and can survive in urban ecosystems which increases the probability of contact with domestic animals and people.Because of these characteristics, studies have been realized of these two species in Mexico to investigate their spatial habitat use, nutritional ecology, biogeography and evolution, among others (González-Pérez et al., 1992; Aranda et al., 1995; Guerrero et al., 2002; Hidalgo-Mihart et al., 2004a; Hidalgo-Mihart., 2004b; Hidalgo-Mihart et al., 2006; Hidalgo-Mihart., 2009; Arroyo-Cabrales and Carranza-Castañeda, 2009; Hernández-Camacho and López González, 2009). Of particular importance, Hidalgo-Mihart et al. (2004b) found that coyotes in tropical environments reduced their home range and group size to exploit open, poorly managed landfills where urban generated resources are in high abundance. These adaptations presumably produce conditions that favor increased parasitism because, not only are population densities more concentrated and group size of coyotes greater, but other mammalian hosts associated with landfills, such as rats and domestic dogs may serve as possible primary and intermediate hosts. However, these conjectures remain to be verified as we found no studies concerning parasitic communities in coyote in Mexico and there are only very few and incidental registries concerning gray fox nematofauna(Pineda-López, 1984; Lamothe et al, 1997).

Studies realized in the United States and Canada, showed that these wild canids can carry a great variety of species of nematodes, generally with little host specificity typical of the carnivores but also several species with zoonotic potential such as *Ancylostomacaninum*, *Uncinariastenocephala*, *Dirofilariaimmitis,* *Toxocara*sp, *Strongyloidesstercoralis*, among others, (Buechner, 1944; Wobeser and Rock, 1973; Simmons et al., 1980; Butterworth and Beverly-Burton, 1981; Seesee et al., 1983; Dyer, 1984; Rogers, 1984; Conti, 1984; King et al., 1984; Measures and Anderson, 1985; Pence et al., 1988; Anderson, 1992; Davidson et al., 1992; Radomski and Pence, 1993; Marquardt et al., 2000; Bush et al., 2001, Pozio et al., 2001; Gompper et al., 2003) (Table 1). Only a few of these species that have received much attention, one of these is *Dirofilariaimmitis* in coyotes (Weinmann and García, 1980; King et al., 1984; Sacks, 1998; Foster et al., 2003; Sacks and Caswell-Chen, 2003; Lee et al., 2010) because the incidence of infection has increased at a world-wide level. Nevertheless is unknown, as for other lesser studied species, the situation of even this well-known parasite in the case of Mexico and this is serious epidemiological problem of public concern.

Several important factors in zoonosis have not been studied sufficiently in Mexico. Two of primal importance is the effect and prevalence of zoonosis in urban ecosystems and the role of domestic and feral dogs in the transmission of diseases with wild canids. The main effect of urban growth in the ecology of disease infections (micro and macroparasites) and in the communities of wildlife is clearly in the alteration of the parasite-host cycle already established that serves as a regulating factor in host populations and by behavior modification of natural hosts as response to dogs presence (intraguild competition), and collaterally to increase the opportunity of contact with human being (Tompkins and Begon, 1999; Daszak et al., 2000; Macpherson, 2005; Ditchkoff et al., 2006; Bradley et al., 2006). The domesticated dog'spresence may be the single most important factor affecting the parasite-host cycle related to nematode zoonosis, and not wild canids. These dogs in Mexico have two sources, as in other regions of the world; first, domestic dogs associated with residential area but with freedom of movement beyond urban boundaries. The second source is those dogs that do not have owner, that survive, feed and reproduce without direct human intervention and can be considered as stray or feral dogs that freely move within the fragments of natural habitat that surround the cities. Both types of dogs constitute an important factor within parasite transmission and that is quite probable that they are the direct factor responsible for the nematode zoonotic diseases in humans in Mexico. Several studies in Mexico have studied the prevalence and abundance of parasitic communities in stray and feral dogs (Eguía-Aguilar et al, 2005, Ponce-Macotela et al 2005; Cantó et al, 2010), and report that the prevalence of some species of zoonotic nematodes as *Ancylostomacaninum* or *Toxocaracanis* are considerably higher that registered in coyotes or gray foxes (Table 1). These results suggest a high potential and progressive dominance of the effects of domestic dogs in the composition of the community of carnivores in remnant fragments of natural habitat (Vanak and Gompper, 2009) around cities through parasite communities modification, which has a direct relation with the loss of diversity of terrestrial vertebrates (potential predators and prey) and in the composition of native parasite communities.

Table 1. *Toxocaracanis, Ancylostomacaninum* and *Dirofilariaimmitis* comparative prevalence values in coyote, gray fox, and domestic dogs in Canada, United States and Mexicoasexamplesof wild and domestic canids zoonotic nematodes,(-) information no available (Buechner, 1944; Seah et al., 1975; WeinmannandGarcía, 1980; Simmons et al., 1980;King andBohning, 1984; Pence et al., 1988; Gommper et al., 2003; Foster et al., 2003; Sacks and Caswell-Chen, 2003; Eguía-Aguilar et al., 2004: Ponce-Macotela et al., 2005, Cantó et al., 2010; Hernández-Camacho et al., 2010; Lee et al., 2010)

Country	Zoonotic nematode	Coyote (*Canislatrans*) Prevalence (%)	Gray Fox (*Urocyoncinereoargenteus*) Prevalence (%)	Dog (*Canisfamiliaris*) Prevalence (%)
Canada	*Toxocaracanis*	-	-	43.2-52.6
	Ancylostomacaninum	-	-	9.5
	Dirofilariaimmitis	-	-	-
United States	*Toxocaracanis*	1.4	-	1.6-60
	Ancylostomacaninum	10.9-33	17	60
	Dirofilariaimmitis	37.3-93	1.8-2.7	12.5
Mexico	*Toxocaracanis*	-	0.8	13.3-93
	Ancylostomacaninum	-	1.6	23-62.5
	Dirofilariaimmitis	-	-	7.7-20

Baseline Information, Management and Control Strategies

Considering the deficiencies and limitations of the basic information of communities of parasitic nematodes of carnivores in Mexico, especially for these two must common species of wild canids and the high potential for contact with domestic and feral dogs, the following information is considered indispensable for future programs of zoonoses management:

1) Community education of parasite risks to animal pets. This aspect is the basis for the success of the remaining proposal strategies, yet is perhaps the most complicated to carry out. However, it is essential that the public understands the biological and ecological responsibility that entails ownership of a pet and that certain measures of sanitary control and hygiene (vaccination, periodic anthelminthic treatments, the pick-up of pets feces in public places) must be obligatory to avoid future health problems.
2) Identification of zoonotic risk zones. This is important to locate potential zoonotic transmission zones of diseases by means of monthly scheduled parasitic inspections of domestic animals and the analysis of wildlife feces in order to generate a general view on the situation of both domestic and wild population in the vicinity of the urban zones to permit the establishment of management practices to minimize human health effects.
3) Campaigns of eradication and sterilization of stray and feral dogs. The population size of stray and feral dogs in some developing countries can exceed the human population density. Control of these animals eliminates an important source of transmission through feces of these animals in urban, suburban, rural and natural areas.
4) Campaigns of wildlife anthelmintic treatment. This proposal, similar to the proposals of vaccination against rabies or parvovirus in the populations of american wildlife (Harrenstien et al, 1997; Steelman et al 1998; Steelman et al, 2000), could be implemented in extreme cases. This considering that we have limited information about the native communities of parasites in Mexican carnivores, specifically nematodes. Such a proposal must also acknowledge that parasites play an important role as regulators in the host populations, and have an important indirect effect at trophic level in the ecosystems.

Conclusion

Maintenance of Parasite Diversity in Mexican Ecosystems

Although parasites are often considered negative diversity and species dominance area inversely related, organisms that must of being eradicated, is necessary to expand the vision towards the indispensable role parasites play in the stability of the natural systems. There are many examples of ill-planned programs to the control or eradication of unwanted organisms that have totally unplanned and disastrous results. Examples come from the introduction of alien species, the use of chemical insecticides to control insects pests (plant parasites), or programs to favor "desirable" organisms by eliminating predators. All these give ample warning of the danger of simplistic solutions to managing complex ecological systems.

Because the study of wild mammals parasites is still in the exploration phase in Mexico (Pérez-Ponce de León, 2001), we must consider this before implementing any strategy so as to minimize the effect on the animal diversity of one of the mega diversity countries of the world.

References

Anderson, R.C. (1992). *Nematode parasites of vertebrates.Their development and transmission*.Canadá.CAB International.

Aranda, M., N. López-Riveraand L. López-De Buen. (1995). Hábitos alimentarios del coyote (*Canislatrans*) en la sierra del Ajusco, México. *Acta Zoológica Mexicana (nueva serie)65:* 89-99.

Arroyo-Cabrales, J. and O. Carranza-Castañeda. (2009). Los cánidos prehistóricos mexicanos antes de la llegada del perro. *Arqueobios 3(1):* 34-45.

Bradley, C. A. and S. Altizer.(2006). Urbanization and the ecology of wildlife diseases.*Trends in Ecology and Evolution 22(2):* 95-102.

Bertoni-Ruiz, F., L. García-Prieto, D. Osorio-Sarabiaand V. León-Règagnon. (2005). A new species of *Gnathostoma* (NEMATODA:GNATHOSTOMATIDAE) in *Procyonlotorhernandezii* from Mexico. *Journal of Parasitology 91(5):* 1143-1149.

Buechner, H. K. (1944). Helminths parasites of the gray fox.*Journal of Mammalogy 25(1)*: 185-188.

Butterworth E. W.and M. Beverly-Burton. (1981). Observations on the prevalence and intensity of Capillaria spp. (Nematoda: Trichuroidea) in wild carnivore from Ontario, Canada. *Proceedings of The Helminthological Society of Washington 48(1):* 24-37.

Bush, A. O., J. C. Fernández, G. W. Eschand J. R. Seed. (2001). *Parasitism.The diversity and ecology of animal parasites*.New York, Cambridge UniversityPress.

Cantó, G. J., M. P. García, A. García, M. J. Guerrero and J. Mosqueda. (2010). The prevalence and abundance of helminths parasites in stray dogs from the city of Queretaro in central Mexico. *Journal of Helminthology* [DOI: 10.1017/S0022149X10000544].

Conti, J. A. (1984). Helminths of foxes and coyotes in Florida.*Proceedings of the Helmintological Society of Washington 51(2):* 365-367.

Cruz-Aubry, J. E. (2002). Zoonosis en la República Mexicana. *Revista de Endocrinología y Nutrición 10(4):* 206-208.

Davidson, W. R., M. J. Appel, G. L. Doster, O. E. Baker and J. F. Brown. (1992). Diseases and parasites of red foxes, gray foxes and coyotes from commercial sources selling to fox-chasing enclosures.*Journal of Wildlife Diseases 28(4):* 581-589.

Daszak, P., A. A. Cunningham and A. D. Hyatt. 2000. Emerging infectious diseases of wildlife. Threats to biodiversity and human health.*Science 21:* 443-449.

Díaz Camacho, S. P., K. Willms, J. G. Rendón-Maldonado, M. C. De la Cruz-Otero, F. Delgado-Vargas, L. Robert, S. Antuna, V. León-Règagnonand Y. Nawa. (2009). Discovery of an endemic area of *Gnathostomaturgidum* infection among opposums, *Didelphisvirginiana*, in Mexico.*Journal of Parasitology 95(3):* 617-622.

Ditchkoff, S. S., S. T. Saalfeldand C. J. Gibson.(2006). Animal behavior in urban ecosystems: modifications due to human-induced stress. *Urban Ecosystems 9:* 5-12.

Dobson, A. and J. Foufopoulos. (2001). Emerging infectious pathogens of wildlife. *Philosophical Transactions of The Royal Society of London 356:* 1001-1012.

Dyer, W. G. (1984). *Paragonimuskellicotti*Ward, 1908 (TREMATODA: PARAGONIMIDAE) from red and gray foxes of Southern Illinois. *Transactions of the Illinois Academy of Science 77 (1-2)*: 33-34.

Eguía-Aguilar, P., A. Cruz-Reyes and J. J. Martínez-Maya.(2005). Ecological analysis and description of the intestinal helminthes present in dogs in Mexico City.*Veterinary Parasitology 127(2):* 139-146.

Foster, G. W., M. B. Main, J. M. Kinsella, L. M. Dixon, S. P. Terreland D. J. Forrester. (2003). Parasitic helminthes and arthropods of coyotes (*Canislatrans*) from Florida, U. S. A. *Comparative Parasitology 70(2):* 162-166.

González-Pérez, G. E., V. M. Sánchez-Bernal, Luis, I. Iñiguez-Dávalos, E. Santana-C.and T. K. Fuller. (1992). Patrones de actividad del coyote (*Canislatrans*), la zorra gris (*Urocyoncinereoargenteus*) y el tlacuache (*Didelphis virginiana*) en la sierra de Manantlán, Jalisco. *Anales del Instituto de Biología, Serie Zoología 63(2):* 293-299.

Gompper. M. E., R. M. Goodman, R. W. Kays, J. C. Ray, C. V. Fiorelloand S. E. Wade. (2003). A survey of the parasites of coyotes (*Canislatrans*) in New York based on fecal analysis. *Journal of Wildlife Diseases 39(3):* 712-717.

Guerrero, S., M. H. Badii, S. S. Zalapaand A. E. Flores.(2002). Dieta y nicho de alimentación del coyote, zorra gris, mapache y jaguarundi en un bosque tropical caducifolio de la costa sur del estado de Jalisco, México. *Acta Zoológica Mexicana (nueva serie)86:* 119-137.

Harrenstien, L. A., L. Munson, E. C. Ramsay, C. F. Lucash, S. A. Kaniaand L. D. N. Potgieter. (1997). Antibody responses of red wolves to canine distemper virus and canine parvovirus vaccination.*Journal of Wildlife Diseases 33(3):* 600-605.

Hernández-Camacho, N. and C. A. López-González. (2009). Efectos de la fragmentación del hábitat en la ecología espacial de la zorra gris (*Urocyoncinereoargenteus*) en el Parque Nacional El Cimatario, Querétaro, México. Pp. 153-172. *In: Perspectivas en Zoología Mexicana.*A. J. Sanchez, M. G. Hidalgo Mihart, S. L. Arriaga Weiss, W. M. Contreras Sanchez (comp.).

Hernández-Camacho, N., R. Pineda-López, C. A. López-González and R. W. Jones. (2010). Nematodes parasites of the gray fox (*Urocyoncinereoargenteus*Schreber, 1775) in the seasonally dry tropical highlands of central Mexico. ParasitologyResearch. [DOI 10.1007/s00436-010-2191-5].

Hidalgo-Mihart, M. G., L. Cantú-Salazar, A. González-Romero andC. A. López-González. (2004a). Historical and Present Distribution of Coyote (*Canislatrans*) in Mexico and Central America.*Journal of Biogeography31:* 2025–2038.

Hidalgo-Mihart, M. G., L. Cantú-Salazar, C. A. López-González, E. C. FernandezandA.González-Romero. (2004b). Effect of a landfill on the home range and group size of coyotes (*Canislatrans*) in a tropical deciduous forest. *Journal of Zoology (London) 263:* 1-9.

Hidalgo-Mihart, M. G., L. Cantú-Salazar, C. A. López González,P. G. Martínez-Gutiérrez, E. C. Fernández and A. González-Romero. (2006). Coyote Habitat Use in a Tropical Deciduous Forest of Western Mexico.*Journal of Wildlife Management 70 (1):* 216-221.

Hidalgo-Mihart, M. G., L. Cantú-Salazar, S. E. Carrillo-Percásteguiand C. A. López González.(2009). Daily Activity Patterns of Coyotes (*Canislatrans*) in a Tropical Deciduous Forest of Mexico.*Studies in Neotropical Fauna and Environment 44 (2):* 77-82.

King, A. W. and A. M. Bohning. (1984). The incidence of heartworm *Dirofilariaimmitis* (Filaroidea) in the wild canids of Northeast Arkansas. *TheSouthwesternNaturalist 29(1):* 89-92.

Lamothe, A. R., L. García Prieto, D. Osorio Sarabia and G. Pérez Ponce de León. (1997). *Catálogo de la colección nacional de helmintos. Universidad Nacional Autónoma de México.*México. Instituto de Biología. Comisión Nacional para el conocimiento y uso de la biodiversidad.

Lamothe-Argumedo, R. (2006). Gnathostomiasis ocular humana.Casos registrados en todo el mundo entre 1937 y 2005. *Revista Mexicana de Oftalmología 80(4):* 185-190.

Lee, A. C. Y., S. P. Montgomery, J. H. Theis, B. L. Blagburnand M. L. Eberhard.(2010). Public health issues concerning the widespread distribution of canine heartworm disease.*Trends in Parasitology 26(4):* 168-173.

León-Règagnon, V., D. Osorio-Sarabia, L. García-Prieto, R. Lamothe-Argumedo, F. Bertoni-RuizandA. Oceguera-Figueroa.(2005). New hosts records of the nematode *Gnathostoma* sp. in Mexico. *Parasitology International 54:* 51-53.

López-Durán, M. M. (2006). Las zoonosis en los sistemas de producción animal de las áreas urbanas y periurbanas de América Latina. *Revista de SanidadMilitar 60(5):* 352-358.

Macpherson, C. N. L. (2005). Human behavior and the epidemiology of parasitic zoonoses.*International Journal of Parasitology 35:* 1319-1331.

Marquardt, W. C., R. S. Demareeand R. B. Grieve. (2000). *Parasitology Vector Biology.2ndedition.*San Diego, California: Academic Press.

Martínez-Meyer, E. (2005). Climate change and biodiversity: some considerations in forecasting shift in species: potential distributions. *Biodiversity Informatics 2:* 42-55.

Martínez-Salazar, E.A. and V. León-Règagnon. (2005). Confirmation of *Gnathostomabinucleatum*Almeyda-Artigas 1991, advanced three stage larvae in Tres Palos Lagoon, México, by morphological and molecular data. *The Journal of Parasitology 91(4):* 962-965.

McCallum, H. and A. Dobson.(2002). Disease, habitat fragmentation and conservation.*Proceedings of the Royal Society of London Bulletin 269:* 2041-2049.

Measures, L. N. and R. C. Anderson.(1985). Centrarchid fish as paratenic hosts of the giant kidney worm, *Dioctophymarenale* (Goeze, 1782), in Ontario, Canada.*Journal of Wildlife Diseases 21(1):* 11-19.

Patz, J. A., T. K. Graczyk, N. Geller and A. Y. Vittor.(2000). Effects of environmental change on emerging parasitic diseases.*International Journal of Parasitology 30:* 1395-1405.

Patz, J. A, P. Daszak, G. M. Tabor, A. Alonso-Aguirre, M. Pearl, J. Epstein, N. D. Wolfe, A. M. Kilpatrick, J. Foufopoulos, D. Molyneux, D. J. Bradleyand Members of the Working Group on Land Use Change and Disease Emergence. (2004). Unhealthy Landscapes: policy recommendations on land use change and infectious disease emergence. *Environmental Health Perspectives 112(10):* 1092-1098.

Pence D. B., F. F. Knowlton and L. A. Windberg.(1988). Transmission of *Ancylostomacaninum* and *Alariamarcianae* in coyotes (*Canislatrans*).*Journal of Wildlife Diseases 24(3):* 560-563.

Pérez-Ponce de León, G. (2001). The diversity of digeneans (Platyhelminthes: Cercomeria: Trematoda) in vertebrates in Mexico. *ComparativeParasitology 68(1):* 1-8.

Pineda López. R. (1984). Primer registro de *Dioctophymarenale* (Goeze, 1782) Mollet-Meygret, 1802 (NEMATODA: DIOCTOPHYMIDAE) en Tabasco, México. *Anales del Instituto de Biología de la Universidad Nacional Autónoma de México. 55 (2):* 307-310 pp.

Ponce-Macotela, M., G. E. Peralta-Abarca and M. N. Martínez-Gordillo. (2005). Giardia intestinalis and other zoonotic parasites: prevalence in adult dogs from the southern part of Mexico City. *VeterinaryParasitology131(1-2):* 1-4.

Pozio, E., D. B. Pence; G. La Rosa, A. Casulliand S. E. Henke. (2001). *Trichinella* infection in wildlife of the Southwestern United States.*Journal of Parasitology 87(5):* 1208-1210.

Radomski, A. A. and D. B. Pence.(1993). Persistence of a recurrent group of intestinal helminthes species in a coyote population from southern Texas.*Journal of Parasitology 79(3):* 371-378.

Reyes-Novelo, E., H. Ruiz-Piña, J. Escobedo-Ortegón, I. Rodríguez-Vivas, M. Bolio-González, A. Polanco-Rodríguez and P. Manríquez-Saide. (2011). Situación actual y perspectivas para el estudio de las enfermedades zoonóticas emergentes, reemergentes y olvidadas en la Península de Yucatán, México. *Tropical and Subtropical Agroecosystems 14:* 35-54.

Rogers, E. D., (1984). Helminths parasites of the gray fox (*Urocyoncinereoargenteus*). *M. C. Dissertation*. Auburn University, Alabama. 81 pp.

Rojas-Molina, N., S. Pedraza-Sanchez, B. Torres-Bibiano, H. Meza-Martínez and A. Escobar-Gutierrez. (1999). Gnathostomosis, anemergingfoodbornezoonoticdisease in Acapulco, Mexico.*Emerging Infectious Diseases 5(2):* 264-266.

Sacks, B. N. (1998).Increasing prevalence of canine heartworm in coyotes from California.*Journal of Wildlife Diseases 34(2):* 386-389.

Sacks, B. J. and E. P. Caswell-Chen.(2003). Reconstructing the spread of *Dirofilariaimmitis* in California coyotes.*Journal of Parasitology 89(2):* 319-323.

Seah, S. K. K., G. Hucaland C. Law. (1975). Dogs and intestinal parasites: a public health problem. *Canadian Medical Association (112)17:* 1191-1194.

Seesee, F. M., M. C. Steiner and D. E. Worley.(1983). Helminths of the coyote (*Canislatrans*Say) in Montana.*Journal of Wildlife Diseases 19(1):* 54-56.

Simmons, J. M., W. S. Nicholson, E. P. Hill and D. B. Briggs. (1980). Ocurrence of *Dirofilariaimmitis* in gray fox (*Urocyoncinereoargenteus*) in Alabama and Georgia.*Journal of Wildlife Diseases 16:* 225-228.

Steelman, H. G., S. E. Henke and G. M. Moore.(1998). Gray fox response to baits and attractants for oral rabies vaccination. *Journal of Wildlife Diseases 34(4):* 764-770.

Steelman, H. G., S. E. Henkeand G. M. Moore.(2000). Bait delivery for oral rabies vaccine to gray foxes.*Journal of Wildlife Diseases 36(4):* 744-751.

Tompkins, D. M.and M. Begon. (1999). Parasites can regulate wildlife populations. *Parasitology Today 15(8):* 311-313.

Vanak, A. T. and M. E. Gompper. (2009). Dogs *Canisfamiliaris* as carnivores: their role and function in intraguild competition. *Mammal review 39:* 265-283.

Weinmann, C. J. and R. García.(1980). Coyotes and canine heartworm in California.*Journal of Wildlife Diseases 16(2):* 217-221.

Wobeser, G. and T. W. Rock. 1973. *Capillaria hepatica* (NEMATODA: TRICHURIDAE) in a coyote (*Canislatrans*). *Journal of Wildlife Diseases 9:* 225-226.

In: Nematodes
Editors: F. Boeri and J. A. Chung
ISBN: 978-1-61470-784-4

Chapter XI

Conservation Medical Strategy of Endemic Host-Parasite Relationships-Case Model of Japanese Rodents (Mammalia) and Their Nematodes

Mitsuhiko Asakawa[1]* and Hideo Hasegawa[2]
[1] Wild Animal Medical Center, Graduate School of Veterinary Medicine, Rakuno Gakuen University, Ebetsu, Hokkaido 069-8501, Japan.
[2] Department of Biology, Faculty of Medicine, Oita University Hasama, Yufu, Oita 879-5593, Japan.

Abstract

We are investigating alien rodents that occur within Japanese Islands for their parasitic nematodes, because some alien nematodes derived from such alien hosts could become highly pathogenic agents to human and livestock, and because the parasites potentially could switch their hosts to and having a negative impact on native host-parasite relationships that have evolved as a result of adaptive radiation or co-speciation between the native rodents and parasitic nematodes. We are generalizing the patterns of host-parasite associations, and providing 4 types: alien rodents and alien nematodes such as *Callosciurus erythraeus* and *Brevistriata callosciuri*, and *Myocastor coypus* and *Strongyloides myopotami*; alien rodents and native nematodes such as *Rattus rattus* and *Heligmosomoides kurilensis* shifted from *Apodemus speciosus*; native rodents and alien nematodes such as *Apodemus argenteus* and *Heligmosomoides polygyrus* shifted from *Mus musculus*; native rodents and native nematodes such as *Apodemus* spp. and *Heligmosomoides* spp.

Hence, aberrant host-parasite relationships due to the presence of alien rodents and/or nematodes comprise the 1st, 2nd and 3rd groups. These relationships should be discriminated from the native host-parasite relationships within Japanese natural

* To whom correspondence should be addressed. E-mail: askam@rakuno.ac.jp

ecosystem. Some potential strategies including monitoring survey for reduction of risk affected by the aberrant relationships were discussed.

Keywords: Alien rodents, alien parasitic nematodes, Japan, natural ecosystems, zoogeography.

Introduction

Because an understanding of the ecology and/or zoogeography of the parasitic helminths that are present in terrestrial vertebrates is one of the essential conservation biological and medical tools for maintaining healthy host-parasite relationships as a part of the Japanese natural ecosystem, we have been investigating parasitic fauna since 1981 (Asakawa et al., 1983, 1991, 1993 and so on), and especially since 2004, the project has been performed as one of conservation medical strategy of Japanese fauna (Asakawa et al., 2004; Zamoto et al., 2004; Tsuji et al, 2006; Haukisalmi et al. 2008; Asakawa, 2010; Onuma and Asakawa, in preparation).

Namely, during this time we have been able to observe some aberrant cases of host-parasite relationships provoked by the presence of alien hosts and/or parasites, so we have to act for conservation of the ecological health. In the present paper, using cases of rodent hosts and their some parasitic nematodes (Nematoda) due to the presence of much data derived from not only our investigations but also those of other authors, we aim to provide an overview and generalized discussion of host-parasite relationships in the Japanese Islands, mainly based on Asakawa (2005) and newly data after then. Moreover, potential strategies for the reduction of risk affected by aberrant relationships will be discussed.

Rodent Fauna of the Japanese Islands

A total of 29 rodent species, including 8 alien species, have been recorded within the Japanese territory (Table 1) (Abe et al. 2005; Ohdachi et al. 2009). However, those deemed as accidental cases of exotic rodents derived immediately from pet animals (e.g., *Cynomys* sp.) are excluded here because they have not yet been recognized as naturalized in Japan, although they could become tentative hosts for parasitic nematodes.

All of the native rodents are divided into 3 groups with respect to their locality: the 1) Endemic species group, including *Sciurus lis, Pteromys momonga, Petaurista leucogenys, Eothenomys* spp., *Microtus montebelli, Apodemus speciosus, A. argenteus*, and *Glirulus japonicus*; the 2) Eurasian species group, including *Tamias sibiricus, Pteromys volans,Myodes (= Clethrionomys)* spp., *Apodemus peninsulae, A. agrarius*, and *Micromys minutus*; and the 3) Ryukyu species group, including *Diplothrix legata, Tokudaia* spp., and *Mus calori.*

Table 1. Checklist of rodent species in Japan (from Abe, 2005; Ohdachi et al., 2009)

Sciuridae
- *Pallas's squirrel, *Callosciurus erythraeus* (Pallas)
- [*]Eurasian red squirrel, *Sciurus vulgaris* Linneaus
- Japanese squirrel, *S. lis* Temminck
- [*]Siberian chipmunk, *Tamias sibiricus* (Laxmann)
- Eurasian flying squirrel, *Pteromys volans* (Linneaus)
- Japanese flying squirrel, *P. momonga* Temminck
- Japanese giant flying squirrel, *Petaurista leucogenys* (Temminck)

Muridae
- northern red-backed vole, *Myodes rutilus* (Pallas)
- gray red-backed vole, *M. rufocanus* (Sundevall)
- king red-backed vole, *M. rex* (Imaizumi)
- Smith's red-backed vole, *Eothenomys smithii* (Thomas)
- Anderson's red-backed vole, *E. andersoni* (Thomas)
- Japanese grass vole, *Microtus montebelli* (Milne-Edwards)
- *muskrat, *Ondatra zibethicus* (Linnaeus)
- large Japanese field mouse, *Apodemus speciosus* (Temminck)
- Korean field mouse, *A. peninsulae* (Thomas)
- small Japanese field mouse, *A. argenteus* (Temminck)
- striped field mouse, *A. agrarius* (Pallas)
- harvest mouse, *Micromys minutus* (Pallas)
- *house mouse, *Mus musculus* (Linnaeus)
- Ryukyu mouse, *M. calori* Bonhote
- *roof rat, *Rattus rattus* (Temminck)
- *Norway rat, *R. norvegicus* Berkenhout
- *Polynesian rat, *R. exulans* (Peale)
- Ryukyu long-furred rat, *Diplothrix legata* (Thomas)
- Amami spiny rat, *Tokudaia osimensis* (Abe)
- Okinawa spiny rat, *T. muenninki* (Johnson)

Gliridae
- Japanese dormouse, *Glirulus japonicus* (Schinz)

Myocastoridae
- *nutria, *Myocastor coypus* (Molina)

*: alien species in Japan.
[*]: native species in Hokkiado I., but escaped or released pet animals found in Japan.

Nematode Records from Japanese Rodents

Using the previous references (mainly, Hasegawa and Asakawa 2003) and our own data (Asakawa unpubl.), here we present relatively recent nematode records and investigations that have been performed since the 1960s.

Sciuridae: In Hokkaido I., *Dirofilaria formasciurorum* from *S. vulgaris* and *Brevistriata skrjabini* (syn. *B. bergerardi*) from *T. sibiricus* have been recorded in the 1980s. More recently, *B. skrjabini* and *Strongyloides robustus* have been obtained from *S. vulgaris* in the Shiretoko Peninsula located in the north-eastern part of the island (Ono et al. 2005, unpubl.). On the other hand, there have been 2 oxyurid species, *Citellina petrovi* from *P. volans*, and

Sypharista kamegaii from *Petaurista leucogenys* obtained in Honshu I., but there is no nematode record from *Sciurus lis*.

Although *S. vulgaris* and *T. sibiricus* are native to Hokkaido I., both sciurids have become an alien species on the Japanese Is. including Hokkaido I. due to the pet trade. Up to now, these sciurids have not been investigated, although *C. erythraeus*, which appears on Izu-Ohshima and Fukue-jima Is., has been examined helminthologically, and *Brevistriata callosciuri, Gongylonema neoplasticum*, and *Strongyloides* cf. *robustus* have been recorded (Matsudate et al. 2003; Shingaki et al. 2004).

Muridae: Among the species described, redescribed, or newly recorded from Japan since the 1960s, three nematode families of zoogeographical interest are mentioned herein: Heligmosomidae, Heligmonellidae, and Oxyuridae. Further, slightly different nematode compositions exist with the parasite genus or species level with respect to the host subfamily level (Microtinae and Murinae). However, it should be noted here that there are exceptions. For example, *Mammanidula hokkaidensis* (Heligmonellidae) which inhabits the mammary gland and Cowper's gland (Figure 1), of host individuals and shows a broad host range, has been recorded from *Myodes* spp., *Apodemus* spp., and *Sorex* spp. (Insectivora).

Figure 1. *Mammanidula hokkaidensis* which inhabits Cowper's gland (left) and mammary gland (right; 5mm mesh) of *Myodes rufocanus*.

Most of the nematodes recorded from native Japanese microtids belonging to the genera *Eothenomys* and *Microtus* are endemic species to the Japanese Is. Meanwhile, the nematodes recorded from the genus *Myodes* that are restricted to Hokkaido within the Japanese Is. are also shared by the continental vole population. For example, *Heligmosomum (Paraheligmosomum) yamagutii* (Heligmosomidae) has been found not only from *Myodes* spp. in Hokkaido, but also from *M. rufocanus* in the Scandinavia Peninsula. From *M. rutilus mikado, H. (P.) mixtum* (Heligmosomidae) and *Syphacia petrusewiczi* (Oxyuridae) have been recorded. *Yatinema japonicum* (Heligmonellidae) was described as a new genus and/or species, *Heligmosomum (Paraheligmosomum) hasegawai* (Heligmosomidae), has been described. They are host-specific to two species of the Japanese *Eothenomys*. *Morganiella cricetuli* (Heligmonellidae), recorded from *E. smithii* in Shikoku I., is of special geographical interest because it had been previously recorded only from the striped hamster (*Cricetulus barabensis*) in China. *Heligmosomoides protobullosus* (Heligmosomidae) (Figure 2), *Heligmosomum (Heligmosomum) halli*, and *Carolinensis minutus* (Heligmonellidae) have been recorded from *Microtus montebelli*, and *H. protobullosus* seems to be endemic to Japan. Among the microtines occuring in the Japanese Is., *Ondatra zibethicus* is the only alien microtin. However, no helminthological investigation of *O. zibethicus* has been performed up to now.

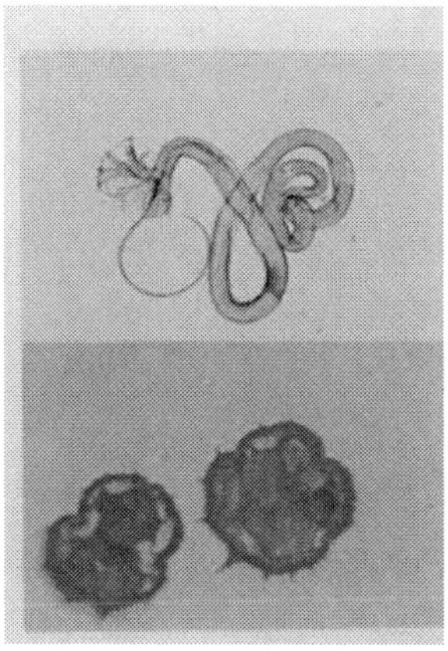

Figure 2. Male whole body (top) and cross-section of its mid-body (bottom) of *Heligmosomoides protobullosus* from *Microtus montebelli.*

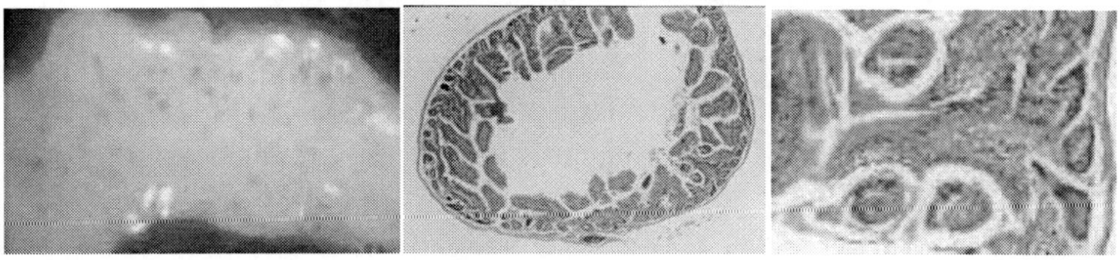

Figure 3. Naked eye observation on small intestine (left) and cross-section of its intestinal wall (middle and right) of *Heligmonoides speciosus* parasitizing *Apodemus speciosus*.

Japanese native murines consist of all species belonging to the genera *Micromys, Apodemus, Diplothrix, Tokudaia,* and *Mus caroli*. *Syphacia vandenbrueli*, which had previously been known as the pinworm of *Micromys minutus* in Europe, was recorded from the Japanese population of the mice in Kyushu I. and Honshu I. as well. Among over 20 nematode species recorded from the Japanese *Apodemus, Heligmosomoides kurilensis, H. desportesi, H. neopolygyrus, Heligmonoides speciosus* (Figure 3), *Syphacia emileromani, S. frederici*, and *S. agraria* are known as biomarkers of zoogeographical epidemiology and environmental status. On the Ryukyu Is., there have been records of *Heligmonoides ikeharai* from *T. osimensis* and of *Heligmonoides ryukyensis* and *Syphacia ohtaorum* from *M. caroli*. Although *Globocephalus longemucronatus*, a common parasitic nematode of the wild boar (*Sus scrofa*), has been obtained from *D. legata*, there is no record of a host specific nematode from the native rats.

There have been many reports and medical/veterinary publications on these parasites that include nematodes from alien murines (*Rattus norvegicus, R. rattus*, and *Mus musculus*) because they harbor many zoonotic helminths. *Orientostrongylus ezoensis* (Heligmonellidae), *Nippostrongylus brasiliensis* (Heligmonellidae), and *Syphacia muris* have been found from *Rattus* spp. captured in wild conditions. Further, *Heligmosomoides polygyrus* and *Syphacia obvelata* are host specific to *M. musculus*.

Gliridae: *Glirulus japonicus* is an endangered species endemic to Honshu, Shikoku, Kyushu, and Doh-go Is. Although *Acanthatrium ovatum* (Trematoda) was obtained from the small intestine of *G. japonicus* captured at Mt. Ontake, there has to date been no nematode recorded. Recently, the present author could also not obtain any parasitic helminths from two bodies of dormice carried to an animal hospital in Tochigi Prefectur (Asakawa unpubl.).

Myocastoridae: On Honshu I., there have been recordings of *Strongyloides myopotami* from coypus, *Myocastor coypus,* captured in Gifu and Okayama Prefs. and of *Calodium hepaticum* in Hiroshima Pref. (Matsudate et al. 2003; Asakawa et al. 2009). More detailed monitoring surveys should be required, however, because both are known as zoonotic agents.

Nematode Zoogeography and Aberrant Host-Parasite Relationships

Changes in the global political framework since the mid 1980s have enabled Japanese parasitologists to study the helminths on the Eurasian continent, and at present several international cooperative research projects are progressing. Through such projects, studies on the origin and speciation of the Japanese helminths that parasitize rodents are becoming more intense and focused. Native murids offer many advantages with regard to the study of host-parasite relationships and zoogeography. For example, their geographical distribution is rarely influenced by human activity, and they are easily collected. Moreover, much taxonomical, zoogeographical, phylogenic, and ecological information on these murids has been accumulated.

Hence, as a part of a zoogeographical research project, an analysis was made on the parasite fauna of the Japanese Microtinae and Murinae, and several endemic host-parasite relationships have been detected (Hasegawa and Asakawa 2003). However, some aberrant host-parasite relationships between Japanese munines and the heligmosomids have been observed in accidental cases of *Heligmosomoides kurilensis* from *Rattus rattus* on Oki Is. (Sakata et al., 2005) and *H. polygyrus* from *Apodemus argenteus* on Hokkaido I. (Asakawa et al. 1994).

Four Types of Host-Parasite Relationships between Rodents and Nematodes

According to the rodent fauna found in Japan (Table 1), there are alien and native host species (abbreviated as "H1" and "H2," respectively) (Table 2). Further, the parasitic nematodes are also divided into alien and native types (abbreviated as "P1" and "P2," respectively) (Table 2).

Table 2. Abbreviations of native/alien rodents and parasitic nematodes (from Asakawa, 2005)

Host rodents (H) / Parasitic nematodes (P)		
alien (H1)	—	alien (P1)
	×	
native (H2)	—	native (P2)

Table 3. Four types of host-parasite relationships between rodents and parasitc nematodes and their some examples accorded with each type of the relationship (from Asakawa, 2005)

Host-Parasie relationships	(Abbreviations of Tab. 2)
H1-P1	
	Callosciurus erythraeus —*Brevistriata callosciuri*
	Myocastor coypus —*Strongyloides myopotami*
	Rattus norvegicus —*Orientostrongylus ezoensis*
	R. norvegicus —*Nippostrongylus brasiliensis*
	Mus musculus —*Heligmosomoides polygyrus*
H1-P2	
	Rattus rattus —*Heligmosomoides kurilensis*
H2-P1	
	Apodemus argenteus —*H. polygyrus*
H2-P2	
	Eothenomys andersoni —*Yatinema japonicum*
	Microtus montebelli —*Heligmosomoides protobullosus*
	Apodemus specuosus —*H. kurilensis*
	A. argenteus —*Heligmosomoides desportesi*
	A. peninsulae —*Heligmosomoides neopolygyrus*
	etc

Consequently, the host-parasite relationships can be summarized into four patterns (Table 3), namely 1) relationships between alien rodents and alien parasitic nematodes (abbreviated as "H1-P1"), 2) relationships between alien rodents and native parasitic nematodes ("H1-P2"), 3) relationships between native rodents and alien parasitic nematodes ("H2-P1"), and 4) relationships between native rodents and native parasitic nematodes ("H2-P2"). In addition, some examples of these according to each type of relationship in Japan are presented in Table 3.

Probable impacts for endemic host-parasite relationships caused by the presence of alien hosts and/or helminths

H1-P1 type: The first cases of relationships between alien rodents and alien parasitic nematodes (e.g., *Callosciurus erythraeus-Brevistriata callosciuri*, *Myocastor coypus-Strongyloides myopotami*, *Rattus norvegicus-Orientostrongylus ezoensis*, *R. norvegicus-Nippostrongylus brasiliensis*, *Mus musculus-Heligmosomoides polygyrus*) indicated that intimate relationships in the places of origin (e.g., South-east Asia, South America) shifted directly into Japan and have been preserved constantly.

Up to now, these alien nematodes have not been known to parasitize Japanese native rodents (H2). However, a constant monitoring survey and an experimental infection in a laboratory with the nematodes to the native rodents should be performed in the future in order to observe whether or not the intimate host-parasite relationship shifts to the H1-P2 or H2-P1 types and/or is newly formed.

H1-P2 type: The accidental case of *Heligmosomoides kurilensis* (P2) obtained from *Rattus rattus* (P1) (Sakata et al. 2005) belongs to this type. Recently, after the immigration of coypus (H1), alien rodent nematode (*Calodium hepaticum*) (P2) that was likely native or old parasitized to coypus in Japan (type H1-P2). These helminths (P2) are native species which comprise part of the natural ecosystem in Japan. However, because coypus uses small waterways (vacant niches for Japanese endemic rodents) and transport these native nematodes to new sites, there seems to be the possibility that a new public or wild animal health problem will be provoked. Hence, even native nematodes could be pathogenic agents using alien hosts as new "vehicles."

H2-P1 type: An example of this type could be obtained from our survey, in a case in which *A. argenteus* (H2) had *Heligmosomoides polygyrus* (P1) in Hokkaido (Asakawa et al. 1994). This nematode species is a common parasite of *M. musculus*, which is an old alien rodent in Japan, and the nematode is well known to be pathogenic to wild and/or captive rodent hosts. In the UK, the nematode species has been obtained from the small intestine of two captive common dormice (*Muscardinus avellanarius*). The total numbers of nematode was ca. 600 to 900 in each animal (Asakawa et al., 2006). It was surmised that the dormice became infected through the intake of diet contaminated with infective larvae likely derived from the feces of house mice. Hence, some host-shifts to type H2-P1 could be dangerous to wild animal health.

H2-P2 type: Except for accidental cases caused by artificial disturbances to the essential habitat of the host species, the H2-P2 type belongs to the natural ecosystem formed by the evolutionary process in Japan. Further, this type of host-parasite relationship should be regarded as an aim of conservation biology, and a top priority to be considered when designing potential strategies is recognition that an endemic host-parasite relationship creates one type of natural biodiversity.

Hence, we must, as potential strategies for the risk reduction of the natural ecosystem in Japan, discriminate the host-parasite relationships derived from alien species (H1-P1, H1-P2 and H2-P1 types) and those from this endemic one (H2-P2 type), and perform monitoring surveys with respect to each type.

Conclusion

We have been investigating nematodes of native and alien rodents that occur within the Japanese Islands because some nematodes derived from alien hosts could possibly become highly pathogenic agents to human and livestock, and because the parasites potentially could also cross species barriers and thus have a negative impact on native host-parasite relationships that have evolved as a result of adaptive radiation or co-speciation between the native rodents and parasitic nematodes. In this paper we have generalized patterns of the native and alien nematode fauna formation, providing an overview of four types of host-parasite relationships: 1) alien rodents and alien nematodes between *Callosciurus erythraeus* and *Brevistriata callosciuri*, between *Myocastor coypus* and *Strongyloides myopotami,* between *Rattus norvegicus* and *Orientostrongylus ezoensis*, between *R. norvegicus* and *Nippostrongylus brasiliensis*, and between *Mus musculus* and *Heligmosomoides polygyrus*; 2) alien rodents and native nematodes between *Rattus rattus* and *Heligmosomoides kurilensis*, shifted from *Apodemus speciosus*; 3) native rodents and alien nematodes between *Apodemus argenteus* and *Heligmosomoides polygyrus*, shifted from *Mus musculus*, and between captive *Muscardinus avellanarius* and *H. polygyrus*; and 4) native rodents and native nematodes between *Apodemus* spp. and *Heligmosomoides* spp. etc. Hence, aberrant host-parasite relationships due to the presence of alien rodents and/or nematodes in Japan comprise groups 1), 2) and 3), and, further, potential strategies for the reduction of risk affected by these relationships should be performed for each group.

Acknowledgements

The present survey is was supported in partly by both Grant-in-Aid (No. 23570120) from the Japan Society for the Promotion of Science, and by the Support Project to Assist Private Universities in Developing Bases for Research from the Ministry of the Education, Science and Culture of Japan, respectively.

References

Abe, H. (ed.) 2005. A Guide to the Mammals of Japan. Tokai University Press, Kanagawa, Japan, 206 pp.

Asakawa, M. 2005. Perspectives of host-parasite relationships between rodents and nematodes in Japan. *Mammal Study,* 30: S95-S99.

Asakawa, M. 2010. Ecotourism with utilization of wild animals - Its impact on conservation medicine and risk assessment in Hokkaido, Japan. In: (Eds. Krause, A. and Weir, E.) Ecotourism: Management, Development and Impact, Nova Science Publishers, Inc., New York: 227-240.

Asakawa, M., Baba K., Fukumoto, S-I., Kudo, N. and Abe, H. 1994. Faunal study on the parasitic nematodes of rodents in southern part of Hokkaido and northern part of Honshu,

Japan. *Bulltin of Biogeographical Society of Japan,* 49: 51-59 (in Japanese with English abstract).

Asakawa, M., Hasegawa, H., Onuma, M., Tatsushima, T. and Ohbayashi, M. 1992. Parasitic nematodes of rodents on the offshore islands of Hokkaido. *Japanese Journal of Parasitology*, 41: 40-41.

Asakawa, M., Onuma, M, Tatsushima, T., and Ohbayashi, M. 1991. A morphological study of the taenid larva, armathyridium, from the Japanese wood mice, *Apodemus argenteus*, on Naka-jima Island of Toya Lake, Hokkaido. *Journal of Rakuno Gakuen University, Natural Science,* 16: 9-13 (in Japanese with English summary).

Asakawa, M., Sainsbury, A. W. and Sayers, G. 2006. Nematodosis caused by *Heligmosomoides polygyrus* (Dujardin, 1845) (Nematoda: Trichostrongyloidea: Heligmosomidae) in the common dormouse (*Muscardinus avellanarius*). *Veterinary Record.* 158: 667-668.

Asakawa, M., Sato, M., Sone, K., Tatsuzawa, S. and Oda, S. 2009. Further helminthological survey on alien rodents, coypu (*Myocastor coypus*: Myocastoridae) in Aichi and Hyogo Prefectures, Jappan. *Journal of Rakuno Gakuen University, Natural Science,*32: 291-292. (In Japanese with English summary).

Asakawa, M., Tenora, F. and Borkovcova, M. 2004. Notes to the Nematoda of the genus *Heligmosomum* Railliet et Henry, 1909 (Nematoda, Heligmosomidae) parasitizing Rodentia in Japan. *Acta Univ. Agric. Silvic. Mendel. Bruno*, 52: 23-26.

Asakawa, M., Yokoyama, Y., Fukumoto, S.-I. and Ueda, A. 1983. A study of the internal parasites of *Clethrionomys rufocanus bedfordiae* (Thomas). *Japanese Journal of Parasitology*, 32: 399-411.

Hasegawa, H. and Asakawa, M. 2003. Parasitic helminth fauna of terrestrial vertebrates in Japan. In: (Eds. Otsuru, M., Kamegai, S. and Hayashi, S.) Progress of Medical Parasitology in Japan, Volume 7, Meguro Parasitological Museum, Tokyo: 129-145.

Haukisalmi, V., Asakawa, M. and A. Gubanyi. 2008. The status of the genus *Hokkaidocephala* Tenora, Gulyaev and Kamiya, 1999 (Cestoda: Anoplocephalidae), parasites of the endemic Japanese field mice (*Apodemus* spp.). *Zootaxa,* 1925: 62-68.

Matsudate, H., Miyoshi, Y., Tamura, N., Murata, K., Maruyama, S., Kimura, J., Nogami, S., Maeda, K., Fukumoto, Y., Akasako, R. and Asakawa, M. 2003. A survey of the parasitic helminths of alien rodents (Belly-banded squirrel *Callosciurus erythraeus* and nutria *Myocastor coypus*) in Japan. Japanese Journal of Zoo and Wildlife Medicine, 8: 63-67 (in Japanese with English abstract).

Ohdachi, S. D., Ishibashi, Y., Iwasa, M. A. and Saitoh, T. (eds.) 2009. The Wild Mammals of Japan, Shokadoh, Koyo, Japan: 544 pp.

Ono, Y., Ueyama, T., Sashika, M., Murakami, T., Tsukada, H., Masuda, Y., Okada, H., Taniyama, H., Ohtaishi N. and Asakawa, M. 2005. Preliminary report of mammalian helminthes in Shiretoko Peninsula, Japan. In (Ed. IX IMC Organizing Committee) Proceedings of the 9th International Mammalogical Congress, Hokkaido University, Sapporo, Japan: 337 pp..

Onuma,M., and Asakawa, M. in preparation. First detection of larval taenid species (*Taenia taeniaeformis*) from two individuals of Ryukyu long-furred rats, *Diplothrix lagata* (Thomas, 1906) on Okinawa Islands, Japan.

Sakata, K., Takada, Y., Kageyama, M., Tenora F. and Asakawa, M. 2005. Parasitic helminthes obtained from the genus *Apodemus* (Muridae: Rodentia) collected on the Oki

Islands, Shimane Pref., Japan, with a new host record of *Heligmosomoides kurilensis* from *Rattus rattus*. *Biogeography,* 7: 97-102.

Shingaki, T., Sakata, K., Takada, Y., Tamura, N., Tsuji, M., Hasegawa, H. and Asakawa, M. 2004. A preliminary survey on the parasitic helminths of alien and endemic rodents on Fukue-jima I., Goto Is., Japan. In: The 10th Meeting of Japanese Society of Zoo and Wildlife Medicine (Program and Abstracts), The University of Tokyo: 90 (in Japanese).

Tsuji, M., Zamoto, A., Kawabuchi, T., Kataoka, T., Nakajima, R., Asakawa, M. and Ishihara, C. 2006. *Babesia microti*-like parasites detected in Eurasian red squirrels (*Sciurus vulgaris orientis*) in Hokkaido, Japan. *Journal of Veterinary Medical Science,* 68: 643-646.

Zamoto, A., Tsuji, M., Kawabuchi, T., Wei, Asakawa, M. and Ishihara, C. 2004. U.S.-type Babesia microti isolated from small wild mammals in eastern Hokkaido, Japan. *Journal of Veterinary Medical Science*, 66: 919-926.

In: Nematodes
Editors: F. Boeri and J. A. Chung
ISBN: 978-1-61470-784-4

Chapter XII

Infection by Anisakid Nematodes in Two Commercially Important Fish Species from Baja California, Mexico (East Pacific)

María Amparo Rodríguez-Santiago[1*] and Jorge Adrián Rosales-Casián[2]

[1] Facultad de Ciencias Marinas, Universidad Autónoma de Baja California, Km 106 carretera Tijuana-Ensenada, C.P. 22800 Ensenada, BC, México.

[2] División de Oceanología, Departamento de Ecología Marina, Centro de Investigación Científica y de Educación Superior de Ensenada BC, carretera Ensenada-Tijuana No. 3918, C.P. 22860 Ensenada, BC, México.

Abstract

Nematode parasites of marine fishes have been recognized as an important component of global biodiversity, but also some species represent a potential risk for human health. The aim of the present study was to assess the changes with time in abundance and prevalence of anisakid nematodes in two commercially important fish species (the red rockfish *Sebastes miniatus*, and the ocean whitefish *Caulolatilus princeps*) from Baja California, Mexico. A total of 301 fish were analyzed (210 of *S. miniatus* and 91 of *C. princeps*) from February to December 2005. The nematode species were identified by the traditional morphological methods. Three nematode species belonging to the genera *Anisakis*, *Hysterothylacium* and *Pseudoterranova* (Anisakidae) were found, which showed certain specificity for host organ. The most abundant species was *Hysterothylacium* sp. (*S. miniatus* = 37 ind/host, and *C. princeps* = 15.5 ind/host), and the parasitic prevalence was higher in *Anisakis* sp. (*S. miniatus* = 91.8% and *C. princeps* = 95%). The group of nematodes showed a significant change in abundance

* corresponding author e-mail: amparor@cicese.mx

with the time in *S. miniatus* (KW, $H = 13.05$, $p = 0.023$), and *C. princeps* (KW, $H = 23.37$, $p = 0.000$), with a marked decrease during winter. Larval stages of *Anisakis* and *Hysterothylacium* were particularly important due to their high abundance and prevalence and because they represent a human health risk (anisakiasis). In this respect, this chapter also provides an illustration of the species found, its location within the host and some recommendations to prevent infection in people.

Introduction

Nematodes of the family Anisakidae are one of the most common parasites of marine organisms, and part of their life cycle can be present in a wide variety of crustaceans, fishes, cephalopods and marine mammals (Navone et al. 1998; Purivirojkul 2009). The occurrence of anisakid larvae on these organisms is of great importance for human health (Ancillo et al. 1997, Audicana et al. 2002, Eguia et al. 2003, Gomez et al. 2004), and for economic (Inoue et al. 2000) and ecological reasons (Konisi and Sakurai 2002; Klimpel et al. 2004, 2007; Podolska et al. 2006). Infection in humans by these parasites (anisakiasis) can be accidental by ingesting raw marine fish and invertebrates infested with third stage larvae (Anderson 1992; Szostakowksa et al. 2002; Doupé et al. 2003). This disease is characterized by acute stomach infections, allergic reactions or anaphylaxis. In this respect, although the infection by anisakid species on marine fish has been widely reported in several studies worldwide (Van Thiel 1962), information about nematodes in commercially important fish species from the Mexican Pacific coasts is still scarce at present time. Some of these studies include pelagic (Alejo-Plata 2003) and benthic species (Castillo-Sánchez et al. 1998).

In the northwestern coasts of Mexico the ocean whitefish *Caulolatilus princeps* Jenyns, 1842 (Malacanthidae) and the red rockfish *Sebastes miniatus* Jordan and Gilbert 1880 (Scorpaenidae) are two common target fish species that are caught all year by the coastal commercial and the recreational fishing (Siri-Chiesa and Moctezuma-Hernández 1989; Elorduy-Garay and Ruiz-Córdova 1998; Rosales-Casián and Gonzáles-Camacho 2003; Rodriguez-Santiago and Rosales-Casián 2008). The ocean whitefish inhabits in rocky reefs from 10 to 150 m depth (Hammann and Rosales-Casián 1990) and its distribution extends from Vancouver Island in British Columbia (Canada) to Peru, including the Galapagos Islands (Dooley 1978). Moreover, the red rockfish is found from Prince William Sound, south Alaska to central Baja California, and inhabits rocky reefs, kelp forests and canyons at depths of 15–467 m, but commonly at 50-150 m (O'Connell et al. 1992; Love et al. 2002). However, in spite of their economical importance, no one study describing nematodes in *S. miniatus* and *C. princeps* at Baja California has been realized.

Therefore, the aim of the present chapter was to identify (by the traditional morphological methods) and to describe the nematodes of both fish species *S. miniatus* and *C. princeps* from the coasts of San Quintín, Baja California, Mexico. Likewise, data on prevalence and abundance of these parasites throughout the year were presented as well as some recommendations to avoid infections by these nematodes.

Monitoring and Identification of Nematodes

The two fish species were sampled bimonthly from February to December 2005, and individuals were obtained from the sport-fishing catches at the coast of San Quintín, Baja California, México (30° 24’ 58’’ N; 115° 58’ 51’’ W). San Quintín is located at 310 km south of the California border (USA) (Figure 1), and six kilometers of dirty road to El Molino Viejo (Old Mill), the site of departure and arrival of boats.

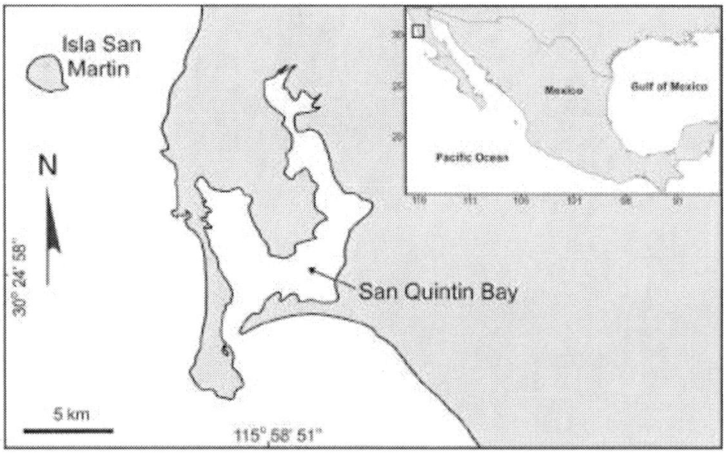

Figure 1. Location of the study area at San Quintin Bay, Baja California, Mexico.

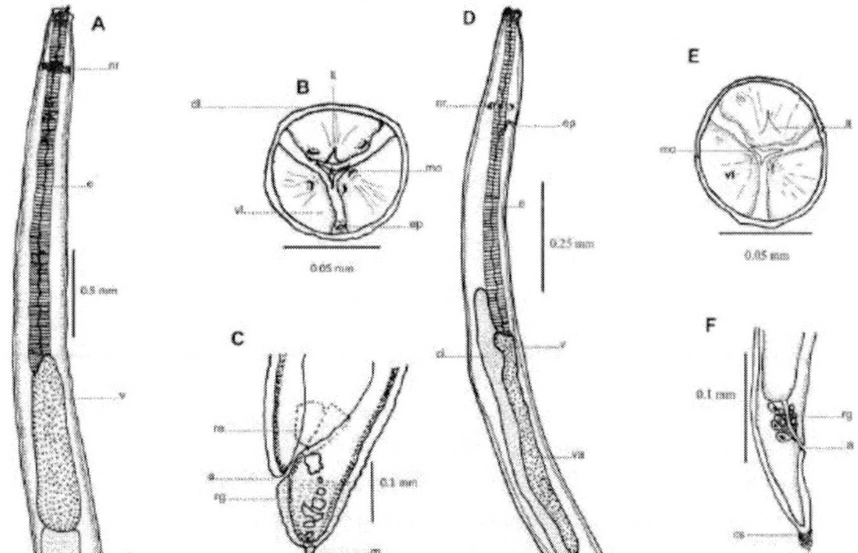

Figure 2. Larval stage of *Anisakis* sp. (A-C) and *Hysterothylacium* sp. (D-F), parasites of *Caulolatilus princeps* from San Quintin Bay, Mexico. A and D) anterior region, lateral view; B and E) cephalic end, apical view; C and F) caudal end, lateral view. (lt) larval tooth; (e) esophagus; (v) ventriculus; (i) intestine; (nr) nerve ring; (ep) excretory pore; (rg) rectal glands; (a) anus; (re) rectum; (m) mucron, (dl) dorsal lip, (lv) ventral lip, (mo) mouth; (ic) intestinal cecum; (va) ventricular appendix; (cs) caudal spines.

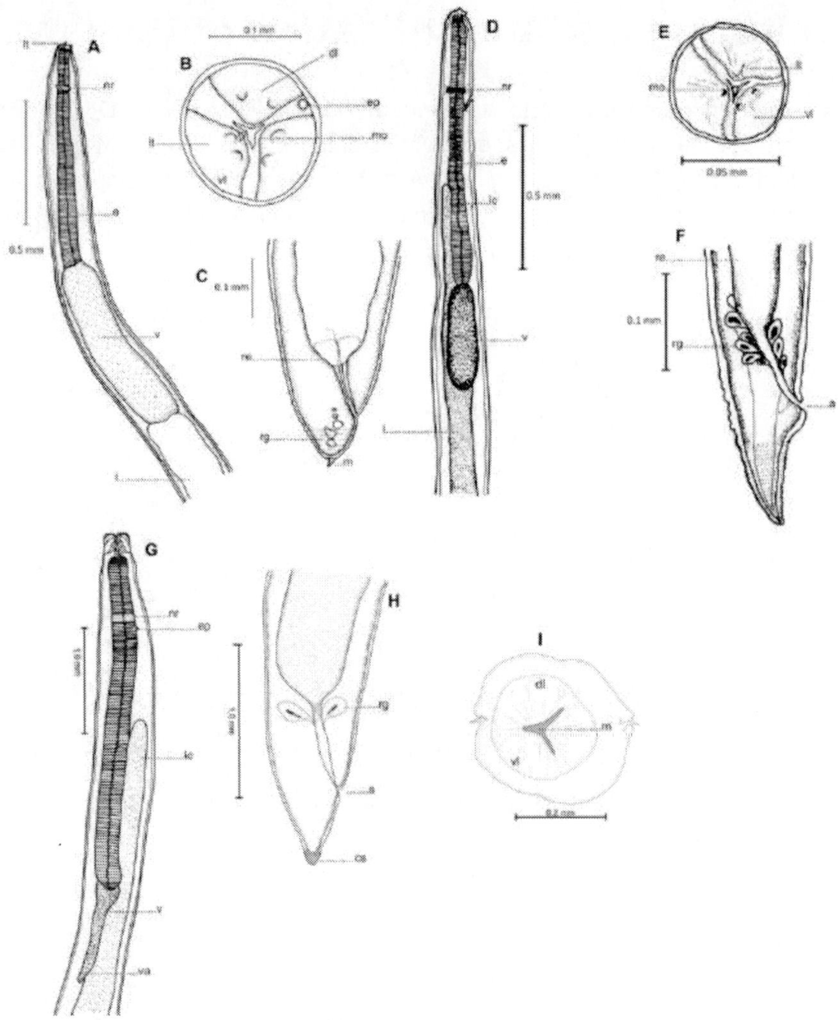

Figure 3. Larval stage of *Anisakis* sp. (A-C), *Pseudoterranova* sp. (D-F) and *Hysterothylacium* sp. (G-I), parasites of *Sebastes miniatus* from San Quintin Bay, Mexico. A, D and G) anterior region, lateral view; B, E and I) cephalic end, apical view; C, F and H) caudal end, lateral view. (lt) larval tooth; (e) esophagus; (v) ventriculus; (i) intestine; (nr) nerve ring; (ep) excretory pore; (rg) rectal glands; (a) anus; (re) rectum; (m) mucron, (dl) dorsal lip, (vl) ventral lip, (mo) mouth; (ic) intestinal cecum; (va) ventricular appendix; (cs) caudal spines.

This activity is guided to many pelagic and bottom fish species (Rosales-Casián and Gonzalez-Camacho 2003; Rodríguez-Santiago and Rosales-Casián 2008), from surface to 150 m depth, and as far as 50 km from Punta Entrada, a rocky point outside of the bay. A total of 91 specimens of the Ocean whitefish *C. princeps* and 210 of the red rockfish *S. miniatus* were revised. Fish specimens were dissected, and organs stored in plastic bags on ice. In the laboratory, internal organs (gills, liver, spleen, intestine, pyloric cecum, heart, gonads, and digestive tract) were examined under a stereoscopic microscope, and all parasites were removed. Identification of parasites was made using morphological data and identification keys (Yamaguti 1961; Anderson et al. 1974-1983).

The examined specimens of *C. princeps* ranged from 370-510 mm total length (TL) and those of *S. miniatus* from 280-610 mm TL. A total of 2,575 nematodes were found in *C.*

princeps belonging to the species *Anisakis* sp. Railli and Henry (Figure 2A-C), 1912 and *Hysterothylacium* sp. Railli and Henry (Figure 2D-F). These nematodes showed certain specificity for host organ and were found in the third larval stage (L3), only. The nematode *Anisakis* sp. was found only in mesentery, whereas larvae of *Hysterothylacium* sp. were found in mesentery, stomach, intestine, and cecum (Table 1).

The individuals of *S. miniatus* ranged from 280-610 mm TL, and a total of 10,323 nematodes were found belonging to the species *Anisakis* sp. (Figure 3A-C), *Pseudoterranova* sp. Railli and Henry (Figure 3D-F) and *Hysterothylacium* sp. (Figure 3G-I).

The nematodes found in *S. miniatus* showed less organ specificity than those in *C. princeps*. They were found in almost all internal organs of the host (coelomic cavity, stomach wall, intestine, mesentery and intestinal cecum) (Table 1).

Table 1. Characterization of nematode infections of *Caulolatilus princeps* and *Sebastes miniatus* from the coasts of San Quintin, Baja California, Mexico. Total number of nematodes per taxa (TNP), average prevalence (%P), and mean abundance (MA) of parasites (± SE). Localization (L) in the host body; G = gills, IC = intestinal cecum, S = stomach, I = intestine, M = mesentery, CC = coelomic cavity, Stw = stomach wall

Parasites	TNP	%P	MA	L
C. princeps				
Anisakis sp.1	1113	95.0 ± 5.5	11.8 ± 2.5	M
Hysterothylacium sp.1	1462	90.4 ± 10.0	15.5 ± 3.1	M,S,I,IC
S. miniatus				
Anisakis sp.2	3,942	91.8 ± 3.8	15.0 ± 3.9	CC, Stw, I, M
Hysterothylacium sp.2	6,373	60.1 ± 14.1	37.9 ± 18.0	M, S, I
Pseudoterranova sp.	8	2.9 ± 1.5	0.2 ± 0.2	S, M, I, IC

Abundance and Prevalence of Nematodes

Abundance (number of parasites per host) and prevalence (% of number of parasite hosts/number of revised hosts) of nematodes was determined according to Margolis et al. (1982). In *C. princeps* the most abundant nematode species was *Hysterothylacium* sp. (1,462 individuals) and presented an overall mean abundance of 16.5 ± 3.5 ind/host, whereas *Anisakis* sp. (1,113 individuals) showed an overall mean abundance of 12.4 ± 4.7 ind/host). The abundance of *Hysterothylacium* sp. was higher in June (21.6 ind/host) and lower in February (7.8 ind/host). Similarly, the species *Anisakis* sp. showed the highest abundance in April (16.4 ind/host) and lowest in February (5.5 ind/host) (Figure 4a). The mean abundances of *Anisakis* sp. showed significant changes (Kruskall-Wallis, H = 11.05, $p < 0.05$) with the time.

In *S. miniatus* the most abundant nematode was *Hysterothylacium* sp. with a total of 6,373 individuals and an overall mean abundance of 45.4 ± 26.7 ind/host (Figure 4c), followed by *Anisakis* sp. (total = 3,942 individuals, overall mean abundance = 14.4 ± 3.9 ind/host). The abundance of *Hysterothylacium* sp. was higher in August (113.9 ind/host) and lower in April (13.7 ind/host), while *Anisakis* sp. showed the highest abundance in June (24.1

ind/host) and lowest in August (8.2 ind/host) (Figure 4c). The abundance of *Pseudoterranova* sp. was below 1.0 ind/host when it was present (June-August). The mean abundances of *Anisakis* sp.2 showed significant changes (Kruskall-Wallis, H = 21.14, $p < 0.05$) with the time.

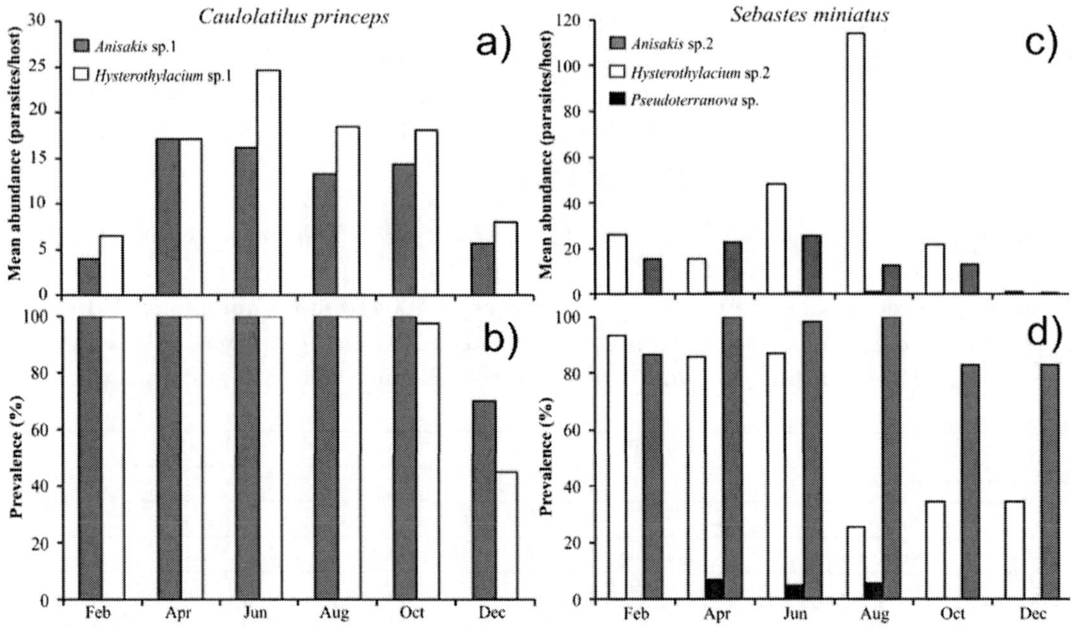

Figure 4. Mean abundance and prevalence of nematodes in *Calulatilus princeps* and *Sebastes miniatus* from San Quintin Bay, Mexico.

Nematodes in *C. princeps* showed in general higher prevalences than those of *S. miniatus* (Figure 4b, d). From February to October, larvae of *Anisakis* sp. (of *C. princeps*) showed a prevalence of 100% and a decrease (72%) in December (Figure 4b). Similarly, *Hysterothylacium* sp. exhibited prevalences of 100% from February to August, followed by a slight decrease in October (95%), and an abrupt diminution in December (50%). In *S. miniatus*, *Pseudoterranova* sp. was only present during spring (6%) and summer (5.7%). Prevalence of *Anisakis* sp. was higher in summer (100%) and lower in autumn (82%), while in *Hysterothylacium* sp. prevalence was higher during spring (86.4%) and lower in summer (25.7%) (Figure 4d).

According to Spearman rank correlation analyses, there were no relationships between abundance or prevalence of nematodes with the size or weight of the hosts (*C. princeps* and *S. miniatus*), or water temperature. However, the highest abundance of nematodes occurs between June and August when surface water temperature reach the highest values (Figure 5). Also the lower abundance and prevalence occurred in winter (February and December) when temperature was the lower of the year.

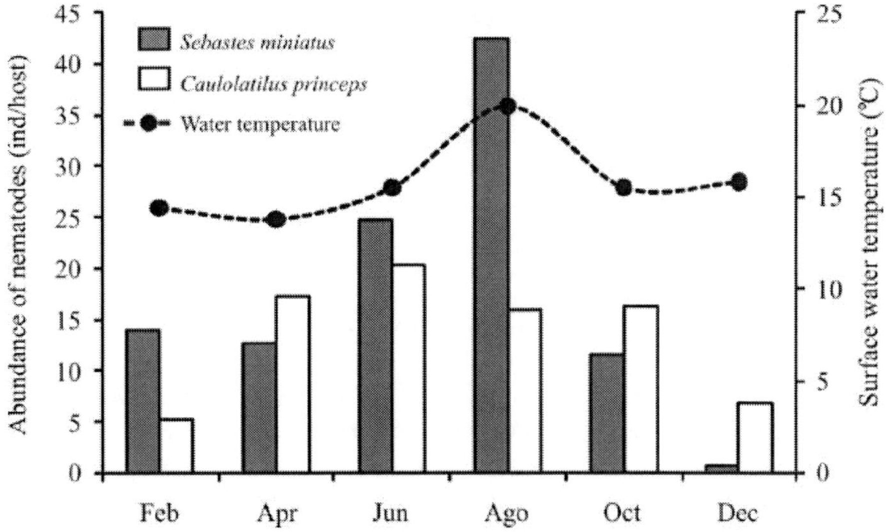

Figure 5. Interannual changes in nematode abundance in *Sebastes miniatus* and *Calulatilus princeps* and in the surface seawater temperature.

Conclusion

The ecological studies of nematodes in commercially important fish species the vermilion rockfish (*S. miniatus*) and the ocean whitefish (*C. princeps*) are important because are appreciated by their white meat, texture and flavor. These two fish species are abundant in the commercial fishing and sport-fishing from Baja California (México) and California (USA). In San Quintín, Baja California (México), these fish are captured all year by tourists or by the commercial fishing and the catch is transported to all big cities of Baja California state, and even the products may reach the California (USA) markets. In the area there is not a parasitological control of captured fishes, and the life cycle and food habits of these fishes are important for recruitment, development and parasite transmissions (Love et al. 2000).

The parasitological record of nematodes for *S. miniatus* and *C. princeps* was represented by 5 species. Two genera of nematodes (*Anisakis* sp. and *Hysterothylacium* sp.) were shared between the two hosts. In the present study, the red rockfish (*S. miniatus*) constitutes a new host for *Hysterothylacium* sp. and the San Quintín coast is registered as a new geographical record for the species of *Pseudoterranova* sp. Respect to the ocean whitefish (*C. princeps*), it is registered as new host for *Anisakis* sp. and *Hysterothylacium* sp., and San Quintín constitutes a new geographic locality for these species too.

The presence of these nematode species in the red rockfish and the ocean whitefish, may constitute a health problem, particularly by eating raw or inadequately cooked fish (Jong-Yil 1986). In our country, a common practice is to prepare "ceviche" that is raw meat of fish with lemon juice. This food is commonly prepared with *S. miniatus* and *C. princeps* by its characteristic white meat. However, given the high abundance of nematodes that cause anisakiasis; this could be an infection risk.

Recommendations

It's important to avoid the consumption of raw or inadequately cooked fish. Also, is priority to increase the parasite investigations about others commercial fishes from the California's region, and to maintain a sanitary control of the fish. A program about health education about parasite risk infections will be also important.

References

Alejo-Plata, M. C., Cerdenares-Ladrón de Guevara, G. and González-Medina, G. (2003). Larvas de *Anisakis* sp. (Nematoda: Anisakidae), presentes en el pez dorado (*Coryphaena hippurus*), en las costas de Oaxaca, México. *Ciencia y Mar*, 7, 45-49.

Ancillo, M., Cabalerro, M. T., Cabanas, R., Contreras, J., Baroso, J. A. M., Barranco, P. and Serrano, M. C. L. (1997). Allergic reactions to *Anisakis simplex* parasitizing seafood. *Annals of Allergy, Asthma and Immunology*, 79, 246-250.

Anderson, R. C., Chabaud, A. G. and Willmott S. (1974-1983) CIH keys to the nematode parasites of vertebrates. *Commonwealth Agricultural Bureaux*, Farnham Royal, Bucks: England, vol. 1-10.

Anderson, R. C. (1992). Nematode parasites of vertebrates. Their development and transmission. C.A.B. International, Cambridge, U. K. 578.

Audicana, M. T., Ansotegui, I. J., Corres, L. F. and Kennedy M. W. (2002). *Anisakis simplex*: dangerous-dead and alive?. *Trends in Parasitology*, 18, 20–25.

Castillo-Sánchez, E., Rosales, A. and Pérez-Ponce de León, G. (1998). Helmintos parásitos de (Osteichthyes: Paralichthydae) en el estero de Punta Banda, bahía de Todos Santos y bahía de San Quintín, Baja California, México. *Ciencias Marinas*, 24, 443-462.

Dooley, J. K. (1978). Systematics and biology of the tilefishes (Periformes: Branchisoterigidae and Malacanthidae) with description of two new species. NOAA Technical Report. *NMFS Circular*, 411, 25-27.

Doupé, R. G., Partridge, G. J. and Lymbery, A. J. (2003). Visible implant fluorescent elastomer tags as pedigree markers for applied aquaculture: an evaluation using black bream *Acanthopagrus butcheri*. *Aquaculture Research*, 34, 681-683.

Eguia, A., Aguirre, J. M., Echevarria, M. A., Martinez-Conde, R. and Ponton J. (2003). Gingivostomatitis after eating fish parasitized by *Anisakis simplex*: a case report. Oral Surgery, *Oral Medicine, Oral Pathology, Oral Radiology and Endodontics*, 96, 437-40.

Elorduy-Garay, J. F. and Ruiz-Córdova, S. S. (1998). Age, growth, and mortality of *Caulolatilus affinis* (Osteichthyes: Branchiostegidae) from the Southern Gulf of California. *Pacific Science*, 52, 259-272.

Gómez, A. A., Ancillo, A. M., Serrano, M. C., Parga, J. M. S., Daschner, A., Caballero, M. T., Barranco, P. and Cabanas R. (2004). *Anisakis simplex* only provokes allergic symptoms when the worm parasitizes the gastrointestinal tract. *Parasitology Research*, 93, 378–384.

Hammann, G. M. and Rosales-Casián, J. A. (1990). Taxonomía y estructura de la comunidad de peces del Estero de Punta Banda y Bahía de Todos Santos, B.C., México. In: Rosa-

Vélez, J. and González-Farías, F. (eds) Temas de Oceanografía Biológica en México, Universidad Autónoma de Baja California, Ensenada, 337.

Inoue, K., Oshima, S. I., Hirata T. and Kimura, I. (2000). Possibility of anisakid larvae infection in farmed salmon. *Fish Science*, 66, 1049–1052.

Jong-Yil, C., Yeon-Myung C., Woon-Mok, S. and Soon-Hyung, L. (1986). Larval anisakids collected from the yellow corvina in Korea. *Journal of Parasitology*, 24, 1-11.

Klimpel, S., Kellermanns, E., Palm, H. W. and Moravec, F. (2007). Zoogeography of fish parasites of the pearlside (*Maurolicus muelleri*), with genetic evidence of *Anisakis simplex* (s.s.) from the Mid-Atlantic Ridge. *Marine Biology*, 52, 725–732.

Klimpel, S., Palm, H. W., Ruckert, S., Piatkowski, U. (2004). The life cycle of *Anisakis simplex* in the Norwegian Deep (northern North Sea). *Parasitology Research*, 94, 1–9.

Konisi, K. and Sakurai, Y. (2002). Geographic variations in infection by larval *Anisakis simplex* and *Contracaecum osculatum* (Nematoda, Anisakidae) in walleye pollock Theragra chalcogramma stocks off Hokkaido, Japan. *Fish Science*, 68, 532-542.

Love, M. S., Caselle, E. and Snook, L. (2000). Fish assemblages around seven oil platforms in the Santa Barbara Channel area. *Fishery Bulletin*, 98, 96-117.

Love, M. S., Yoklavich, M. and Thorsteinson, L. (2002). The rockfishes of the Northeast Pacific. University of California Press, Los Angeles, 405.

Margolis, L., Esch, G. W., Holmes, J. C., Kuris, A. M. and Schad. G. A. (1982). The use of ecological terms in parasitology (report of an ad hoc committee of the American Society of Parasitologists). *Journal of Parasitolology*, 68, 131-133.

Navone, G. T., Sardella, N. H. and Timi, J. T. (1998). Larvae and adults of *Hysterothylacium aduncum* (Rudolphi, 1802) (Nematoda: Anisakidae) in fishes and crustaceans in the South West Atlantic. *Parasite,* 5, 127-136.

O'Connell, V. A., Gordon, D. A., Hoffmann, A., and Hepler, K. (1992). Northern range extension of the vermilion rockfish (*Sebastes miniatus*). *California Departament of Fish and Game*, 78, 173.

Podolska, M., Horbowy, J. and Wyszynski, M. (2006). Discrimination of Baltic herring populations with respect to *Anisakis simplex* larva infection. *Journal of Fish Biology*, 68, 1241–1256.

Purivirojkul, W. (2009). An investigation of larval ascaridoid nematodes in some marine fish from the Gulf of Thailand. *Kasetsart J.* (Natural Science), 43, 85–92.

Rodríguez-Santiago, M. A. and Rosales-Casián, J. A. (2008). Abundance and size composition of vermilion rockfish, *Sebastes miniatus* (Jordan and Gilbert 1880), from sport fishing catches of San Quintín, Ensenada, Baja California, México. *Bulletin of the Southern California Academy of Science*, 107, 25-32.

Rosales-Casián, J. A. and Gonzáles-Camacho, J. R. (2003). Abundance and importance of fish species from the artisanal fishery on the Pacific coast of northern Baja California. *Bulletin of the Southern California Academy of Science*, 102, 51-65.

Siri-Chiesa, M., Moctezuma-Hernández, P. (1989). La pesca de Baja California. Universidad Autónoma de Baja California, México, 211.

Szostakowska, B., Myjak, P. and Kur, J. (2002). Identification of anisakid nematodes from the southern Baltic Sea using PCR-based methods. *Molecular and Cellular Probes*, 16, 111–118.

Van Thiel, P. H. (1962). Anisakiasis. *Parasitology*, 52, 16–17.

Yamaguti, S. (1961). Systema Helminthum. III. Nematodes. Interscience Publishers Inc, New York, USA, 1261.

In: Nematodes
Editors: F. Boeri and J. A. Chung
ISBN: 978-1-61470-784-4

Chapter XIII

Confocal Scanning Laser Microscopy of *Lagochilascaris Minor* Eggs

***Mara Silvia Carvalhaes**[*1] **and Toshie Kawano**[2]*
[1] Institute of Tropical Pathology and Public Health (IPTSP), Federal University of Goiás (UFG), Postal Box 1031, 74605-050 Goiânia, Goiás, Brazil.
[2] Sector of Parasitology, Butantan Institute, São Paulo, Brazil, in memorium

Introduction

Lagochilascaris minor (Leiper, 1909) is a nematode only rarely encountered in animals and man that causes purulent abscesses in the subcutaneous tissue of the ear, neck, jaw, orbit, mastoid and retropharyngeal tissue of humans. Sometimes, *L. minor* invades the pulmonary tissue and central nervous system in fatal cases (Palheta Neto et al, 2002; Roig et al, 2010; Guimaraes et al, 2010).

The extraordinary capacity of *L. minor* to migrate across different human tissues can also be observed in animal models of the disease such as mice and cats (Semerene et al, 2004, Barbosa et al, 2005). In mice orally inoculated with *L. minor* infective eggs, hatched larvae can soon be observed migrating in the intestinal tract. After the process of hatching, third stage larvae (L3) migrate through intestinal mucosa reaching vessels and hepatic parenchyma and disseminate to other tissues such as lungs, skeletal muscles, subcutaneous tissues, renal parenchyma and pancreas (Semerene et al, 2004; Volcan et al, 1992). In cats inoculated with skinned carcasses of infected mice, L3 migrate through the esophagus, pharynx, trachea, rhino and oropharynx, and cervical lymph nodes (Campos et al, 1992).

The genus *Lagochilascaris* (Leiper, 1909) was taxonomically reviewed by Sprent (1971) who confirmed *L. buckleyi* from *Felis concolor* and *L. major* from the stomach, oesophagus

[*] Email: maracarvalhaes@bol.com.br

and trachea of a cat. Nowadays, we have five species recognized in the genus: *L. minor*, from cat and man; *L. major* from cat and lion; *L. turgida*, from opossums; *L. bucleyi*, from cougar; and *L. sprenti* from opossums (Sprent, 1982). The eggs of *L. minor* are report in the literature, as spheroid to ovoid, with maximum diameter 67 μm and minimal diameter 59 μm, with 15-26 pits on the outer circumference (Volcan et al, 1992). A difference exists in the measurement of the diameter and number of pits among several authors. The morphology of the eggs of *L. minor* was evaluated in all of the works described in the literature for common optic microscopy. Here, we evaluated the morphology of the parasite eggs for confocal microscopy without coloration, in different periods of culture in formalin .

Material and Methods

Parasite eggs were collected from the faeces of *Felis domesticus e*xperimentally infected with a human isolate of *L. minor*. Faeces from infected animals were submitted to Hoffman's method and kept in culture in formalin solution (1%) at room temperature for different periods (1 to 120 days). Always, cultures were submitted to Faust's method for optimal recovery of eggs free from faecal debris (Oliveira et al, 2002). Egg suspensions were exhaustively washed with phosphate buffered saline (pH 7.4).

The quantification of the number and measurement of pits was performed using a confocal microscope. The fixed parasites were analyzed under a confocal microscope (Laser Scanning Microscopy,LSM 510 META, Zeiss) at 488 nm (exciting) and 505 nm (emission) as described by (Moraes et al, 2009). A minimum of ten specimens were assessed of each culture in each different time of incubation. The measurement and number of pits were counted with the Zeiss LSM Image Browser software. Measurements were expressed as means and standard errors. Data were analyzed Student's t tests.

Results and Discussion

By confocal microscopy, the eggs of *L. minor* presented maximum diameter of 70.5 μm and minimal diameter of 44.0 μm (Figure 1A), with 12-25 pits on the outer circumference (Figure 1B), and each pit showed 14.0-5.0 μm of diameter (Figure 1 C). We can't observe significant difference among eggs collected in different days of culture. A difference exists in the number of pits in the total egg, that was smaller in day 1 of culture, when compared to other periods, and it varied from 83.0 to 125.0 pits/egg (Figure 1D). In the literature, the eggs of *L. minor* are report with maximum diameter 67 μm and minimal diameter 59 μm (Volcan et al, 1992), a little smaller than the ones measured by confocal microscopy. The number of pits on the outer circumference described in the literature is of 15-26 pits (Volcan et al, 1992), similar to that found by confocal microscopy. This first analyze of *L. minor* egg by confocal microscopy permitted a tridimensional image of the egg with the enumeration of total pits per egg (Figure 2 and 6), with the visualization of the larvae into the egg (Figure 3, 4 and 6), and of the larvae leaving the egg (Figure 5). It is interesting to stand out that the eggs of *L. minor* presented natural fluorescence in the confocal microscopy.

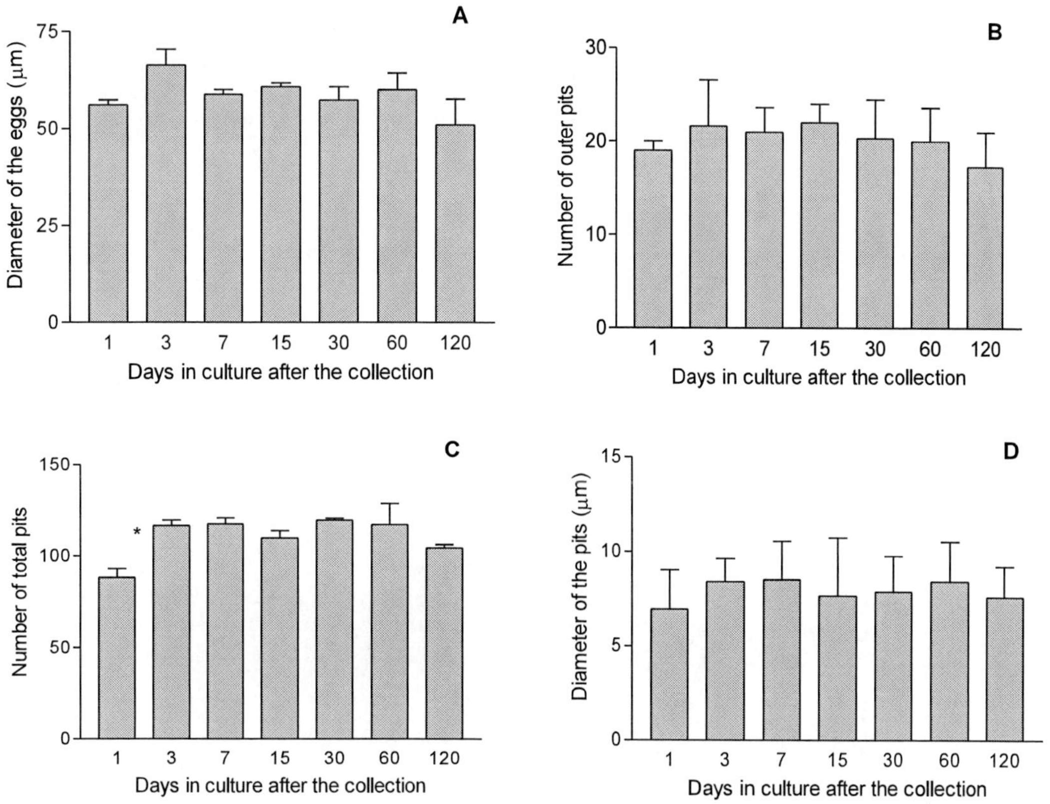

Figure 1. Analyze of *Lagochilascaris minor* eggs (ten/day in culture) by confocal microscopy. A) Diameter of the eggs; B) Number of out pits; C) Number of total pits; D) Diameter of the pits. The measurement and number of pits were counted with the Zeiss LSM Image Browser software. Data are show as means ± standart errors (* $p \leq 0.05$).

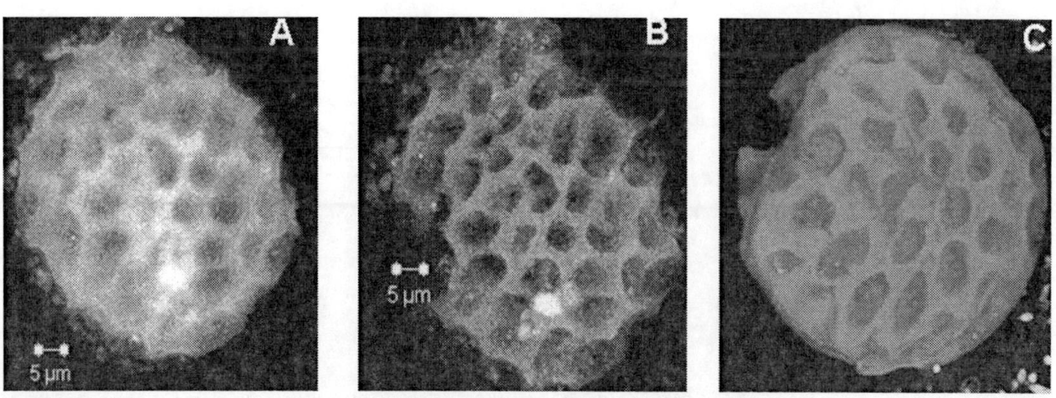

Figure 2. Tridimentional image of *Lagochilascaris minor* eggs denoting general morphology. A,B,C- Fluorescent confocal microscope image, scale of gray color, showing general view with pits (scan mode stack).

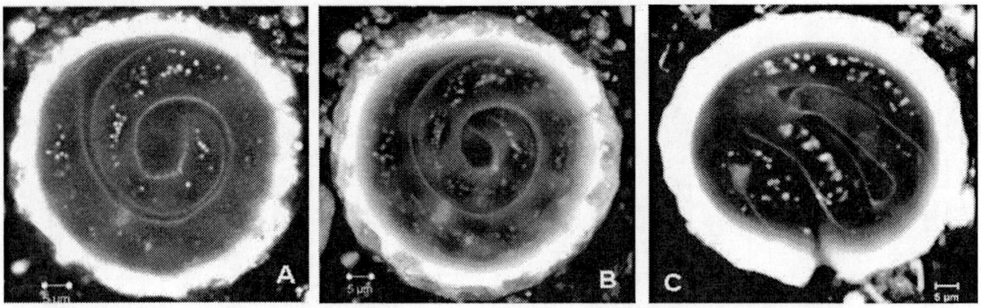

Figure 3. Confocal image of *Lagochilascaris minor* eggs denoting viable larvae into the eggs. A,B,C- Fluorescent confocal microscope image, scale of gray color, showing one larvae into the egg (scan mode plane).

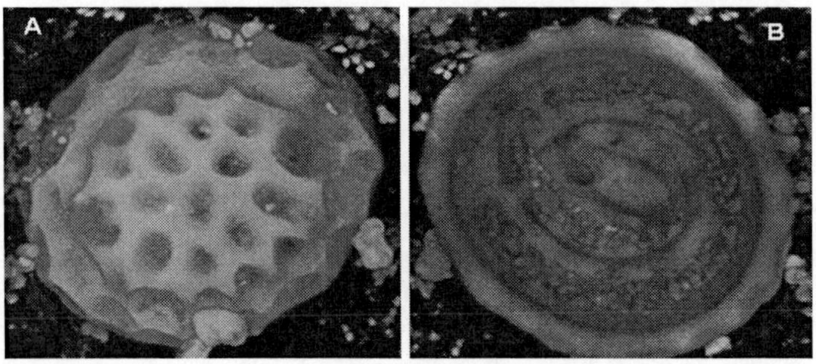

Figure 4. Confocal image of *Lagochilascaris minor* eggs denoting general view of the egg (A), with the corresponding larvae into the egg (B). Fluorescent confocal microscope image, scale of gray color. Scan mode stack (A) and plane (B).

Microscopy has a long and distinguished history in the study of helminth parasites and has made a singularly outstanding contribution to understanding how these complex animals organize their lives and relate to their hosts.

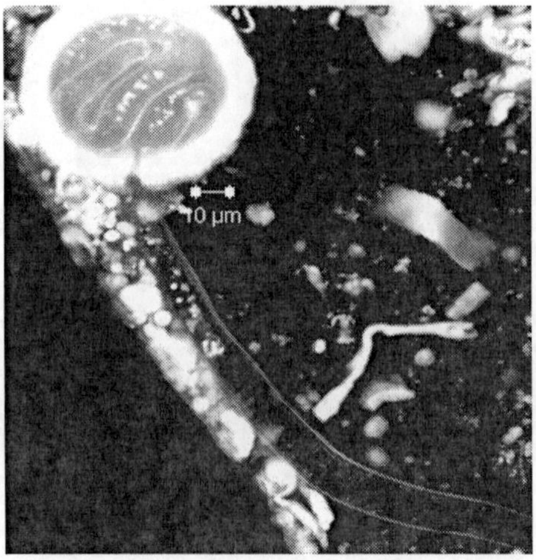

Figure 5. Confocal image of *Lagochilascaris minor* egg with a larvae leaving. Fluorescent confocal microscope image, scale of gray color (scan mode plane).

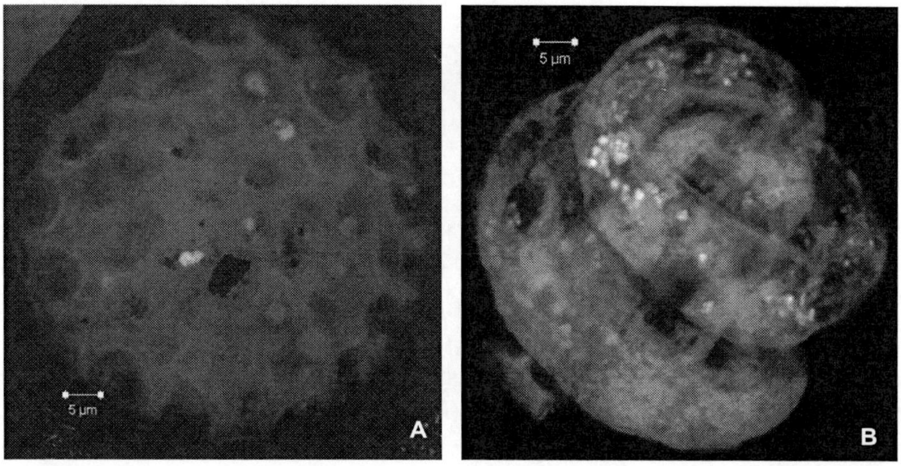

Figure 6. Confocal image of *Lagochilascaris minor* eggs denoting general view of the egg (A), with a detail of the corresponding larvae into the egg (B). Confocal microscope image in scale of gray color. Scan mode stack (A) and plane (B).

Parasite surfaces have understandably demanded most of the attention of microscopists, largely as a result of the pioneering studies using transmission electron microscopy. Their findings focused the attention of physiologists and immunologists on the tegument and cuticle of helminths and in doing so helped unravel the complex molecular exchanges that are fundamental to understanding host–parasite interactions. Scanning electron microscopy succeeded in augmenting these data by revealing novel microtopographical features of the host–parasite relationship, as well as proving invaluable in helminth taxonomy and in assessing the efficacy of test substances in drug screens.

Control of helminth parasites has never been more critical: problems of drug resistance demand urgent action to identify exploitable targets for new generation anthelmintics. In this regard, confocal scanning laser microscopy may be essential in detect the action of potential chemotherapeutic agents against *L. minor* eggs, with fantastic sensibility.

Acknowledgments

The authors thank Ms Denise Vilarinho Tambourgi and Mr. Alexsander S. Souza for his excellent technical assistance with confocal Microscopy. This study received financial support from the Conselho Nacional de Pesquisa (CNPq, Brazil), Fundação de Amparo á Pesquisa do Estado de São Paulo (FAPESP, Brazil), and Fundação de Apoio á Pesquisa do Estado de Goiás (FAPEG, Brazil).

References

Barbosa C.A., Barbosa A.P., Campos D.M. Gato doméstico (*Felis catus domesticus*) como possível reservatório de *Lagochilcascaris minor* Leiper (1909). *Rev. Patol. Trop*. 34: 251-253, 2005.

Campos D.M.B., Freire-Filha L.G., Vieira M.A., Paçõ J.M., Maia M.A. Experimental life cycle of *Lagochilascaris mino*r Leiper 1909. *Rev. Inst. Med. Trop. São Paulo* 34: 277-87, 1992.

Fraiha H, Leão RNQ, Costa FSA. Lagochilascaríase humana e dos animais domésticos. *Zoon Rev Inst.* 1:25-33,1989.

Guimaraes V.C., Barbosa A.P., Camargo L.A., Siqueira P.H., Silva Filho J.C., Castro V.L.S., Barbosa M.A., Campos D.M.B. *Lagochilascaris minor* otomastoiditis in child: case repost. *Int. Arch. Otorhinolaryngol.* 14:373-376, 2010.

Leiper RT. A new nematoda worm from Trinidad: *Lagochilascaris minor*. *Proc Zool Soc Lond.* 4:742-743, 1909.

Moraes J., Silva M.P., Ohlweiler F.P., Kawano T. *Schistosoma mansoni* and other larval trematodes in Biomphalaria tenagophila (Planorbidae) from Guarulhos, São Paulo State, Brazil. *Rev. Inst. Med. Trop. de São Paulo* 51, 77–82, 2009.

Oliveira J. A, Barbosa C. L., Vieira M. A, Paço J. M., Carrijo A. S., Fiorini M. P., Campos, D. M. B. Isolado de *Lagochilascaris minor*: procedimentos para obtenção de ovos infectantes. *Rev. Patol. Trop.*, 31: 121 – 128, 2002.

Palheta-Neto, Leão R.N.Q., Fraiha-Neto H. Tomita S., Tavares de Lima M.A.M., Pezzin-Palheta A.C. Contribuition to the study of human lagochilascariasis. *Rev. Bras. Otorrinolaringol.* 68: 101-105, 2002.

Roig J.L.R., Forteza J.L.R.C., Granato L., Serafini D.P. Ostomastoiditis with right retroauricular fistula by Lagochilascaris minor. *Braz. J. Otorhinolaryngol.* 76: 407, 2010.

Semerene A.R., Lino-Junior R.S., Oliveira J., Magalhães A.V., Stefani M.M., Barbosa A.P., Campos D.M.B. Experimental lagoshilascariosis: histopathological study of the inflammatory response to larval migration in the murine model. *Mem. Inst. Oswaldo Cruz*, 99: 393-398, 2004.

Sprent J.F.A. Speciation and development in the genus Lagochilascaris. *Parasitology* 62:71-112, 1971.

Sprent J.F.A. Ascaridoid nematodes of South American mammalis, with a definition of a new genus. *J. Helminthol.* 56: 275-295, 1982.

Volcan G.S., Medrano C.E., Payares G. Experimental heteroxenous cycle of Lagochilascaris minor in White mice and cats. *Mem. Inst. Oswaldo Cruz* 87: 525-532, 1992.

In: Nematodes
Editors: F. Boeri and J. A. Chung

ISBN: 978-1-61470-784-4

Chapter XIV

Pochonia chlamydosporia: Biological, Ecological and Physiological Aspects in the Host-Parasite Relationship of a Biological Control Agent of Nematodes

R. H. Manzanilla-López[*1], *I. Esteves*[2] *and M. M. Finetti-Sialer*[3]
[1] Plant Pathology and Microbiology Department,
Rothamsted Research, Harpenden, Herts, UK.
[2] IMAR-CMA, Department of Life Sciences,
Faculty of Sciences and Technology,
University of Coimbra, Coimbra, Portugal.
[3]CNR-IGV, Consiglio Nazionale delle Ricerche,
Istituto di Genetica, Vegetale, Bari, Italy.

Abstract

The nematophagous fungus *Pochonia chlamydosporia* (Clavicipitaceae) can parasitize eggs of nematodes including cyst (*Globodera* spp., *Heterodera* spp.), root-knot (*Meloidogyne* spp.), false-root knot (*Nacobbus* spp.) and reniform (*Rotylenchulus reniformis*) nematodes. Its potential as a biological control agent has been the subject of numerous studies aimed at understanding the micro-ecological conditions, including the

* Plant Pathology and Microbiology Department, Rothamsted Research. Harpenden, Herts. AL5 2JQ, UK. rosa.manzanilla-lopez@rothamsted.ac.uk

tritrophic (*e.g.*, plant, fungus, nematode) and host-parasite relationships that allow the fungus to thrive in the soil and rhizosphere environment. In order to effectively exploit the fungus in regulating plant endo-parasitic nematodes, a careful selection of fungal isolates (biotypes) appropriate for both host plant and nematode is essential. *Pochonia* survives in soil in the absence of a nematode host and, although it behaves as a saprophyte, research evidence points to a physiological 'switch' from the saprophytic to the parasitic stage that is triggered by nutrition. Fungal chlamydospore-based formulations for application to soil as an inoculum have been shown to be viable commercial formulations. The basic biology and physiology of the fungus is herein reviewed to provide new insights into the host-parasite interaction and commercial production methodology.

I. Introduction

The nematophagous species of the genus *Pochonia* Batista & Fonseca mainly parasitize eggs of sedentary plant-parasitic nematodes including cyst (*Heterodera*, *Globodera*), root-knot (*Meloidogyne*), false root-knot (*Nacobbus*) and reniform (*Rotylenchulus reniformis*) nematodes, with the exception of *P. microbactrospora*, which was isolated from a rotifer (Zare *et al.*, 2001). *Pochonia* has been isolated worldwide from forest, grassland, savannah, arable and garden soils, as fungal spores and as a parasite of eggs of various invertebrates such as molluscs (Domsh & Gams, 1972). *Pochonia chlamydosporia* and its teleomorph (*Cordyceps chlamydoporia*) have been reported to grow on slug eggs (Zare *et al.*, 2001). Other potential hosts of *P. chlamydosporia* include helminth eggs (Fassatiová & Lysek, 1982; Araújo *et al.*, 2009) and eggs of other animal-parasitic nematodes (Braga *et al.*, 2010; Frassy *et al.*, 2010; Silva *et al.*, 2010) in which the ovicidal activity of the fungus has been demonstrated.

Pochonia chlamydosporia (Clavicipitaceae) is one of the most studied biocontrol agents of plant endo-parasitic nematodes. Several factors have contributed to target this fungus for nematode biocontrol. They include the saprotrophic behaviour and permanence in soil in the absence of both plant and nematode hosts, its worldwide distribution, natural occurrence in nematode suppressive soils, laboratory culturing, access to isolates (fungal collections) and effectiveness in nematode control. These factors, and the need to find alternatives to nematicides, have lead to the screening and testing of numerous isolates to find potential biocontrol agents against *Heterodera*, *Globodera*, *Meloidogyne*, *Nacobbus* and *Rotylenchulus* as shown by the increasing number of *Pochonia* native isolates reported (Hidalgo-Díaz *et al.*, 2000; Olivares-Bernabeu & Lopez-Llorca, 2002; Bourne *et al.*, 2004; Wang *et al.*, 2005; Flores-Camacho *et al.*, 2007; Franco-Navarro *et al.*, 2009; Nagesh *et al.*, 2007). However, only a few isolates are commercially available. In this chapter, we will present aspects of the biology of the fungus, ecological and physiological factors of the tritrophic interaction (plant, fungus and nematode), and how they affect the effectiveness of *P. c.* var. *chlamydosporia* as a biological control agent and its use as a commercial biopesticide. Relevant information on other species and varieties of *Pochonia* is included for comparison.

II. Morphology and Vegetative Structures (Zare *et al*., 2001)

Colonies on agar are white or whitish to ochre yellow and cottony with the same colour on the reverse side and can reach 15-40 mm diam. in 10 days. Conidiophores are prostrate and little differentiated from vegetative hyphae, or erect and differentiated, verticillate or solitary. Conidia adhere to form globose heads or chains, subglobose, ellipsoidal to rod-shaped, isodiametric-polyhedric or falcate with blunt ends, depending on the fungus subspecies, with size varying between 3-4 μm by 1.5-2.0 μm (Gams, 1988). Dictyoclamydospores with diameter varying between 15 to 30 × 10-20 μm are usually formed on short stalks and produced on the surface of the colony or submerged in the agar, and crystals absent (Zare *et al.*, 2001). Uninucleate conidia and crystals can be present in some isolates of *P. c.* var. *chlamydosporia* (Manzanilla-López, unpublished).

The diagnosis of the genus, the description of *P. chlamydosporia* teleomorph (*Cordyceps chlamydosporia*) as well as a taxonomic key to *Pochonia* species, can be found in Zare *et al.* (2001). *Pochonia chlamydosporia* has two varieties: *P. chlamydosporia* (Goddard) Zare & W. Gams var. *chlamydosporia* and *P. chlamydosporia* var. *catenulata* (Kamyschko ex Barron & Onions) Zare & W. Gams; the latter variety produces conidia in chains and partly in heads (Zare *et al.*, 2001).

Aerial and Submerged Conidia

It has been reported that *P. chlamydosporia* does not produce chlamydospores and blastospores in liquid culture and that the fungus produces conidia directly on phialides in liquid culture that are not different in size and shape from those produced in solid culture. The pathogenicity of different propagules to nematodes including conidia and blastospores produced under different nutritional conditions has not been assessed (Mo *et al.*, 2005).

Resistant Structures

Most species of *Pochonia* produce dictyochlamydospores or at least some irregularly vegetative hyphae, although these may be absent in some strains. The term 'dictyochlamydospores' refers to a stalked, hyaline, thick walled, multicellular resting stage (spores), which can be formed either in large numbers in the aerial mycelium or submerged in the agar. The shape of the dictyochlamydospores is variable from subglobose, isodiametric, to less differentiated, and without a stalk. *Pochonia bulbillosa* can form intercalary swollen hyphal cells rather than well differentiated dictyochlamydospores. Some isolates have either lost the capacity to form dictyochlamydospores or, if they do occur, they have yet to be seen (Zare *et al.*, 2001).

Molecular Studies and Classification

The genus *Pochonia* is included in the order Hypocreales (Clavicepitaceae). The genus is closely related to *Verticillium*, a genus into which *Pochonia* species were previously classified. Zare *et al.* (2000) used the ITS regions of nrDNA (including parts of the small and large nuclear rDNA genes and the central 5.8S rDNA gene) to analyse the phylogenetic relationships among taxa in *Verticillium*, particularly the section Prostrata. Molecular studies supported the distinctness of the two genera. They found at least four clades (A-D) that are also morphologically separable (*i.e.*, conidia and resting structures). The species of clade D (*i.e.*, *Pochonia*) consisted of species parasitic on cyst nematodes.

Chemotaxonomical studies support the generic segregation of Gams and Zare (2001) based on the exclusive presence of pochonins in species of the genus *Pochonia* (Stadler *et al.*, 2003). Minisatellite PCR fingerprinting used to segregate species and strains of *Pochonia* pointed towards the existence of cryptic species. The genes encoding for production of secondary metabolites have so far been rarely included in molecular genetic studies (Stadler *et al.*, 2003).

III. Biology and Life Cycle

The fungus inhabits the soil as a saprophyte where it occurs as hyphae, mycelium, conidia and chlamydospores. *Pochonia* can survive in its saprophytic stage in the absence of both the plant and nematode hosts. In the rhizosphere, the fungus colonizes the roots of host plants and some *Pochonia* species have an 'endophytic behaviour' that may be beneficial to the host plant's defence against other soil borne pathogens (Bordallo *et al.*, 2002; Maciá-Vicente *et al.*, 2009a, b; Manzanilla-López *et al.*, 2011a). As a parasite of eggs of sedentary plant endo-parasitic nematodes, colonization of the root surface by the fungus is closely linked to egg production and may also be related to changes in root exudation due to nematode infection of the roots (Bourne & Kerry, 1999). Root colonization has been reported to be greater as numbers of juveniles increased in the plant root and parasitism of *Globodera pallida* and *G. rostochiensis* females can occur as they burst through the root cortex (Tobin *et al.*, 2008). When egg masses emerge on the root surface, the fungus colonises the gelatinous matrix and parasitizes the eggs (de Leij *et al.*, 1992). The eggshell of potential nematode hosts has a similar basic structure but differs in thickness as shown by transmission electronic microscopy in *Meloidogyne javanica* (0.6 μm) and *Rotylenchulus reniformis* (0.3 μm) eggs and the layers that compose the eggshell: vitelline layer (10-40 nm), chitin layer (50-400 nm) and the lipid layer. The eggshell of *M. incognita* is rich in protein (50% dry weight) and chitin (30% dry weight). The chitin and lipid layers are made probably of chitin-protein, and lipo-protein complexes respectively (Bird & McClure, 1976). Hence fungal penetration may show differences between hosts.

Cyst and root-knot nematode females can be parasitized by vegetative hyphae in the rhizosphere without the production of specialized infective structures apart from an appressorium. The fungus can reduce the number of mature females and their fecundity as a result of infection. Differences in biology and embryogenesis time between nematode hosts can affect egg parasitism. Root-knot nematode eggs are laid inside a gelatinous matrix that is pushed outside the root galls leaving the egg-mass exposed into the rhizosphere. Eggs within the gelatinous matrix of *Meloidogyne* spp. and *Nacobbus* spp. can be found in different stages of embryogenesis, *e.g.*, from segmentation to fully developed second-stage juvenile (J2). The embryogenesis period varies according to nematode species but it is shorter at higher temperatures and can last around 10 days at 25°C. The eggshell of *Heterodera*, *Globodera* and *Meloidogyne* has different thickness according to species; thus, the *H. avenae* eggshell is half the thickness of *G. pallida*, which is thicker than that in *Meloidogyne* (Lopez-Llorca & Robertson, 1992). In contrast to root-knot nematodes, cyst nematode eggs are not directly exposed to the rhizosphere since they remain inside the encysted female until they develop into J2. Therefore, it is considered that the time that the fungus has to infect and kill root-knot nematode eggs is shorter than for cyst nematode eggs (Kerry, 1997). However, other anatomical and physiological factors should be considered during the infection process, *e.g.*, eggshell thickness, nutrient availability for the fungus and enzymatic activity within fungal isolates. For example, infection tests of eggs of root-knot and cyst nematodes using conidia inoculated in a yeast extract liquid medium have shown that within 48 h differences in percentage of egg parasitism occur depending on the biotype of cyst or root-knot nematode (Esteves, 2007; Mutua *et al.*, 2009; Ward *et al.*, 2010).

Appressorium formation and egg infection must be preceded by environment signalling, a process that is still largely unknown (Lopez-Llorca *et al.*, 2002a). Thigmotropic responses (*i.e.*, towards hydrophobic and hydrophilic surfaces) and adhesion process mediated by glycoproteins, and enzymes (*e.g.*, serine proteases) are involved in pre-penetration events (Lopez-Llorca *et al.*, 2002a). The formation of appressoria in *P. suchlasporia* (= *Verticillium suchlasporium*) has been studied with *H. avenae* and *G. rostochiensis* (Lopez-Llorca & Claugher, 1990) using scanning electron microscopy. Vegetative hyphae can penetrate the eggshell by forming appressoria at the point of egg infection, a process followed by the development of a single, very narrow, penetration hypha that perforates the eggshell. Once penetration is accomplished, an infection bulb is formed inside the egg. Vesicular 'lomasome-like' structures and vacuoles occur in the infection bulb and the pre-apressorium hypha. Penetration of the eggs can also occur through the germ tube of conidia attached to the outside of the eggshell (Lopez-Llorca & Robertson, 1992). A similar process has been observed in *P. c.* var. *chlamydosporia* conidia infecting *M. incognita* eggs (Manzanilla-López, unpublished). The presence on infected eggs of low electron-density areas in contact with hyphae and conidia of *P. suchlasporia* suggests that enzymes are secreted *in situ* by the fungus to weaken the eggshell before penetration. Once the egg contents have been destroyed the eggshell is degraded from inside, even ruptured (post-infection 'break-outs'). Infection bulbs develop into trophic hyphae that destroy the body contents of the nematode (Lopez-Llorca & Robertson, 1992).

Egg mortality caused by fungal parasitism can effectively reduce the primary inoculum, or pool of eggs, available in the soil to produce the next generation of J2 to infect the plant. The number of nematode generations is an important factor that should be accounted for when assessing fungus-plant-nematode interactions in a particular crop since increasing nematode population and root galling cannot be prevented in the short term by the fungus, although multiplication of the nematode can be reduced significantly. Results from a long term field study to evaluate the effect of *P. chlamydosporia* and other biocontrol agents on slow decline of black pepper (*Piper nigrum* L.), emphasised the importance of the density-dependant multiplication of root-knot nematodes to suppress nematode populations through time following application of biological control agents alone or in combination (Eapen *et al.*, 2009). Viaene and Abawi (2000) reported that the percentage of *M. hapla* egg masses colonized by *P. chlamydosporia* (= *Verticillium chlamydosporium*) increased with increasing fungus application levels (*i.e.*, 2 or 4 *M. hapla* eggs cm^{-3} soil). However, in *M. hapla* infested soils at population levels equivalent to the threshold damage level (8 eggs cm^{-3} soil), no difference in colonization could be detected when 5,000 or 10,000 chlamydospores cm^{-3} soil were added to soils.

Endophytic Behaviour

Studies on root colonization have been made until recently using mainly classical approaches to estimate root colonization per cm^2 and rhizosphere colonization percentage (Bourne *et al.*, 1996; Hidalgo-Díaz *et al.*, 2000). Recently developed microscopy and molecular tools have made it possible to quantify more accurately the fungus and conduct studies on root colonization and endophytic behaviour of the fungus at the cellular level under laboratory conditions (López-Llorca *et al.*, 2002b; Maciá-Vicente *et al.*, 2009a, b). As an endophyte, *P. chlamydosporia* can complete its life cycle inside barley roots (Lopez-Llorca *et al.*, 2002b) and the hyphae can penetrate the root hairs of barley, although more commonly it is the epidermal cells that are so affected. The resulting hyphal network can be intra- and intercellular with hyphae not reaching the vascular cylinder.

Other species, such as *P. rubescens*, apparently induce changes in the host plant that increase root length of barley seedlings (*Hordeum vulgare*) and reduce *Gaeumannomyces graminis* var. *tritici* root colonization and peroxidase activity (Monfort *et al.*, 2005; Lopez-Llorca *et al.*, 2008). *Fusarium equiseti* and *P. chlamydosporia* were inoculated on barley roots to evaluate their root population dynamics and to monitor their effect on *G. graminis* under non-axenic condition. Results showed that *P. chlamydosporia* had an early plant growth-promoting effect in fresh shoot weight and also in length and weight of roots. The fungus was capable of remaining as an endophyte for at least two months under non-axenic conditions and to reduce the occurrence of other fungi in the root cortex (Maciá-Vicente *et al.*, 2009b). However, a single inoculation of the endophytic population of the fungus may be not enough to maintain high population levels of the fungus since they gradually decreased after inoculated plants were transferred into the soil. It is likely that the long-term endophytic root colonization depends on an equilibrium between the plant defences and the fungal rhizosphere competence of *Pochonia* (Maciá-Vicente *et al.*, 2009b).

Root colonization and anatomical host responses have been studied in plants with different root anatomy and rhizosphere properties, including tomato (Dicotyledonea) and barley (Monocotyledonea). Epidermis penetration was facilitated by appressoria in both hosts and cell wall modifications included papillae and appositions, although long lignituber ensheathing hyphae were only found in tomato. Compounds commonly associated with plant resistance in other fungus-plant interactions, such as phenolics (including lignin), protein deposits and callose, were present in cell wall papillae. Cell wall modifications did not prevent fungal growth, but colonization in tomato roots was restricted in comparison with barley (Bordallo *et al.*, 2002). A more recent approach to investigate the spatial root colonization and host responses included the combination of real-time PCR, specific primers, molecular beacon probes (based on the *vcp1* gene), and the genetic transformation of a

Pochonia isolate with the green fluorescent protein (*GFP*) gene, which allowed the quantification and study of the fungus within the roots using laser confocal microscopy. Results demonstrated the root endophytic behaviour of the fungus and the eliciting of plant defence reactions against endophyte colonization so as to prevent fungal growth within tissues (Maciá-Vicente *et al.*, 2009a).

Host growth can be affected by the fungus as noticed on several crops under glasshouse and field conditions. Increased growth under glasshouse conditions occurred with a single application of a native isolate of *P. chlamydosporia* (5×10^3 chlamydospore g^{-1} soil) to reduce the numbers of *R. reniformis* on cotton roots and soil (Wang *et al.*, 2005). Longer roots and shoots have been reported in potato plants inoculated with *P. c.* var. *chlamydosporia* in the absence of nematodes. However, a negative effect occurred on plant growth when the fungus and *M. incognita* were inoculated simultaneously on plants (Manzanilla-López *et al.*, 2011a). The application of *P. c.* var. *chlamydosporia* to banana plants under field conditions in Minas Gerais (Brasil) has shown a gradual reduction in *Meloidogyne* spp., *Radopholus similis*, *Helicotylenchus* sp. and *Pratylenchus* sp. populations. Although the parasitic effect of the fungus on eggs of sedentary plant-parasitic nematodes is well known, there is no information about how the control of migratory plant-parasitic nematodes is affected by the fungus. It has been proposed that, since the fungus is a root endophyte, it can affect migratory endo-parasitic nematodes such as *Pratylenchus* by producing a toxin or by inducing banana resistance (Freitas *et al.*, 2009). There are also a few reports on the effect of the fungus on the juvenile stages of nematodes (Braga *et al.*, 2010). Mukhta & Pervaz (2003) reported that the action of the fungal filtrate of *P. chlamydosporia* on J2 of *M. javanica* caused 13.25% mortality, and *P. c.* var. *chlamydosporia* caused 12% mortality on J2 of *Heterodera schachtii* (Ayatollahy *et al.*, 2008). The red pigment produced by *P. rubescens* (= *Verticillium suchlasporium*) was reported to reduce the mobility of the J2 of *G. rostochiensis* (Lopez-Llorca & Boag, 1993).

IV. Ecology

Pochonia-Rhizosphere Interactions and Control of Plant-Parasitic Nematodes

Plants and their rhizodeposits are an important source of C and N for soil microbiota and to maintain some entomopathogenic and nematophagous fungi in a saprophytic stage in soil (Bruck, 2010). Fungal isolates may differ in their capacity to colonize soil, rhizosphere and eggs and may behave as poor or good colonizers in these habitats. However, one isolate from the same or different nematode biotypes (*i.e.*, from the original nematode host) can dominate over other isolates when sharing the same habitat (Manzanilla-López *et al.*, 2009a, b). Fungal biotypes can also show a host preference at the infraspecific level, a feature that is related to differences in egg parasitism associated with specific amino acid polymorphisms in the fungal serine protease VCP1 (Morton *et al.*, 2003a). Plants also differ in their host status, not only for the fungus, but also for the nematode. These differences affect the degree of control exerted by the fungus on the nematode.

Kerry (2000) pointed out a series of factors that should be considered in order to enhance the performance of a biological control agent, such as *Pochonia*, in a tritrophic interaction (*i.e.*, plant, fungus, nematode). They can be summarised as ten key factors:

1. Isolates differ in their potential as biological control agents due to their ability (*i.e.*, competence): (a) to colonize the rhizosphere of plants and cultivars; (b) to produce chlamydospores *in vitro*; and (c) to infect nematode eggs (*i.e.*, percentage of parasitism).
2. Nematode control is dependent on fungal rhizosphere colonization. A good control of nematodes can be achievable if fungal isolates are capable of proliferating in the rhizosphere and nematode control will be poor if fungal isolates are incapable of colonizing the rhizosphere.
3. Good hosts (plants). Plant species differ in their root exudates and rhizodeposits, as well as in their ability to support *P. chlamydosporia* growth in their rhizosphere, and in their susceptibility to infection by nematodes. Crops and plant species that can support more than 200 colony forming units (CFU) cm^{-2} of root are considered to be good hosts for the fungus. Good hosts for *Pochonia chlamydosporia* include beans, cabbage, crotalaria, kale, pigeonpea, potato, pumpkin and tomato (de Leij, 1992; Bourne *et al.*, 1996).
4. Plant susceptibility to nematode attack. Plant species differ in their nematode host status, *e.g.*, if plants can or cannot be parasitised by nematodes. The result of nematode interaction with the host (*i.e.*, resistant, susceptible, tolerant) will determine nematode reproduction, as well as the extent of damage and yield loss caused to the host. *Pochonia chlamydosporia* hosts can be good or bad (poor) hosts for the nematode. In general, fungal control can be higher in plants that are poor hosts for the nematode but also support extensive rhizosphere fungal growth.
5. Plant tolerance to nematode attack. The fungus does not reduce initial plant invasion by nematodes (*i.e.*, juvenile stages), therefore to minimize yield losses fungal isolates should be applied in plant species that are tolerant to nematode attack.
6. Gall size. Plant susceptibility and tolerance to nematode infection and reproduction can result in large root galls in susceptible hosts. Eggs retained inside large galls cannot be exposed to *P. chlamydosporia* parasitism (Atkins *et al.*, 2003a) and 'escape' fungal infection. Therefore, application of the fungus to a relatively poor host of the nematode, which produces small galls so that most egg masses are exposed in the rhizosphere, might provide more effective control by the fungus (Kerry, 1997).
7. Fungus application rate. Increasing the fungus application rate can increase the amount of fungus in the soil but not necessarily in the rhizosphere as may be the case in a poor host for the fungus where growth in the rhizosphere is not increased by higher applications to the soil (de Leij *et al.*, 1992).
8. Egg development. Immature eggs (*i.e.*, root-knot nematodes) are more susceptible to parasitism by the fungus than those containing a J2 (*i.e.*, cyst nematodes). However, egg susceptibility to fungal isolates depends on host preference (biotype) and origin of isolates (see point 10).
9. Temperature. The fungus is less effective at temperatures above 30°C. *Pochonia* species showed optimum growth at 25°C but isolates can differ in their optimal

temperature with those obtained from cyst nematodes having lower optimal temperatures than those from root-knot nematodes (Bourne & Kerry, 2000).

10. Nematode species. Root-knot and cyst nematodes differ in their susceptibility to fungal isolates, the latter showing nematode host preference.

These basic principles and their combination can be integrated with crop rotation schemes and other control strategies as part of an integrated pest management (IPM) approach and many examples are available in the literature (Viaene & Abawi, 2000; Atkins *et al.*, 2003a; Verdejo-Lucas *et al.*, 2003; Van Damme *et al.*, 2005). The combined use of poor nematode hosts and the soil application of the fungus may be a useful strategy for the management of root-knot nematodes, especially in organic production. Application of a mixture of *P. c.* var. *catenulata* (10 g m^{-2}) with cattle manure in a cropping sequence of beans (*Vigna ungiculata sesquipedalis* cv. Lila), Chinese cabbage (*Beta vulgaris cicla* cv. PK-7) and tomato (*Solanum lycopersicum* cv. Campbell-28) reduced *M. incognita* soil infestations in a field trial following a heavily infested red beet crop. However, as an equivalent of 100 kg ha^{-1} of colonised medium was used to control the nematode, such a strategy requires more extensive evaluation (Atkins *et al.*, 2003a).

Establishment and reproduction of the fungus following an adequate initial level of fungal inoculum applied to soil, can build up levels found in naturally suppressive soils (*i.e.*, 2-8 × 10^3 CFU g^{-1} soil; Kerry *et al.*, 1993) under proper management conditions, and eventually produce long-term efficacy in the control of nematodes (Viaene & Abawi, 2000; Atkins *et al.*, 2003a; Van Damme *et al.*, 2005). Stirling and Smith (1988) proposed that levels of fungal parasitism increase as inoculum densities increase, and that *P. chlamydosporia* may be adapted to take advantages of the exponential increase in food supply that becomes available with second and subsequent generations of the root-knot nematode populations, nematode levels that result from susceptible crops being grown in consecutive growing seasons. Under such conditions, populations of the fungus can reach the point where it is suppressive to nematodes.

Ecophysiology (Habitat, Niche, Soil Web)

Pochonia chlamydosporia is a saprotroph in soil, a facultative parasite of invertebrates and a hyperparasite of other fungi. It is one of the most important parasites responsible for the natural control of both cereal and beet cyst nematodes. One study has shown that numbers of CFU in soils in which cereal and beet cyst nematode multiplication was suppressed were greater (1.7 – 7.5 and 4.3 – 5.1 × 10^3 CFG g^{-1} soil, respectively) than in those in which the nematode multiplied (<10^{-3} CFU g^{-1} soil). Since the fungus survived long enough to kill nematode eggs and females that develop on roots of spring-sown crops, pre-cropping applications of the fungus can be an alternative to control these pests (Kerry *et al.*, 1993). There are several reports on the effectiveness and limitations of *Pochonia* to control root-knot nematodes (de Leij & Kerry, 1991; de Leij *et al.*, 1993; Campos & Campos, 1997; D'Angieri & Campos, 1997; Viaene & Abawi, 2000; Verdejo-Lucas *et al.*, 2003; Lopes *et al.*, 2007), *Nacobbus aberrans* (Flores-Camacho *et al.*, 2007) and *Rotylenchulus reniformis* (Wang *et al.*, 2005). It is also considered that fungi adapted to different mutualism and parasitism interactions, such as *Pochonia*, are likely to have developed different strategies, including chemical agents (*i.e.*, secondary metabolites) to overcome their host's defence or act as signals in their natural habitat (Stadler *et al.*, 2003).

Tritrophic Interactions: Soil, Plant and Nematode Hosts

Increasing the amount of fungus applied to soil may not necessarily have an effect on the extent of colonization in the rhizosphere as shown by a tenfold increase in chlamydospore application from 10^3 to 10^4 chlamydospores g^{-1} soil having no effect on colonization (de Leij *et al.*, 1992). Growth of the fungus in the rhizosphere differs between plant species and colonization by the fungus is restricted to the rhizosphere and rhizoplane without detrimental effects on a range of crop species. *Pochonia chlamydosporia* var. *catenulata* did not affect plant germination and growth of maize (*Zea mays* L.) or beans (*Phaseolus vulgaris* L.) and analysis of tissue concentrations indicated that the fungus did not accumulate in aerial parts and there were no phytotoxic effects (García *et al.*, 2008). Not long ago it was considered that the fungus did not spread into root tissues. However, it has now been found that, although species of *Pochonia* can colonize the epidermis of some species of Graminae and Solanaceae, they do not invade the vascular cylinder in comparison to plant pathogenic fungi (Maciá-Vicente *et al.*, 2009a).

Multitrophic Interactions

Most studies on *P. chlamydosporia* ecology have focused on the tritrophic interaction between the fungus, the plant and nematode hosts and little is known about the interaction of the fungus with other soil microbia. *Pochonia chlamydosporia* is a mycoparasite of oomycetes, several species of *Puccinia* (rust fungi), and an antagonistic to *Pythium aphanidermatum* in soil (Lumsden *et al.*, 1982), *Fusarium oxysporum* and *Rhizoctonia solani* (Leinhos & Buchenauer, 1992; Jacobs *et al.*, 2003; Monfort *et al.*, 2005). *Pochonia chlamydosporia* also has antagonistic effects against *Phytophthora capsici* when applied as a root dip (Sutherland & Papavizas, 1991).

Different species of nematophagous fungi, such as *Paecilomyces lilacinus* and *Monographella cucumerina*, can occur with *Pochonia* in nematode infested soils (Gaspard *et al.*, 1990; Olivares-Bernabeu & Lopez-Llorca, 2002; Atkins *et al.*, 2004; Manzanilla-López *et al.*, 2009a). The potential control of *Meloidogyne incognita* using an isolate of *P. chlamydosporia* and *Pasteuria penetrans* was tested by de Leij *et al.* (1992). The two biological control agents tended to complement each other, thus providing up to 92% of nematode control 14 weeks after nematode inoculation. Puertas *et al.* (2006) have reported the compatibility of *P. c.* var. *catenulata* with *Rhizobium* sp., *Trichoderma harzianum* and *Glomus clarum* to control *M. incognita* on *Vigna unguiculata* subsp. *sesquipedalis* cv. Lina and *Solanum lycopersicum* (= *Lycopersicon esculentum*). The parasitic activity of *P. c.* var. *catenulata* was enhanced in combined applications with the other microbia. Application of *P. chlamydosporia* and *Hirsutella rossiliensis*, alone or together did not reduce final population densities of *M. hapla* on lettuce when the initial population density of the nematode was at its damage threshold (*i.e.*, 8 eggs cm^{-3} soil; Viaene & Abawi, 2000).

In a study to assess potential interactions between nematophagous fungi and their response to chemical pesticides to control potato cyst nematodes, the compatibility of *P. chlamydosporia* to compete with *Paecilomyces lilacinus*, *Monographella cucumerina* (= *Plectosphaerella cucumerina*) and the soil fungi *Rhizoctonia solani*, *Chaetomium globosum*, *Fusarium oxysporum*, *Penicillium bilaii* and *Trichoderma harzianum* was tested *in vitro*. *Pochonia chlamydosporia* radial growth was the least susceptible to growth inhibition by the other fungi at 20°C, and inhibited *P. lilacinus, T. harzianum* and *R. solani*. *Pochonia chlamydosporia* was the most effective fungus in inhibiting growth of the saprophytic fungi and also the least susceptible to being inhibited by them. It was also the most tolerant to nematicides (Jacobs *et al.*, 2003).

Kok *et al.* (2001) demonstrated that the egg masses of *M. fallax* or *M. hapla* from tomato or potato supported higher bacterial population densities than those in the rhizosphere soil, and that most fungi identified from egg masses showed some type of antagonistic interaction with *P. chlamydosporia*. O'Flaherty *et al.* (2003) investigated the impact of *M. incognita* on the size and diversity of the rhizosphere population of heterotrophic bacteria, and compared the effects of *P. chlamydosporia* and aldicarb on tomato and cabbage. It was found that tomato roots infested with the nematode supported a different bacterial population from roots on which infestation was controlled either by the fungus or nematicide. Cabbage did not show this effect but its rhizosphere population was affected by the Vc10 (= Pc10) treatment of *P. chlamydosporia* and modifed the rhizosphere for other microorganisms.

The fungus does not affect abundance of other nematodes and microbes as inferred from total counts of mycophagous, bacteriophagous and predatory nematodes on different plant species (*i.e.*, tomato and cabbage) or application of the fungus and *M. incognita* (Tahseen *et al.*, 2005).

Physical and Chemical Soil Parameters

Environmental factors are known to affect the survival and proliferation of *P. chlamydosporia* in soil (Bourne & Kerry, 2000). When released in soil, *P. chlamydosporia* is exposed to fluctuating abiotic factors which can have a great impact on establishment in the field and compromise its efficacy in controlling nematodes. Soil texture and structure influence nematode activity and the spread of micro-organisms. In general, the mean numbers of CFU that proliferated from chlamydospores of *P. chlamydosporia* and *P. c.* var. *catenulata* have been reported to be greater in organic that in mineral soils (Kerry *et al.*, 1993) and just a few applications of the fungus in organic soil may be enough to obtain adequate and lasting control (Viaene & Abawi, 2000; Puertas & Hidalgo-Díaz, 2009). Tolerance to environmental-stress conditions, such as water availability, temperature and microclimate, can determine the success of an ecologically competent biocontrol agent (Magan, 2001). An effective biocontrol agent should be able to withstand drying and persist over periods of little or even no rainfall, common conditions of many sub-tropical, tropical and Mediterranean regions. Strains of the fungus can tolerate low values of free water that can reduce but not arrest fungal growth. This may allow the fungus to cope water stress in agricultural fields (Olivares-Bernabeu & Lopez-Llorca, 2002). The minimum water potential for survival appears to be –8 to –10 MPa at 40°C, whereas optimal growth occurs between –1.0 to –2.0 MPa, at 25 to 30°C (Bourne & Kerry, 2000) but isolates can differ in their optimal temperature. Also, *P. chlamydosporia* isolates have shown differences in growth and tolerance to osmotic and matric water stress (Esteves *et al.*, 2009a). Under water stress, isolates accumulated a combination of different polyhydroxyalcohols (polyols) which are important in osmoregulation. The quantities of the major polyols accumulated in colonies of *P. chlamydosporia* depended on the solute used to generate the water stress. Whereas arabitol and glycerol were the main polyols accumulated in osmotically modified media with ionic and non-ionic solutes, respectively, erythritol was the main sugar accumulated in media amended with polyethylene glycol (PEG 8000). The versatility to accumulate different polyols and perhaps different sugars under different stress conditions might have the advantage of allowing the fungus to survive in soil when environmental conditions are less favourable or highly variable. When water is freely available, *P. chlamydosporia* is able to produce conidia and chlamydospores but chlamydospore production seems to be repressed by an increase in water stress (Esteves *et al.*, 2009a).

V. Physiology of Parasitism

Host Signalling and Penetration Events

Pochonia chlamydosporia produces an appressorium when in contact with the surface of a nematode egg (Segers, 1996). This specialised infection structure enables the fungus to become firmly attached to the egg and facilitates penetration. In this process, mechanical and enzymatic forces are known to be involved. Enzymes are secreted to help destroy the outer vitelline membrane of the eggshell and expose the chitin layer, allowing the infection peg to penetrate the egg. The appressorium focuses the mechanical force and enzymatic degradation into a small area, thereby promoting host penetration (Lopez-Llorca *et al.*, 2002a). Appressoria of *P. chlamydosporia* are produced either at the ends of long hyphae from lateral branches or at the tip of the conidia germ tube (Segers *et al.*, 1996; Lopez-Llorca & Robertson, 1992). Substrate hydrophobicity, appressorium formation and protease production are factors known to be involved in early parasitism of nematode eggs by *P. chlamydosporia* (Lopez-Llorca *et al.*, 2002a) but further research is needed to understand which stimuli are involved in adhesion and attachment, formation of appressoria and penetration of the nematode eggshell. One study on the formation of appressoria on artificial surfaces by *P. rubescens*, *P. chlamydosporia* and *Lecanicillium lecanii*, showed that the three fungi formed appressoria less often on hydrophilic materials than on hydrophobic surfaces such as polysterenes. *Pochonia chlamydosporia* was the most effective fungus, forming most appressoria on all materials except polyvinyl chloride (Lopez-Llorca, *et al.*, 2002). No tropism towards plant roots has been observed in *Pochonia* (Bordallo *et al.*, 2002). Appressoria of *P. rubescens* on nematode eggs have mucilaginous substances between the surface of the appressoria and the eggshell and these could function as a fungal adhesive to assist in eggshell penetration by the fungus. The labelling of spore walls of ungerminated conidia, germlings and hyphae with FITC-labelled lectin Con A suggested the early secretion of substances containing sugars (α-mannoside and α-glucoside groups) and their accumulation after appresorium differentiation on nematode eggs (Lopez-Llorca *et al.*,

2002a). The PCCHI44 chitinase from *P. chlamydosporia* has a glycoprotein nature due to two potential N-glycosilation sites. A conservative Ser/Thr-rich domain was identified in PCCHI44 which can be a potential site for o-mannosylation and that may contribute to enhance the enzyme's affinity for chitin (Mi *et al.*, 2010). Nematode gelatinous matrix and other excretions from root-knot, false root-knot and reniform nematodes also contain glycoproteins (Sharon & Spiegel, 1993; Manzanilla-López *et al.*, 1998; Agudelo *et al.*, 2004) important in semiochemical signalling for the nematode to find its plant host. Glycosilation of nematode proteins has been considered to be involved in host-parasite relationship and in protection against micro-organisms (Agudelo *et al.*, 2004).

The Switch from Saprophytic to Parasitic Phase

Appressoria formation is thought to be a nutritional response in many plant and entomopathogenic fungi, with conducive conditions being provided by nutrient depletion, low levels of complex nitrogen sources such as low concentrations of yeast and low C:N ratios (Blakeman & Parbery, 1977; Jackson & Schisler, 1992; St. Leger *et al.*, 1989). Low C:N ratios influence egg pathogenicity by *P. chlamydosporia* (Segers, 1996), whereas production of conidia seems to be favoured by growth in media containing relatively high C:N (Mo *et al.*, 2005). The change, or 'switch', from the saprophytic to the parasitic phase of the fungus may therefore be related to nutrients (including C and N) that are either released by plants into their rhizosphere or available in nematode eggs. A recent analysis of serine proteases enzyme production as related to nutrients available for the fungus in presence/absence of nematode eggs has been conducted using a VCP1 assay, in order to investigate the effect of different nutrients on genes expression in early stages of parasitism. VCP1 was found to be repressed by glucose and stimulated by ammonium chloride but there was also evidence of an increase in VCP1 mRNA when glucose and eggs were present. VCP1 expression was higher in more alkaline pH conditions (Ward *et al.*, 2011).

Enzymatic Activity

Enzymatic assays and zymography of extracts of conidia and germlings of *P. chlamydosporia* and *P. rubescens* have shown proteolytic activity very similar to serine proteases VCP1 and P32 and that these proteases occur in the absence of nematode eggs. Particular extracellular enzymes secreted by *P. chlamydosporia* are thought to play an important role in the infection process of eggs as they enable the fungus to degrade the host's major barrier to infection, the nematode eggshell. Specific proteases and chitinases have been isolated and purified and their activity against the nematode eggshell demonstrated. They are considered to be involved in the infection process and serve as virulence factors (Huang *et al.*, 2004; Morton *et al.*, 2004; Yang *et al.*, 2007; Mi *et al.*, 2010; Palma-Guerrero *et al.*, 2010). Proteolytic activity has been reported to show the earliest and highest activity within the other extracellular enzymes tested for *P. chlamydosporia* isolates. In these assays, chitinolytic activity was shown to be lower than proteolytic or lipolytic activities and amylase activity could not be detected. The higher proteolytic activity observed supports the importance of proteases and chitinases in the degradation of the eggshell necessary to assist fungal penetration (Olivares-Bernabeu & Lopez-Llorca, 2002). Other proteases have been reported for *P. rubescens* (P32), serine protease (VCP1) for *P. chlamydosporia*, as well as chitinases/chitosanases (CHI43) (Lopez-Llorca *et al.*, 2008; Palma-Guerrero *et al.*, 2010). An endochinase (CHI43) from *P. chlamydosporia* was reported to be involved in the breakdown of *G. pallida* eggshells (Tikhonov *et al.*, 2002) and Mi *et al.* (2010) cloned the gene *pcchi44* encoding an extracellular endochinase from *P. chlamydosporia*. The purified recombinant chitinase PCCHI44 exhibited activity under a broad range of temperatures (30-60°C) and pH (4.0 to 8.0) with the highest activity at pH 6.0. The enzyme can damage *M. incognita* and *Bombyx mori* (Lepidoptera) eggs (Mi *et al.*, 2010).

Chitosan and Chitosanases

Chitosan is a partly deacetylated form of chitin, and its derivatives display antibiotic activity against microorganisms. Chitosan is not toxic to plants and it can enhance plant resistance in seeds and reduce disease caused by fungal pathogens (Palma-Guerrero *et al.*, 2007). The conidia of *P. chlamydosporia* are surrounded by a material that probably helps to bind chitosan to the conidia surface and enter the fungal cells in an energy-dependent process. Once inside fungal cells chitosan causes membrane disorganization and damage (Palma-Guerrero *et al.*, 2007). Bioassays to study the effect of chitosan on colony growth, and spore germination of root pathogenic (*Fusarium oxysporum* f.sp. *radicis-lycopersici*, *Gaeumannomyces graminis* var. *tritici*, *Pythium ultimum*, *Rhizoctonia solani*) and biocontrol fungi, including entomopathogenic (*Beauveria bassiana*, *Lecanicillium psalliotae*), mycoparasitic (*Trichoderma harzianum*, *T. atroviride*) and several isolates of *Pochonia* spp., showed that the growth of the fungi was slightly affected by chitosan at 1-2 mg ml^{-1} in *P. chlamydosporia* (isolate Pc123) but that it produced a significant growth increase (111%) in *P. rubescens* at 2 mg ml^{-1} (Palma-Guerrero *et al.*, 2007). Plant pathogens and mycoparasitic fungi were more affected by chitosan than nematophagous or entomopathogenic fungi with spores being more sensitive to chitosan than the hyphae. The lower effect of chitosan on *P. chamydosporia* was considered to be possibly related to the presence in the fungus of better adapted chitosan degrading enzymes (*e.g.*, chitinase, deacetylases, deaminases, *etc.*) – hence the potential use of chitosan in combination with *P. chlamydosporia* for biological control of plant pathogens and pests, including nematodes (Hallmann *et al.*, 1999; Kokalis-Burelle *et al.*, 2004). Extracellular proteins from culture filtrates of *P. chlamydosporia*, grown either with colloidal chitin or chitosan as main carbon and nitrogen sources, were also investigated by Palma-Guerrero *et al.* (2010) using proteomic techniques. In this study, chitinolytic and chitosanolytic activities were analyzed using FITC-chitin or FITC-chitosan as substrates, which allowed detection of a much higher endochitinolytic activity in chitosan in comparison to chitin culture filtrates. Chitosanolytic activity was only detected when chitosan was present in the culture medium. Palma-Guerrero *et al.* (2010) proposed that chitosan induces expression of extracellular hydrolytic enzymes (*e.g.*, chitinases, chitosanases, proteases, *etc.*) involved in egg penetration. Only 14 out of 52 protein spots could be identified (around 27%), emphasising the scarcity of the proteomic data available for *P. chlamydosporia*. Some proteins that were identified were either over-expressed or newly induced in the presence of chitosan. Pathogenesis proteins included a chitinase, a mannose, an exoglucanase and an alkaline serine protease. Entomopathogenic fungi can transform chitin in the host cuticle to chitosan using chitin deacetylases, chitosan being subsequently degraded by fungal chitosanases during fungal penetration (Nahar *et al.*, 2004). The nematode eggshell consists of several layers, including a chitinous layer composed of a protein matrix where chitin microfibrils are embedded. This layer is the thickest and acts as an important barrier to fungal infection (Bird & Bird, 1991; Palma-Guerrero *et al.*, 2010). Hence, Palma-Guerrero *et al.* (2010) proposed that a mechanism, similar to entomopathogenic fungi, may be present in nematophagous fungi to degrade nematode eggshell chitin microfibrils.

VCP1 Serine Protease

During the egg infection process, a 33 kDa subtilisin-like serine protease (VCP1) is produced by *Pochonia* (Segers *et al.*, 1994). Immunolocalization of this enzyme at the penetration site indicated that VCP1 degraded the vitelline membrane on the surface of the eggshell and exposed the chitin layer (Segers *et al.*, 1996). This enzyme is serologically and functionally related to Pr1, the much studied enzyme produced by the entomopathogenic fungus *M. anisopliae* (Segers *et al.*, 1995). As demonstrated with Pr1 and other pr1-like proteases produced by various nematophagous fungi, VCP1 plays an important role in the infection process. VCP1 is induced by the nematode host and depletion of nutrients *in vitro*. The enzyme is repressed by glucose and N sources easy to metabolize and is induced by nitrate and insoluble proteins (*e.g.*, collagen and chitin) (Segers, 1996; Esteves *et al.*, 2009b). Morton *et al.* (2003a, b) also pointed out that aminoacid polymorphisms in VCP1 might contribute to host preference since: "The VCP1 of *P. chlamydosporia* shows amino acid polymorphisms in root-knot and cyst nematode isolates at positions 171 and 208. The polymorphism at 171 is a replacement of a glutamic acid (Glu) in root-knot nematode isolates by a glutamine (Gln) in cyst nematode isolates. The polymorphism at position 208 is a replacement of a glycine (Gly) in root-knot nematode isolates with an alanine (Ala) in cyst nematode isolates". In the absence of a nematode host, proteases such as VCP1 are also expressed in roots of plants colonised endophytically by *P. chlamydosporia*, thus suggesting that proteolytic enzymes may also play a role in protecting plants from nematode (Lopez-Llorca *et al.*, 2010). Nutrition affects key enzymes involved in the change from saprophytic to parasitic phase and fungal isolates differ in their *in vitro* production of extracellular enzymes. However, it has not been possible to relate such differences with isolates virulence and ability to colonize the rhizosphere (Esteves *et al.*, 2009b).

Proteases of Fungal Origin in Host Roots

When developing endophytically, nematophagous fungi interact with their plant host eliciting defence responses and also promoting plant growth (Maciá-Vicente *et al.*, 2009b). Such responses have been proven *in vivo* in barley root using a *P. chlamydosporia* GFP-expressing transformant and the lipophilic tracker FM4-64 (Maciá-Vicente *et al.*, 2009a). Detection of fungal proteases in roots of barley endophytically colonised by *P. chlamydosporia* and *P. rubescens* has been carried out through specific polyclonal antibodies against proteases VCP1 (*P. chlamydosporia*) and P32 (*P. rubescens*), and one specific polyclonal antiserum to P32 (Lopez-Llorca *et al.*, 2010). The presence of the two fungal proteases was revealed in extracts from roots inoculated with either two of the fungal species. A new serine carboxypeptidase of fungal putative origin, named CP1, from *P. chlamydosporia* was detected from inoculated roots. Hence, nematophagous fungi not only are capable of colonizing roots endophytically but also are able to express in roots some of the proteases that they produce when interacting with their nematode hosts. The presence of proteases and chitinases from nematophagous fungi in the root could also play an important role in plant defence against nematodes and other root pathogens (Lopez-Llorca *et al.*, 2010; Mi *et al.*, 2010).

Secondary Metabolites

Overproduction by fungal biopesticides of mycotoxins or other secondary metabolites with detrimental biological effects to soil biota may pose an environmental risk for their development, large-scale application and marketing. Hence, it is important that secondary metabolites of candidate biocontrol agents can be identified and studied for biological activities (Stadler *et al.*, 2003). *Pochonia chlamydosporia* produces antifungal and phytotoxic compounds that have been tested in *Puccinia coronata* on oat seedlings, *P. recondita* on wheat and *P. sorghi* on cotton seedlings (Leinhos & Buchenauer, 1992). Production of active metabolic compounds by *Pochonia* depend on the nutrition solution and fermentation times. Ethyl acetate extracts of peptone-Czapeck and malt extract culture filtrates reduced rust incidence up to 80%. Different concentrations of monorden (syn. radicicol), were detected during fermentation reaching a maximal concentration at 5 – 7 days of incubation and were correlated with a decrease of pH in culture medium. Monorden has a very broad *in vitro* antifungal spectrum against important plant pathogens including nematodes (Leinhos & Buchenauer, 1992; Khambay *et al.*, 2000; Stadler *et al.*, 2003). Pochonins are antiviral and antiparasitic resorcylic acid lactones (RAL) structurally related to monorden that are found in *P. chlamydosporia* and *P. c.* var. *catenulata* (Stadler *et al.*, 2003). Production and distribution of metabolites in cultures of different species of *Pochonia* was studied by Stadler *et al.* (2003) using High Performance Liquid Chromatography with UV-visual and mass spectrometric detection. Monocillins and monocillin II glycoside were present in *P. c.* var. *catenulata* and the alkaloid pseurotin A in *P. chlamydosporia*. The UV-visual spectra of the characteristic compounds that were found in most strains of *P. suchlasporia* var. *catenata* and *P. bulbillosa* CBS 248.83 resembled those of aurovertin B and citreoviridin A, compounds known to have phytotoxic, antiparasitic, broad-spectrum and antibiotic activities (Stadler *et al.*, 2003). Extracellular enzymatic activity can assist egg penetration but egg death can be also be due to toxic compounds produced by the fungus. Aurovertins are among the major constituents produced by *P. chlamydosporia* and aurovertin-type metabolites (A3 and A4), obtained from the fungus, have proved to be toxic to *Panagrellus redivivus* in *in vitro* assays, although they did not inhibit hatching (Niu *et al.*, 2010). The latter effect can be related to the nematode eggshell preventing external compounds getting into the egg. The chemical structures of aurovertin-like metabolites have been shown to be partly similar to phomalactone, a compound with nematicidal properties (Khambay *et al.*, 2000; Niu *et al.*, 2010). The mechanism of action of aurovertins-type metabolites could be similar to that of aurovertin B, an inhibitor of ATPase-like enzymes, due to chemical structural similarity. A novel compound called aurovertin 1 (A1) was the first natural aurovertin, with one less carbon atom, to be identified from *P. chlamydosporia* (Niu *et al.*, 2010).

Pigments

Attempts have been made to identify and characterise the compounds responsible for culture pigmentation related to secondary metabolism of the fungus (Stadler *et al.*, 2003). Characteristic pigments produced by species such as *P. rubescens* have been only observed in solid culture (Lopez-Llorca *et al.*, 1994) but some isolates of *P. chlamydosporia* produce orange pigments when grown in liquid culture (Manzanilla-López, unpublished). One Spanish strain of *P. chlamydosporia* produced diffusible pigments at about pH 3; similar pigments have been found to be produced in *P. rubescens* (= *P. suchlasporium*) in response to environmental stress (Olivares-Bernabeu & Lopez-Llorca, 2002).

Molecular Analyses

In the genomic era, elucidating gene and protein functions from an organism is a task of particular interest, which may provide a deeper view of the fungus behaviour on different habitats (*i.e.*, soil, rhizosphere, nematode eggs) and metabolic conditions (*i.e.*, as a parasite, a root endophyte or a saprophyte), through a functional, applied approach. Particular attention has been given to genes expressed by a number of parasitic fungi in certain cell or tissue types under particular conditions, such as those observed in host pathological reactions. Most recent studies, however, have focused on entomo-pathogenic species, and data on nemathophagous fungi are in comparison, scarce.

One useful approach for molecular analysis of host-pathogen interactions is the determination of changes in steady state mRNA levels, occurring during the process of host invasion and parasitism. Such transcriptomic approaches may shed light on complex interactions, and several platforms are available today, *i.e.* microarray analysis (if genome data are available), cDNA-amplification fragment length polymorphism (AFLP), allowing comparison of expression profiles to identify novel genes without previous sequence informations (Bachem *et al.* 1998) or, alternatively, whole-genome, open-architecture technologies (Polesani *et al.*, 2008).

Recently, a study of the molecular response of *P. chlamydosporia* to different nutritional conditions, including the presence of nematode eggs, was carried out with two isolates of *P. chlamydosporia*, each isolate having a host preference either for RKN or PCN. Expression patterns obtained using the cDNA-AFLP technique (Rosso *et al.*, 2011) were analysed by principal component and hierarchical cluster analyses, that highlighted groups of different genes induced by nematode eggs that were specifically co-expressed, and related to parasitism. Screening of two distinct isolates and host associations confirmed the utility of a broad approach in categorizing parasitism genes, since differences may also be isolate-specific (Johannesson *et al.*, 2006). Gene expression data also suggested differential regulation in *P. chlamydosporia*, due to structural and functional differences among isolates.

The fungus parasitic behaviour, although appearing as an intermediate trophic condition between the rich and poor media tested, revealed specific metabolic pathways at the beginning of nematode parasitism, as some genes were differentially activated only in the presence of eggs. *Pochonia chlamydosporia* also showed a transcriptional re-programming, as indicated by the changes in expression levels of, *i.e.*, up-regulated genes involved in detoxification, metabolism or cellular transport. Some induced/expressed or repressed genes also suggested a concerted regulation, especially when activated only at 4-8 hours after egg exposure.

Some transcripts active during parasitism were suggestive of a detoxification and cellular protein protection mechanism related, but not limited to, glucose-6-phosphate and in some way linked to a number of transmembrane transport and activation factors.

A second biochemical path showed co-expression of homologues to the GTPase activating protein Rga6, a chromosome segregation protein conserved domain and a hypothetical protein, a GAL-4-like transcription regulator and a transposon polyprotein. GAL4 is a fungal specific activator of the GAL system, a group of enzymes and regulatory elements of a glucose-repressible and galactose-initiated catabolic system (Johnston, 1987). In fungi, the GAL4 domain is involved in several biochemical pathways and cellular processes. Galactose is a filamental growth promoter, and is present in the fungal cell wall, mainly as a glycoprotein component. It is also a good C source for *P. chlamydosporia*, as shown in growth and conidiation assays (Liu & Chen, 2003).

The presence of GAL4 in *P. chlamydosporia* suggests a sensing capability of this fungus for galactose residues, in a path similar to other fungi (Brown *et al.*, 2009). Galactose has been often detected in nematodes, as the cuticle of several species bears D-galactose and N-acetyl-D-galactosamine residues (Forrest & Robertson, 1986). These molecules are targets on the host body surface that are recognized by nematode-trapping fungi (Nordbring-Hertz *et al.*, 1982; Premachandran & Pramer, 1984). Galactosamine residues are also present in the gelatinous matrix of RKN egg masses (Sharon & Spiegel, 1993) and in the cyst wall of *Heterodera schachtii* (Clarke, 1970). The GAL4 transcriptional regulator could be involved in one or more *P. chlamydosporia* cellular metabolic processes, including cell wall formation, but it may also act as a putative metabolic switch or receptor, related to external nutritional stimuli. This latter hypothesis is worth further studies, as GAL4 may be a good candidate to identify the promoter(s) switching *P. chlamydosporia* metabolism towards egg parasitism, in a parallel with other pathosystems (Dufresne *et al.*, 2000, Chung & Liao, 2008).

Other *P. chlamydosporia* clusters of transcript showed regulation of genes involved in cellular proliferation, and DNA replication, recombination or repair, together with a retrotransposable element, probably active in the cellular proliferation and/or in processes recombining metabolic factors. Secondary metabolism and transport transcripts, mainly expressed in rich medium and down-regulated in the poor nutritional conditions or in presence of eggs, showed homologies to a glycosyltransferase, a reducing polyketide synthase (RPS), a patatin-like serine hydrolase, a nuclear pore related protein (nucleoporin) and transcriptional regulators. In particular, the RPS, expressed only in rich nutritional conditions, showed high affinity with a *P. chlamydosporia* radicicol gene cluster, catalysing a resorcilic acid lactone inhibiting the HSP90 chaperone. Polyketides are products associated with the release of several fungal compounds having antibiotic and antifungal properties. However, in *P. chlamydosporia* this transcript appeared related to favourable nutritional conditions rather than to parasitism. Finally, other transcripts included microcycle conidiation, a monoxygenase, a NCS family nucleoside transporter, a possible cystathionin-γ-lyase and a RNA dependent RNA polymerase as well as constitutive products, including a chaperonin

GroEL (HSP60 family), and a Fot5 transposase. The presence of transposable elements, as well as of gene-silencing mechanisms, both having implications in gene expression of important biological functions, require further investigations.

Differences in gene activation and expression also occur between *P. chlamydosporia* isolates. Although similar host stages (*i.e.*, eggs) were colonized, the fact that expression profiles and induced transcriptional changes were only partially similar suggests a differential reaction to the host nematode species. A better understanding of the molecular mechanism that give rise to the colonization of nematodes eggs will greatly assist in the development of *P. chlamydosporia* as a biocontrol agent, through the identification of key factors required to improve its performance in more sustainable pest management strategies.

VI. Limiting Factors for the Development of *P. chlamydosporia* as a Biopesticide

A fundamental difference between biological control agents and chemical agents is that, to be effective, biological control agents (BCA) must survive and proliferate in the environment (Bruck, 2010). These properties are particularly important if the agent is to be used in arable crops in which inundative application rates are impractical (Kerry, 1997). Successful establishment has generally been associated with the addition of an energy source to help the organisms overcome competition from the residual soil microflora (Kerry *et al.*, 1993). The potential benefits of BCA remaining *in situ* over time, in comparison to the faster degradation of a pesticide active ingredient, tend to be overlooked because biological control can be inconsistent, less effective, complicated to use and slower acting than chemicals (Bruck, 2010; Jaronski, 2010). Other limiting factors to adopt microbial control agents in the industrial world include regulatory constraints and availability of cheaper efficacious chemicals (Bruck, 2010). In general, research on the manipulation of natural control, has been limited, under resourced and concentrated mainly on the introduction of specific agents to provide a more rapid control that might be commercially exploitable (Kerry, 1997). However, to improve the chances of success of isolates as potential BCA, selection should not be based on laboratory bioassay results alone and should include research into ecology in the screening process before releasing fungal isolates in the field (Bruck, 2010). It is also likely that successful application of BCA will depend on integration with other control measures according to farming systems (Kerry, 2000).

Nutrition and Mass Production of Fungal Propagules

The growth of mycelia and sporulation of *P. chlamydosporia* on artificial media are important characteristics used to evaluate potential BCA and are influenced by the components of the culture medium and growth conditions (Kerry *et al.*, 1986; Liu & Chen, 2003; Mo *et al.*, 2005). Different carbon and nitrogen sources support *P. c.* var. *chlamydosporia* in solid and liquid culture and the ratio of carbon and nitrogen might be important for parasitic activities (Liu & Chen, 2003). Mo *et al.* (2005) conducted a study to optimize carbon and nitrogen sources (C:N) to assess growth (biomass) and sporulation of *P.*

chlamydosporia in liquid cultures. Optimal carbon and nitrogen sources for mycelia growth were sweet potato (510 mg of dry biomass 100 ml^{-1} culture) and L-tyrosine (640.67 mg 100 ml^{-1} culture), and for sporulation, sweet potato (58 × 10^6 conidia ml^{-1}) and casein peptone (175.5 × 10^6 conidia ml^{-1}). The C:N ratio of 10:1 at pH 3.7 gave the maximum yield of conidia (51.4 × 10^6 conidia ml^{-1}). Regarding biomass production, a C:N ratio of 40:1 at pH 6.8 gave a maximum yield of 925.7 mg 100 ml^{-1}. The optimal pH for sporulation was within the 3.5-4.5 range and 5-6 for growth. It was noticed that after 7 days cultivation the pH was higher than the initial value. Results indicated that the nutritional requirement for growth and sporulation of the fungus are different.

Application and Formulation

Farming systems have a marked effect on the way a BCA is used and applied (Davies *et al.*, 1991). Growers in developed agriculture require an agent that can be applied to arable crops grown in monocultures over large areas using standard application machinery. According to Kerry (1997) “Formulated products with a good shelf-life, low application dosage and seed treatments are preferred. In contrast, crops in subsistence farming systems tend to be grown in mixed stands in relative small areas. Large application rates of an unformulated agent can be mixed into the soil manually as long as the BCA could be produced cheaply and locally”. For a BCA to be combined with other integrated pest management (IPM) strategies, it is important to understand how a BCA interacts with the resident microbiota, including non-target pathogens, chemical pesticides and other potential biocontrol agents. The success of a BCA also depends on its formulation. Nutrient sources can be added to alginate granules to increase sporulation and the period of time during which the pellet can support survival of the fungus (Jacobs *et al.*, 2003). Duan *et al.* (2008) improved the shelf-life of alginate pellets formulation containing mycelium of *P. chlamydosporia* at low temperatures (4 to –20°C), low water activity (a_w = 0.12) and a nitrogen-filtrate atmosphere. The method of mass culturing of *P. chlamydosporia* can also have a marked effect on the subsequent survival and proliferation of the fungus in soil (Kerry, 1997). Inoculum produced in shaken liquid cultures consist mostly of hyphae and conidia, both of which require an energy source to ensure proliferation in soil (Kerry, 1987; Kerry *et al.*, 1993), whereas chlamydospores are produced in large numbers on solid media, can be added to aqueous suspension and rapidly establish the fungus (de Leij & Kerry, 1991; Kerry *et al.*, 1993). Chlamydospores (resting spores) are not produced in liquid culture and, for large scale production and application, they are produced on a substrate (solid phase) inoculated with conidia/chlamydospores (Montes de Oca, 2004). At present, the fungus is applied as chlamydospores on colonised substrates. Application rates vary greatly and so far there has been limited progress in reducing them. Facilities and resources for fungal production can be basic and are not always subjected to regulatory mechanisms, factors which may limit their use and distribution. Other important factors include production energy cost being lower in the tropics than in temperate countries (FAO, 1992), long term storage facilities, shelf life and local distribution. Hence, there is a need to improve the different scales of fungus production (*i.e.*, commercial *vs* local, cottage industry) and formulation.

The capacity of hypocreales species (Jaronski, 2010) to produce infective propagules (*e.g.*, aerial, conidia submerged conidia or blastospores) *in vitro* in large quantities has also been explored for *Pochonia*. Mycelia and conidia granular formulation of *P. chlamydosporia* could remain viable for up to 12 months at 25°C and be biologically active when introduced into soil (Stirling *et al*., 1998; Mo *et al*., 2005). A commercial product (Xianchongbike) made from mycelia and conidia of *P. chlamydosporia* produced in liquid culture reduced root-knot nematodes in field trials in tobacco (Mo *et al*., 2005). Fungal chlamydospore-based products for application to soil have been shown to be commercially viable for *P. c.* var. *catenulata* in Cuba (KlamiC). The fungus is also produced in Portugal (Clamitec, Évora University) in liquid fermentation using micro-filtration and separation of clamydospores. In Cuba and Brasil the fungus is produced in a bi-phasic fermentation process (Montes de Oca, 2004, 2005; Freitas *et al*., 2009). The fungus is also produced commercially in Kenya (Dudutech) and Brazil (Rizotec, Rizoflora Biotecnologia).

Establishment and Monitoring

Introduction of a novel BCA for permanent establishment and long term pest control has been a classic approach towards biological control. However, conservation biocontrol, based on crop environment manipulation, to enhance resident microbial agent populations, should not be ignored (Jaronski, 2010; Pell *et al*., 2009). Population densities of the introduced BCA should be tested and monitored to ensure that it has survived through the period that activity against the nematode target is required. This can be achieved using molecular methods that nowadays are sufficiently discriminating to enable the detection of specific isolates of fungi in soil (Atkins *et al*., 2003b; Sidiqqui *et al*., 2009).

Fungistasis

Conidia and hyphal fragments have low survival rates in soil and are subjected to fungistasis when they are added to soil without a supplementary energy source. Chlamydospores have sufficient food reserves to enable the fungus to establish without the addition of other energy sources (Kerry *et al*., 1993; Mauchline *et al*., 2002). Adding an energy source to alginate granules containing *P. chlamydosporia* markedly improved the establishment of the fungus and fungistasis since this food source enables the fungus to grow through the soil (Lewis & Papavizas, 1985 cited by Kerry *et al*., 1993). Nevertheless, addition to soil of comminutes of 'fermentation biomas', which contained hyphal fragments and conidia that had limited food reserves, resulted in the eventual establishment of *P. chlamydosporia* in soil in appreciable numbers (Kerry *et al*., 1993). A soil can be more receptive to indigenous than to non-indigenous isolates and soil microbiota can be a major determinant in soil fungistasis and affect the ability of nematophagous fungi to proliferate in soil (Monfort *et al*., 2006). Root endophytism provides biocontrol agents with an adaptive advantage by reducing competition with other soil microbiota. Endophytic fungi can easily find nutrients and suitable environmental conditions for multiplication within the root, thus creating a stable source of inoculum to sustain the population of the microorganism and to escape competition and lack of receptivity in the rhizospheric soil (Maciá-Vicente *et al*., 2009).

Organic Amendments

Soil amendments may reduce nematode infestations and also provide an energy source for BCA. Pre-colonized substrates are most effective in establishing nematophagous fungi in soil (Kerry, 1997). However, colonization of soil organic matter by *P. chlamydosporia* is limited. De Leij *et al.* (1992) reported a large difference between the high and low rates of fungal inoculum (*i.e.*, 10^3 and 10^4 spore g^{-1} soil) in the establishment of the fungus in the compost. Differences in soil populations of the fungus in peat compost did not result in significantly different fungal populations on the root surface. However, readily accessible nutrients in soil and addition of organic soil improvers reduce fungal parasitic activity and hence biological control by the fungus (Esteves *et al.*, 2009c).

Nutritional Effects in Parasitic Activity

Nutrition is considered to be a key factor in switching from saprophytic to parasitic phases of the fungus. Mauchline *et al.* (2004) pointed out that differential growth of *Pochonia* isolates indicated great variation in the ability of *P. chlamydosporia* isolates to use root exudates saprophytically, and the qualitative and/or quantitative difference in nutrients available in the rhizosphere of plants plays an important role. The enzyme VCP1 is a useful indicator of the change in state of the fungus (*i.e.*, saprophyctic *vs* parasitic), and research is focusing on the effect of nutrient availability on *vcp1* gene expression (Ward *et al.*, 2010, 2011). The analysis of expression profiles of *P. chlamydosporia* using a cDNA-amplified fragment length polymorphism approach showed that genes expressed during egg parasitism are involved among others in metabolic functions, cellular signal expression and DNA repair, thus being possible that differential expression of parasitism related genes can play a role in fungal signalling and infection (Rosso *et al.*, 2011).

Virulence and Fitness Cost

Increased dependence of the fungus on nematodes as a source of nutrition has been associated with a decrease in saprotrophic competitiveness and is often negatively correlated with their parasitic ability (Siddiqui *et al.*, 2009). These authors reported a negative correlation between population densities in soil of different *P. chlamydosporia* biotypes and their parasitism of *M. hapla* and *G. pallida* eggs in tests using compatible biotypes. The cyst nematode biotype was the most virulent against both nematode eggs, but was a less effective soil and rhizosphere colonizer in comparison to other isolates tested, thereby suggesting that virulence may have a fitness cost (Siddiqui *et al.*, 2009).

Genetic and Functional Diversity

PCR-based DNA fingerprinting techniques can provide a rapid means to examine the genetic variation of fungal isolates from various geographical regions and nematode hosts

(Kerry *et al.*, 2004), and also to monitor biocontrol agents after their release into soil. However, few investigations have been made on population diversity and its effect in the control exerted on nematode populations for individual or combinations of fungal isolates. Nowadays, PCR in combination with classical methods, facilitates identification of fungal varieties (PCR-β-tubulin), intra-specific variation by ERIC-PCR fingerprinting, abundance (qPCR) and associated parasitic ability of the fungus between root-knot and cyst nematode based on a serine protease assay (PCR-VCP1) (Arora *et al.*, 1996; Morton *et al.*, 2003b). One study, using ERIC-PCR fingerprinting profiles, revealed that isolates that occurred together in roots and eggs could show different prevalence or absence in some combinations, thus indicating that root colonization and parasitic abilities of a single isolate may differ when applied in combination with other isolates (Manzanilla-López *et al.*, 2009b).

A biological aspect that has not received much attention is the genetic compatibility/incompatibility between native and introduced fungal biotypes and its significance for biological control. There is evidence from a limited number of fungal biotypes tested *in vitro* that anastomosis may occur within biotype isolates from the same nematode host (*i.e.,* root-knot, potato and cereal cyst nematodes) but fungal biotypes isolated from different hosts appear not to anastomose, thereby supporting the supposition that different biotypes can occur together in the same niche and may compete with each other within the rhizosphere yet are unlikely to exhibit genetic interactions. However, further studies with more isolates are required to confirm incompatibility (Manzanilla-López *et al.*, 2011b).

Epidemiology

Although molecular methods to quantify the fungus in soil and root tissues are becoming increasingly available, there are still limitations in correlating levels of genomic DNA (qPCR) with classic CFU counts to assess fungal abundance. There is limited knowledge of the epidemiology and quantitative relations of the fungus with its hosts. Development of models incorporating multiple variables is critical to get a better understanding of the factors that operate together to affect fungal BCA efficacy (Jaronsky, 2010).

VII. Integrated Pest Management and Future Perspectives

Integrated pest management for plant-parasitic nematodes includes the use, in addition to crop rotation, of resistant cultivars and nematicides, although the former is not always effective or economically viable (Jacobs *et al.*, 2003). As a result, there is a need for effective novel control strategies that can be included in an IPM framework. *Pochonia chlamydosporia* is an important enemy of nematodes in suppressive soils, hence its potential use to control nematode pests (Kerry & Crump, 1998; Kerry *et al.*, 1993; Atkins *et al.*, 2003a; Jacobs *et al.*, 2003; Wang *et al.*, 2005; Tobin *et al.*, 2008). Rhizosphere interactions between *P. chlamydosporia*, plant and nematode are complex and isolates from different nematode hosts should be considered as different biotypes according to their original host. Bio-management strategies for control of nematodes incorporate the use of *P. chlamydosporia* in combination with selected cultivars of host plants that are less susceptible or resistant to the nematode and support extensive growth of the fungus in their rhizosphere. Hosts that release rhizodeposits that could support fungal growth in the rhizosphere and an endophytic behaviour will also need to be considered (Manzanilla-López *et al.*, 2011a). According to Jacobs *et al.* (2003), some potato growers already apply two control measures for PCN, a fumigant in the autumn followed by a granular nematicide in the spring at a cost of *ca* £900 ha^{-1}. Hence separate applications of two biological control agents, such as *Pochonia* spp. and another nematophagous fungus (*e.g.*, *Paecilomyces lilacinus*), may be feasible (Jacobs *et al.*, 2003). Tobin *et al.* (2008) have also pointed out the potential of using *P. chlamydosporia* to control PCN in potato crops grown under commercial field conditions. However, the potential success of such a biological control agent should be based upon the careful selection and combination of the fungal isolate biotype and host plant to be included as break crops or crop rotation against a specific pest species. Although fungal chlamydospore-based products for application to soil as an inoculum have been shown to be commercially viable, there is a need to reduce application dosages and optimise production and formulation methods as well as to conduct more validation tests of fungal products in the field. So far, seed treatment has been ineffective, but recent studies on the endophytic behaviour of the fungus could encourage research on this area. From the ecological perspective it is important to consider soil receptivity to non-native isolates (Monfort *et al.*, 2006; Stirling *et al.*, 1998) as well as to understand how a BCA interacts with the resident microbiota (*e.g.*, competition, predation and parasitism by other microorganisms), including non-target pathogens, and other potencial biocontrol agents as well as their response to chemical pesticides (Jacobs *et al.*, 2003). There is also a need to conduct studies on simultaneous application of different biocontrol agents as well as to build up infrastructure capability to follow the establishment, dynamics and, if possible, the epidemiology of the fungus in the soil and rhizosphere using molecular tools once an isolate is introduced for biocontrol purposes (Atkins *et al.*, 2003b). According to Leinhos and Buchenauer (1992): "The broad range of antagonistic modes of action including antibiosis and hyperparasitism as well as the wide range of susceptible fungi and nematodes may indicate *P. chlamydosporia* as an outstanding candidate for the development of a biological control agent against both pests and diseases".

Proteomics and other –omics will open new research opportunities for the fungus, including genome sequencing of at least some fungal isolates, characterization of structure and function of the genome, discovery of new genes involved in the host-parasite relationship as well a re-evaluation of those involved in virulence, mechanisms involved in signalling and reception of stimuli. It is expected that a better understanding of the fungus-nematode interactions at the molecular level will provide new methods to improve genetic and chemical interactions that could lead to a more efficient and effective use of this biocontrol agent (Manzanilla-López *et al.*, 2011b).

Acknowledgements

Authors would like to dedicate the chapter to Prof. Brian R. Kerry. Rothamsted Research is an institute of the Biotechnology and Biological Research Council (BBSRC), UK.

References

Agudelo, P., Robbins, R.T., Stewart, J. McD. & Murphy, J.B. (2004). Glycoproteins in the gelatinous matrix of *Rotylenchulus reniformis*. *Journal of Nematology* 34: 229-234.

Araújo, J.M., Araújo, J.V., Braga, F.R., Carvalho, R.O., Silva, A.R. & Campos, A.K. (2009). Interaction & ovicidal activity of nematophagous fungus *Pochonia chlamydosporia* on *Taenia saginata* eggs. *Experimental Parasitology* 121: 338-341.

Arora, D.K., Hirsch, P.R. & Kerry, B.R. (1996). PCR-based molecular discrimination of *Verticillium chlamydosporium* isolates. *Mycological Research* 100: 801-809.

Atkins, S.D., Hidalgo-Diaz, L., Kalisz, H., Mauchline, T.H., Hirsch, P.R. & Kerry, B.R. (2003a). Development of a new management strategy for the control of root-knot nematodes (*Meloidogyne* spp.) in organic vegetable production. *Pest Management Science* 59: 183-189.

Atkins, S.D., Hidalgo-Díaz, L., Clark, I.M., Morton, C.O., Montes de Oca, N., Gray, P.A. & Kerry, B.R. (2003b). Approaches for monitoring the release of *P. chlamydosporia* var. *catenulata*, a biological control agent of root-knot nematodes. *Mycological Research* 107: 206-212.

Atkins, S.D., Sosnowska, D., Evans, V.J., Clark, I.M., Hirsch, P.R. & Kerry, B.R. (2004). Investigation of three nematophagous fungi in two potato cyst nematode suppressive soils. *Multitrophic Interactions in Soil and Integrated Control IOBC wprs Bulletin* 27: 1-8.

Ayatollahy, E., Fatemy, S. & Etebarian, H.R. (2008). Potential for biological control of *Heterodera schachtii* by *Pochonia chlamydosporia* var. *chlamydosporia* on sugar beet. *Biocontrol Science and Technology* 18: 157-167.

Bachem, C., Oomen R. & Visser, G. (1998). Trancript Imaging with cDNA-AFLP: A step-by-step protocol. *Plant Molecular Biology Reporter* 16: 157-173.

Bird, A.F. & Bird, J. (1991). *The structure of nematodes*. San Diego, Academic Press.

Bird, A.F. & McClure, M.A. (1976). The tylenchid (Nematoda) egg shell: structure, composition and permeability. *Parasitology* 72: 19-28.

Blakeman, J.P. & Parbery, D.G. (1977). Stimulation of appressorium formation in *Colletotrichum acutatum* by phylloplane bacteria. *Physiological Plant Pathology* 11: 313-325.

Bordallo, J.J., Lopez-Llorca, L.V., Hansson, H.B., Salinas, J., Persmark, L. & Asensio, L. (2002). Colonization of plant roots by egg-parasitic and nematode trapping-fungi. *New Phytologist* 154, 491-499.

Bourne, J.M., Kerry, B.R. & de Leij, F.A.A.M. (1996). The importance of the host plant on the interaction between root-knot nematodes (*Meloidogyne* spp.) and the nematophagous fungus *Verticillium chlamydosporium* Goddard. *Biocontrol Science and Technology* 6: 539-548.

Bourne, J.M. & Kerry, B.R. (1999). Effect of the host plant on the efficacy of *Verticillium chlamydosporium* as a biological control agent of root-knot nematodes at different nematode densities and fungal application rates. *Soil Biology and Biochemistry* 31: 75-84.

Bourne, J.M. & Kerry, B.R. (2000). Observations on the survival and competitive ability of the nematophagous fungus *Verticillium chlamydosporium* in soil. *International Journal of Nematology* 10: 9-18.

Bourne, J.M., Karanja, P.K., Kalisz, H., Karanja, D.K., Mauchline, T.H. & Kerry, B.R. (2004). Incidence and severity of damage caused by *Meloidogyne* spp. and isolation and screening of the nematophagous fungus *Pochonia chlamydosporia* from some of the main vegetable growing areas in Kenya. *International Journal of Nematology* 14: 111-122.

Braga, F.R., Araújo, J.V., Carvalho, R.O., Silva, A.R., Araujo, J.M., Soares, F.E.F., André, G.L.H., Ferreira, S.R. & Queiroz, J.H. (2010). Ovicidal action of a crude enzymatic extract of the fungus *Pochonia chlamydosporia* against cyathostomin eggs. *Veterinary Parasitology* 72: 264-268.

Brown, V., Sabina, J. & Johnston, M. (2009). Specialized sugar sensing in diverse fungi. *Current Biology* 19: 436–441.

Bruck, D.J. (2010). Fungal entomopathogens in the rhizosphere. *Biocontrol* 55: 103-112.

Campos, H.D. & Campos, V.P. (1997). Efeito da época e forma de aplicação dos fungos *Arthrobototrys conocides*, *Arthobotrys musiformis*, *Paecilomyces lilacinus* e *Verticillium chlamydosporium* no controle de *Meloidogyne exigua* do cafeeiro. *Fitopatologia Brasileira* 22: 361-365.

Chung, K.R. & Liao, H.L. (2008). Determination of a transcriptional regulator-like gene involved in biosynthesis of elsinochrome phytotoxin by the citrus scab fungus, *Elsinoë fawcettii*. *Microbiology* 154: 3556–3566.

Clarke, A.J. (1970). The composition of the cyst wall of the beet cyst-nematode *Heterodera schachtii*. *Biochemical Journal* 118: 315-318.

D'Angieri, C.N.F. & Campos, V.P. (1997). Control de *Meloidogyne javanica* em Jaborandi (*Pilocarpus microphyllus*) com *Arthrobotrys conoides*, *Paecilomyces lilacinus* e *Verticillium chlamydosporium*. *Nematologia Brasileira* 21: 23-30.

Davies, K.G., de Leij, F.A.A.M. & Kerry, B.R. (1991). Microbial agents for the biological control of plant-parasitic nematodes in tropical agriculture. *Tropical Pest Management* 37: 303-320.

de Leij, F.A.A.M. (1992). *Significance of ecology in the development of* Verticillium chlamydosporium *as a biological control agent against root-knot nematodes (*Meloidogyne *spp.)*. University of Wageningen, The Netherlands, (Ph.D. Thesis).

de Leij, F.A.A.M. & Kerry, B.R. (1991). The nematophagous fungus, *Verticillium chlamydosporium*, as a potential control agent for *Meloidogyne arenaria*. *Revue Nématologie* 14: 157-164.

de Leij, F.A.A.M., Davies, K.G. & Kerry, B.R. (1992). The use of *Verticillium chlamydosporium* and *Pasteuria penetrans* alone and in combination to control *Meloidogyne incognita* on tomato plants. *Fundamental and Applied Nematology* 15: 235-242.

de Leij, F.A.A.M., Kerry, B.R. & Dennehy, J.A. (1993). *Verticillium chlamydosporium* as a biological control agent for *Meloidogyne incognita* and *M. Hapla* in pots and micro-plot tests. *Nematologica* 39: 115-126.

Domsh, K.H. & Gams, W. (1972). Fungi in agricultural soils. Edited by Longman editions, pp. 290.

Duan, E., Yang, E., Xiang, M. & Liu, X. (2008). Effect of storage conditions on the survival of two potential biocontrol agents of nematodes, the fungi *Paecilomyces lilacinus* and *Pochonia chlamydosporia*. *Biocontrol Science and Technology* 6: 613-620.

Dufresne M., Perfect S., Pellier A.L., Bailey, J. A. & Langin, T. (2000). A GAL4-like protein is involved in the switch between biotrophic and necrotrophic phases of the infection process of *Colletotrichum lindemuthianum* on common bean. *Plant Cell* 12: 1579-1589.

Eapen, J.A., Beena, B. & Ramana, K.V. (2009). Field evaluation of *Trichoderma harzianum*, *Pochonia chlamydosporia* and *Pasteuria penetrans* in a root knot nematode infested black pepper (*Piper nigrum* L.) garden in India. *Journal of Plantation Crops* 37: 196-200.

Esteves, I. (2007). *Factors affecting the performance of Pochonia chlamydosporia as a biological control agent for nematodes*. Cranfield University, United Kingdom, (Ph.D. Thesis), 254 pp.

Esteves I., Peteira, B., Powers, S., Magan, N. & Kerry, B.R. (2009a). Effects of osmotic and matric potential on radial growth and accumulation of endogenous reserves in three isolates of *Pochonia chlamydosporia*. *Biocontrol Science and Technology* 19: 185-199.

Esteves, I., Peteira, B., Atkins, S.D., Magan, N. & Kerry, B.R. (2009b). Production of extracellular enzymes by different isolates of *Pochonia chlamydosporia*. *Mycological Research* 113: 867-876.

Esteves, I., Manzanilla-López, R.H. & Kerry, B.R. (2009c). Can an understanding of the biology and ecology *Pochonia chlamydosporia* improve its performance as a biological control agent? 2nd International Congress of Tropical Nematology (40th Organization of Nematologists of Tropical America and 28th Sociedade Brasileira de Nematologia Meetings). October 4-9, 2009. Maceió, Alagoas State, Brazil. *(www.ontaweb.org).*

FAO Agricultural Services Bulletin 96 (1992). Small-scale processing of microbial pesticides. Rome, Italy, 90 pp.

Fassatiová, O. & Lysek, H. (1982). Ovicidal fungi in soil ecological system. *Acta Universitatis Carolinae-Biologica* 9: 297-334.

Flores-Camacho, R., Manzanilla-López, R.H., Cid del Prado-Vera, I. & Martínez-Garza, A. (2007). Control of *Nacobbus aberrans* (Thorne, 1935) Thorne y Allen, 1944 with *Pochonia chlamydosporia* (= *Verticillium chlamydosporium*) (Goddard) Zare and W. Gams. *Revista Mexicana de Fitopatología* 25: 26-34.

Forrest, J.M. S. & Robertson, W.M. (1986). Characterization and localization of saccharides on the head region of four populations of the potato cyst nematode *Globodera rostochiensis* and *G. pallida*. *Journal of Nematology* 18, 23-23.

Franco-Navarro, F., Vilchis-Martínez, K. & Miranda-Damián, J. (2009). New records of *Pochonia chlamydosporia* from Mexico: isolation, root colonization and parasitism of *Nacobbus aberrans* eggs. *Nematropica* 39: 133-142.

Frassy, L.N., Braga, F.R., Silva, A.R., Araújo, A.R., Ferreira, S.R. & de Freitas, L.G. (2010). Destruição de ovos de *Toxocara canis* pelo fungo nematófago *Pochonia chlamydosporia*. *Revista da Sociedade Brasileira de Medicina Tropical* 43: 102-104.

Freitas, L.G., Dallemole-Giaretta, R., Ferraz, S., Zooca, R.J.F & Podestá, G.S. (2009). Controle biológico de nematoides: Estudo de casos. In Zambolim, L. and Picanço M.C. (Eds). Controle biológico de pragas e doenças: exemplos práticos. Viçosa, MG, UFV/DFP. 41-82.

Gams, W. (1988). A contribution to the knowledge of nematophagous species of *Verticillium*. *The Netherlands Journal of Plant Pathology* 94: 123-148.

García, L., Melchor, G., Arévalos, J. & Hidalgo-Díaz, L. (2008). Phytoxicity testing of the strain IMI SD 187 of *Pochonia chlamydosporia* var. *catenulata* on *Zea mays* L. and *Phaseolus vulgaris* L. *Revista de Protección Vegetal* 23: 38-42.

Gaspard, J.T., Jaffee, B.A. & Ferris, H. (1990). *Meloidogyne incognita* survival in soil infested with *Paecilomyces lilacinus* and *Verticillium chlamydosporium*. *Journal of Nematology* 22: 176-181.

Hallmann, H., Rodríguez-Kábana, R. & Kloepper (1999). Chitin-mediated changes in bacterial communities of the soil, rhizosphere and within roots of cotton in relation to nematode control. *Soil Biology and Biochemistry* 31: 551-560.

Hidalgo-Díaz, L., Bourne, J.M., Kerry, B.R. & Rodriguez, M.G. (2000). Nematophagous *Verticillium* spp. in soils infested with *Meloidogyne* spp. in Cuba: isolation and screening. *International Journal of Pest Management* 46: 277-284.

Huang, X., Zhao, N. & Zhang, K. (2004). Extracellular enzymes serving as virulence factors in nematophagous fungi involved in infection of the host. *Research in Microbiology* 155: 811-816.

Jackson, C.W. & Schisler, D.A. (1992). The composition and attributes of *Colletotrichum truncatum* spores are altered by the nutritional environment. *Applied and Environmental Microbiology* 58: 2260-2265.

Jacobs, H., Gray, S.N. & Crump, D.H. (2003). Interactions between nematophagous fungi and consequences for their potential as biological agents for the control of potato cyst nematodes. *Mycological Research* 107: 47-56.

Jaronski, S.T. (2010). Ecological factors in the inundative use of fungal entomopathogens. *BioControl* 55: 159-185.

Johannesson, H., Kasuga, T., Schaller, R.A., Good, B., Gardner, M.J., Townsend, J.P., Cole, G.T. & Taylor, J.W. (2006). Phase-specific gene expression underlying morphological adaptations of the dimorphic human pathogenic fungus, *Coccidioides posadasii*. *Fungal Genetics and Biology* 43: 545-559.

Johnston, M. (1987). A model fungal gene regulatory mechanism: the *GAL* genes of *Saccharomyces cerevisiae*. *Microbiological Reviews* 51: 458-476.

Kerry, B.R. (1987). Biological control. In R.H. Brown & B.R. Kerry (Eds). *Principles and Practice of Nematode Control in Crops*. Academic Press, Sydney, Australia, pp. 233-263.

Kerry, B.R. (1997). Biological control of nematodes: prospects and opportunities. FAO Corporate Document Repository. Plant Nematode Problems and their control in the Near East Region. FAO Plant Production and Protection Paper, 144. *http://www.fao.docrep/V9978E00.htm*

Kerry, B.R. (2000). Rhizosphere interactions and the exploitation of microbial agents for the biological control of plant-parasitic nematodes. *Annual Review of Phytopathology* 38: 423-441.

Kerry, B.R. & Crump, D.H. (1998). The dynamics of the decline of the cereal cyst nematode, *Heterodera avenae*, in four soils under intensive cereal production. *Fundamental and Applied Nematology* 21: 617-625.

Kerry, B.R., Irving, F. & Hornsey, J.C. (1986). Variation between strains of the nematophagous fungus *Verticillium chlamydosporium* Goddard I. Factors affecting growth *in vitro*. *Nematologica* 32: 461-473.

Kerry, B.R., Kirkwood, I.A., de Leij, F.A.A.M., Barba, J., Leijdens, M.B. & Brookes, P.C. (1993). Growth and survival of *Verticillium chlamydosporium* Goddard, a parasite of nematodes in soil. *Biocontrol Science and Technology* 3: 355-365.

Kerry, B.R., Atkins, S., Mauchline, T., Morton, O. & Hirsch, P. (2004). Variation in *Pochonia chlamydosporia* and its potential as a biological control agent for root-knot nematodes. *Multitrophic Interactions in Soil and Integrated Control IOBC wprs Bulletin* 27: 127-132.

Khambay, B.P.S., Bourne, J.M., Cameron, S., Kerry, B.R., Zaki, M. & Jave, C.S. (2000). A nematicidal metabolite from *Verticillium chlamydosporium*. *Pest Management Science* 56: 1098-1099.

Kok, C.J., Papert, A. & Hok-a-Hin, C.H. (2001). Microflora of *Meloidogyne* egg masses: species composition, population density and the effect on the biocontrol agent *Verticillium chlamydosporium* (Goddard). *Nematology* 3: 729-734.

Kokalis-Burelle, N., Martinez-Ochoa, N., Rodríguez-Kábana, R. & Kloepper, J. (2002). Development of multi-component transplant mixes for suppression of *Meloidogyne incognita* on tomato (*Lycopersicon esculentum*). *Journal of Nematology* 34: 362-369.

Leinhos, G.M.E. & Buchenauer, H. (1992). Inhibition of rust diseases of cereals by metabolic products of *Verticillium chlamydosporium*. *Journal of Phytopathology* 136: 177-193.

Lewis, J.A. & Papavizas, G.C. (1985). Effect of mycelia preparations of *Trichoderma* and *Gliocladium* on populations of *Rhizoctonia solani* and the incidence of damping-off. *Phytopathology* 75: 812-817.

Liu, X.Z. & Chen, S.Y. (2003). Nutritional requirements of *P. chlamydosporia* and ARF18, fungal parasites of nematode eggs. *Journal of Invertebrate Pathology* 83: 10-15.

Lopes, E.A., Ferraz, S., Ferreira, P.A., Freitas, L.G., Dhingra, O.D., Gardiano, C.G. & Carvalho, AS.L. (2007). Potential de isolados de fungos nematófagos no controle de *Meloidogyne javanica*. *Nematologia Brasileira* 31: 74-84.

Lopez-Llorca, L.V., & Claugher, D. (1990). Appressoria of the nematophagous fungus *Verticillium suchlasporium*. *Micron and Microscopica Acta* 21: 125-130.

Lopez-Llorca, L.V. & Robertson, W.M. (1992). Ultrastructure of infection of cyst nematode eggs by the nematophagous fungus *Verticillium suchlasporium*. *Nematologica* 39: 65-74.

Lopez-Llorca, L.V. & Boag, B. (1993). Biological properties of a red pigment produced by the nematophagous fungus *Verticillium suchlasporium*. *Nematologia Mediterranea* 21: 143-149.

Lopez-Llorca, L.V., Moya, M. & Linares, A. (1994). Effect of pH on growth and pigment production of nematophagous and entomogeneous fungi. *Micologia Vegetale Mediterranea* 8: 107-112.

Lopez-Llorca, L., Olivares-Bernabeu, C., Salinas, J., Jansson, H.B. & Kolattukudy, P.E. (2002a). Pre-penetration events in fungal parasitism of nematode eggs. *Mycological Research* 106: 499-506.

Lopez-Llorca, L.V., Bordallo, J.J., Monfort, E. & López-Serna, M.L. (2002b). Use of light and scanning electron microscopy to examine colonisation of barley rhizosphere by the nematophagous fungus *Verticillium chlamydosporium*. *Micron* 3: 61-67.

Lopez-Llorca, L.V., Maciá-Vicente, J.G. & Jansson, H.B. (2008). Mode of action and interactions of nematophagous fungi. In *Integrated Management and Biocontrol of Vegetable and Grains Crops Nematodes*. Eds A. Ciancio & K.G. Mukerji. Springer, pp. 51-76.

Lopez-Llorca, L., Gómez-Vidal S., Monfort E., Larriba, E., Casado-Vela, J., Elortza, F., Jansson, H.B., Salinas, J. & Martín-Nieto, J. (2010). Expression of serine proteases in egg-parasitic nematophagous fungi during barley root colonization. *Fungal Genetics and Biology* 47: 342-351.

Lumsden, R.D., Frias, T.G., Gracia, E.R. & Dunn, M.T. (1982). Biocontrol of *Pythium aphanidermatum* on cucumber by microbial isolates from Mexican soils. *Phytopathology* 72: 1010 [Abstract].

Maciá-Vicente, J.G., Jansson, H.B., Talbot, N.J. & Lopez-Llorca, L.V. (2009a). Real-time PCR quantification and live-cell imaging of endophytic colonization of barley (*Hordeum vulgare*) roots by *Fusarium equiseti* and *Pochonia chlamydosporia*. *New Phytologist* 182: 213-228.

Maciá-Vicente, J.G., Rosso, L.C., Ciancio, A., Jansson, H.B. & Lopez-Llorca, L.V. (2009b). Colonisation of barley roots by endophytic *Fusarium equiseti* and *Pochonia chlamydosporia*: effects on plant growth and disease. *Annals of Applied Biology* 155: 391-401.

Magan, N. (2001). Physiological approaches to improving the ecological fitness of fungal biocontrol agents in fungi as biocontrol agents. In *Fungi as Biocontrol Agents -Progress, Problems and Potential*, pp. 239-251. Eds. T. Butt, C. Jackson & N. Magan. London: CAB International.

Manzanilla-López, R.H., Rowe, J., Gravato-Nobre, M., von Mende, N. & Evans, K. (1998). External secretions from the false root-knot nematode *Nacobbus aberrans* (Thorne, 1935) Thorne & Allen, 1944 (Tylenchida: Pratylenchidae). *Nematologica* 44: 326-330.

Manzanilla-López, R.H., Atkins, S.D., Clark, I.M., Kerry, B.R. & Hirsch, P.R. (2009a). Measuring abundance, diversity and parasitic ability in two populations of the nematophagous fungus *Pochonia chlamydosporia* var. *chlamydosporia*. *Biocontrol Science and Technology* 19: 391-406.

Manzanilla-López, R.H., Clark, I.M., Atkins, S.D. & Kerry, B.R. (2009b). Rapid and reliable DNA extraction and PCR fingerprinting methods to discriminate multiple biotypes of the nematophagous fungus *Pochonia chlamydosporia* isolated from plant rhizospheres. *Letters in Applied Microbiology* 48: 71-76.

Manzanilla-López, R.H., Esteves, I., Powers, S.J. & Kerry, B.R. (2011a). Effects of crop plants on abundance of *Pochonia chlamydosporia* and other fungal parasites of root-knot and potato cyst nematodes. *Annals of Applied Biology* 159: 118-129.

Manzanilla-López, R.H., Clark, I.M., Atkins, S.D., Hirsch, P.R. & Kerry, B.R. (2011b). Exploring competitiveness and variation in the nematophagous fungus *Pochonia chlamydosporia* var. *chlamydosporia* and its significance for biological control. *Bulletin OILB/SROP* 63:37-40.

Mauchline, T.H., Kerry, B.R. & Hirsch, P. (2002). Quantification in soil and the rhizosphere of the nematophagous fungus *Verticillium chlamydosporium* by competitive PCR and comparison with selective plating. *Applied and Environmental Microbiology* 68: 1846-1853.

Mauchline, T.H., Kerry, B.R. & Hirsch, P. (2004). The biocontrol fungus *Pochonia chlamydosporia* shows nematode host preference at the infraspecific level. *Mycological Research* 108: 106-169.

Mi, Q., Yang, J., Ye, F., Gan, Z., Wu, C. Niu, X., Zou, C. & Zhang K.Q. (2010). Cloning and overexpression of *Pochonia chlamydosporia* chitinase gene *pcchi44*, a potential virulence factor in infecting against nematodes. *Process Biochemistry* 45: 810-814.

Mo, M., Xu, C. & Zhang, K. (2005). Effects of carbon and nitrogen sources, carbon-to-nitrogen ratio, and initial pH on the growth of nematophagous fungus *Pochonia chlamydosporia* in liquid culture. *Mycopathologia* 159: 381-387.

Monfort. E., Lopez-Llorca, L.V., Jansson, H.B., Salinas, J., Park, J.O. & Sivasithamparam, K. (2005). Colonisation of seminal roots of wheat and barley by egg-parasitic nematophagous fungi and their effects on *Gaemannomyces graminis* var. *tritici* & development of root-rot. *Soil Biology & Biochemistry* 37: 1229-1235.

Monfort, E., Lopez-Llorca, L.V., Jansson, H.B. & Salinas, J. (2006). *In vitro* soil receptivity assays to egg parasitic nematophagous fungi. *Mycological Progress* 5: 18-23.

Montes de Oca, N. (2004). *Buenas prácticas de fabricación para la obtención de un bionematicida a partir de la cepa Vcc 108 de* Pochonia chlamydosporia *var.* catenulata. Tesis en opción al grado de Doctor en Ciencias Agrícolas. Universidad Agraria de la Habana - UNAH. Centro Nacional de Sanidad Agropecuaria -CENSA. Havana, Cuba.

Montes de Oca, N. (2005). Perspectiva para la fabricación de un bionematicida a partir del hongo *Pochonia chlamydosporia*. *http://www.monografias.com/trabajos26/bionematicida.shtml* (18April 2011)

Morton, C.O., Hirsch, P.R., Peberdy, J.P. & Kerry, B.R. (2003a). Cloning of and genetic variation in protease VCP1 from the nematophagous fungus *Pochonia chlamydosporia*. *Mycological Research* 107: 38-46.

Morton, C.O., Mauchline, T.H., Kerry, B.R. & Hirsch, P.R. (2003b). PCR-based DNA fingerpring indicates host-related genetic variation in the nematophagous fungus *Pochonia chlamydosporia*. *Mycological Research* 107: 198-205.

Morton, O.C., Hirsch, P.R. & Kerry, B.R. (2004). Infection of plant-parasitic nematodes by nematophagous fungi-a review of the application of molecular biology to understand infection process and to improve biological control. *Nematology* 6: 161-170.

Mukhta, T. & Pervaz, I. (2003). *In vitro* evaluation of ovicidal and larvicidal effects of culture filtrate of *Verticillium chlamydosporium* against *Meloidogyne javanica*. *International Journal of Agriculture and Biology* 5: 576-579.

Mutua, G., Kerry, B.R., Hirsch, P., Kimenju, J., Manzanilla-López, R.H. & Ward, E. (2009). The impact of nutrition on parasitism of nematode eggs by *Pochonia chlamydosporia*. British Society for Plant Pathology (BSPP) Meeting. Oxford, UK, 21-22 September 2009, p. 48 [Abstract].

Nagesh, M., Hussaini, S.S., Ramanujam, B. & Rangeswaran, R. (2007). Molecular identification, characterization, variability and infectivity of Indian isolates of the nematophagous fungus *Pochonia chlamydosporia*. *Nematologia Mediterranea* 35: 47-56.

Nahar, P., Ghormade, V. & Deshpande, M.V. (2004). The extracellular constitutive production of chitin deacetylase in *Metharhizium anisopliae*: possible edge to entomopathogenic fungi in the biological control of insect pests. *Journal of Invertebrate Pathology* 85: 80-88.

Niu, X., Wang, Y., Chu, Y., Xue, H., Li, N, Wei, L., Mo, M. & Zhang, K. (2010). Nematode toxic aurovertin-type metabolites from a root-knot nematode parasitic fungus *Pochonia chlamydosporia*. *Journal of Agricultural and Food Chemistry* 58: 828-834.

Nordbring-Hertz, B., Friman, E., & Mattiasson, B. (1982). A recognition mechanism in the adhesion of nematodes to nematode-trapping fungi, p. 83-90. In T. C. Bog-Hansen (Ed.), *Lectins-biology, biochemistry and clinical biochemistry*, vol. 2. Walter de Gruyter, Berlin.

O'Flaherty, S.M., Hirsch, P.R. & Kerry, B.R. (2003). The influence of the root-knot nematode *Meloidogyne incognita*, the nematicide aldicarb and the nematophagous fungus *Pochonia chlamydosporia* on heterotrophic bacteria in soil and the rhizosphere. *European Journal of Soil Science* 54: 759-766.

Olivares-Bernabeu, C.M. & Lopez-Llorca, L.V. (2002). Fungal egg-parasites of plant-parasitic nematodes from Spanish soils. *Revista Iberoamericana de Micología* 19: 104-110.

Palma-Guerrero, J., Jansson, H.B., Salinas, J. & Lopez-Llorca, L.V. (2007). Effect of chitosan on hyphal growth and spore germination of plant pathogenic and biocontrol fungi. *Journal of Applied Microbiology* 104: 541-553.

Palma-Guerrero, J., Gómez-Vidal, S., Tikhonov, V.E., Salinas, J., Jansson, H.B. & Lopez-Llorca, V. (2010). Comparative analysis of extracellular proteins from *Pochonia chlamydosporia* grown with chitosan or chitin as main carbon and nitrogen sources. *Enzyme and Microbial Technology* 46: 568-574.

Pell, J.D., Hannam, J. & Steinkraus, D. (2009). Conservation biological control using fungal entomopathogens. *BioControl* 55: 187-198.

Polesani, M., Desario, F., Ferrarini, A., Zamboni, A., Pezzotti, M., Kortekamp, A. & Polverari, A. (2008). cDNA-AFLP analysis of plant and pathogen genes expressed in grapevine infected with *Plasmopara viticola*. *BMC Genomics*: 9, 142.

Premachandran, D. & Pramer, D. (1984) Role of N-acetylgalactosamine specific protein in trapping of nematodes by *Arthrobotrys oligospora. Applied and Environmental Microbiology* 47: 1358- 1359.

Puertas, A. & Hidalgo-Díaz, L. (2009). Effect of different organic manures over *Pochonia chlamydosporia* var. *catenulata* in the substrate and rhizosphere of tomato plants. *Revista de Protección Vegetal* 24: 162-165.

Puertas, A., de la Noval, B.M., Martínez, B., Miranda, I., Fernández, F. & Hidalgo-Díaz, L. (2006). Interacción de *Pochonia chlamydosporia* var. *catenulata* con *Rhizobium* sp., *Trichoderma harzianum* y *Glomus clarum* en el control de *Meloidogyne incognita*. *Revista de Protección Vegetal* 21: 80 -89.

Rosso, L.C., Finetti-Sialer, M.M., Hirsch, P.R., Ciancio, A., Kerry, B.R. & Clark, I.M. (2011). Transcriptome analysis shows differential gene expression in the saprotrophic to parasitic transition of *Pochonia chlamydosporia*. *Applied Microbiology and Biotechnology* 90: 1981-1994.

Segers, R. (1996). The nematophagous fungus *Verticillium chlamydosporium*: aspects of pathogenicity. Pp. 223: University of Nottingham.

Segers, R., Butt, T., Kerry, B.R. & Peberdy, J.F. (1994). The nematophagous fungus *Verticillium chlamydosporium* produces a chymoeslastase-like protease which hydrolyses host nematode proteins in situ. *Microbiology* 140: 2715-2723.

Segers, R., Butt, T.M., Keen, J.N., Kerry, B.R. & Peberdy, J.F. (1995). The subtilisins of the invertebrate mycopathogens *Verticillium chlamydosporium* and *Metarhizium anisopliae* are serollogicaly and functionally related. *FEMS Microbiology letters* 126: 227-232.

Segers, R., Butt, M.T., Kerry, B.R., Beckett, A. & Peberdy, J.F. (1996). The role of the proteinase VCP1 produced by the nematophagous *Verticillium chlamydosporium* in the infection process of nematode eggs. *Mycological Research* 100: 421-428.

Sharon, E. & Spiegel, Y. (1993). Glycoprotein characterization of the gelatinous matrix in the root-knot nematode *Meloidogyne javanica*. *Journal of Nematology* 25: 585-589.

Siddiqui, I.A., Atkins, S.D. & Kerry, B.R. (2009). Relationship between saprotropic growth in soil of different biotypes of *Pochonia chlamydosporia* and the infection of nematode eggs. *Annals of Applied Biology* 155: 131-141.

Silva, A.R., Araújo, J.V., Braga, F.R., Alves, C.D.F. & Frasy, L.N. (2010). *In vitro* ovicidal activity of the nematophagous fungi *Duddingtonia flagrans*, *Monacrosporium thaumasium* and *Pochonia chlamydosporia* on *Trichuris vulpis* eggs. *Veterinary Parasitology* 172: 76-79.

St. Leger, R.J., Butt, T.M., Goettel, M.S., Staples, R.C. & Roberts, W.D. (1989). Production *in vitro* of appressoria by the entomopathogenic fungus *Metarhizium anisopliae*. *Experimental Mycology* 13: 274-288

Stadler, M., Hans-Volker, T., Katsiou, E. & Hellwig, V. (2003). Chemotaxonomy of *Pochonia* and other conidial fungi with *Verticillium*-like anamorphs. *Mycological Progress* 2: 95-122.

Stirling, G.R. & Smith, L. (1998). Field tests of formulated products containing either *Verticillium chlamydosporium* or *Arthrobotrys dactyloides* for biological control of root-knot nematodes. *Biological* Control 11: 231-239.

Stirling, G.R., Licastro, K.A. West, L.M. & Smith, L.J. (1998). Development of commercially acceptable formulations of the nematophagous fungus *Verticillium chlamydosporium*. *Biological Control* 11: 215-221.

Sutherland, E.D. & Papavizas, G.C. (1991). Evaluation of oospore hyperparasites for the control of Phytophthora crown rot on pepper. *Journal of Phytopathology* 131: 33-39.

Tahseen, Q., Clark, I.M., Atkins, S.D., Kerry, B.R. & Hirsch, P.R. (2005). Impact of the nematophagous fungus *Pochonia chlamydosporia* on nematode and microbial populations. *Communications Applied Biological Sciences* 70: 81-87.

Tikhonov, V.E., Lopez-Llorca, L.V., Salinas, J. & Jansson, H.B. (2002). Purification and characterization of chitinases from the nematophagous fungi *Verticillium chlamydosporium* and *V. suchlasporium*. *Fungal Genetics and Biology* 35: 67-78.

Tobin, J.D., Haydock, P.P.J., Hare, M.C., Woods, S.R. & Crump, D.H. (2008). Effect of the fungus *Pochonia chlamydosporia* and fosthiazate on the multiplication rate of potato cyst nematodes (*Globodera pallida* and *G. rostochiensis*) in potato crops grown under UK field conditions. *Biological Control* 46: 194-201.

Van Damme, V., Hoedekie, A. & Viaene, N. (2005). Long-term efficacy of *Pochonia chlamydosporia* for management of *Meloidogyne javanica* in glasshouse crops. *Nematology* 7: 727-736.

Verdejo-Lucas, S., Sorribas, F.J., Ornat, C. & Galeano, M. (2003). Evaluating *Pochonia chlamydosporia* in a double-cropping system of lettuce and tomato in plastic houses infested with *Meloidogyne javanica*. *Plant Pathology* 52: 521-528.

Viaene, M. & Abawi, G.S. (2000). *Hirsutella rhossiliensis* and *Verticillium chlamydosporium* as biocontrol agents of the root-knot nematode *Meloidogyne hapla* on lettuce. *Journal of Nematology* 32: 85-100.

Wang, K., Riggs, R.D. & Crippen, D. (2005). Isolation, selection, and efficacy of *Pochonia chlamydosporia* for control of *Rotylenchulus reniformis* on cotton. *Phytopathology* 95: 890-893.

Ward, E., Mutua, G., Manzanilla-López, R.H. & Kerry, B.R. (2010). Nutrient regulation of parasitism in *Pochonia chlamydosporia*. Proceedings Organization of Nematologists of Tropical America XLII Annual Meeting. Quito, Ecuador 3-8 October 2010, p. 40. *(www.ontaweb.org).*

Ward, E., Kerry, B., Manzanilla-López, R. & Hirsch, P. (2011). Nutrient regulation of parasitism genes in the nematophagous fungus *Pochonia chlamydosporia*. The 26th Fungal Genetics Conference at Asilomar, USA, 15th-20th March 2011. *http://www.fgsc.net/26thfgc/26FGCProgramAndAbstracts.pdf*

Yang, J., Tian, B., Liang, L. & Zhang, K. (2007). Extracellular enzymes and the pathogenesis of nematophagous fungi. *Applied Microbiology and Biotechnology* 75: 21-31.

Zare, R., Gams, W. & Culham, A. (2000). A revision of *Verticillium* sect. Prostrata I. Phylogenetic studies using ITS sequences. *Nova Hedwigiia* 71: 465-480.

Zare, R., Gams, W. & Evans, H.C. (2001). A revision of *Verticillium* section Prostrata. V. the genus *Pochonia*, with notes on *Rotiferophthora*. *Nova Hedwigia* 73: 51-86.

In: Nematodes
Editors: F. Boeri and J. A. Chung

ISBN: 978-1-61470-784-4

Chapter XV

Biodiversity of Nematodes in the Era of 'Omics'

Punyasloke Bhadury*
Integrative Taxonomy and Microbial Ecology Group, Department of Biological Sciences, Indian Institute of Science Education and Research-Kolkata, Mohanpur Campus, West Bengal, INDIA.

Abstract

Nematodes play important role in terrestrial and marine ecosystems; nevertheless nematode species diversity in ecology has been poorly studied. Taxonomic intractability is one of the main reasons that has resulted in poor understanding of biodiversity of nematodes from different ecological niches. The era of genomics is playing a pivotal role in exploring and understanding nematode biodiversity in different environments ranging from terrestrial to deep-sea ecosystems. In this review paper we will discuss how DNA barcoding and massively parallel sequencing approaches have helped us gain a better understanding of nematode diversity from different environments. Also we will discuss wider applicability of massively parallel sequencing technique at the transcriptome level so as to understand the role of this key metazoan group facilitating major ecosystem processes.

Introduction

Nematodes are the most abundant metazoans in the world comprising of upto 10^8 individuals per square metre (Lambshead 2004). Although their species richness is predicted

* Integrative Taxonomy and Microbial Ecology Group, Department of Biological Sciences, Indian Institute of Science Education and Research-Kolkata, Mohanpur Campus, P.O. BCKV Campus Main Office, Mohanpur-741252, Nadia, West Bengal, INDIA. Email: pbhadury@iiserkol.ac.in

to be over 1 million (Lambshead 2004), the number of described nematode species is estimated to be around 24,000 (Hallan 2007). The discrepancy between 'known' and estimated species richness in nematodes is mainly due to the limitations of light microscopy based on superficial morphological characters. Application of molecular techniques and the development of DNA sequence-based approaches have led to considerable progress in nematode identification over the last decade, revealing unusually large diversity even in environments that are relatively well studied (e.g. Floyd et al. 2002). The application of molecular techniques and the development of DNA sequence-based approaches have revolutionized the field of biological sciences. Techniques including polymerase chain reaction dependent denaturing gradient gel electrophoresis (PCR-DGGE), clone library and sequencing approach and terminal-restriction fragment length polymorphism (T-RFLP) are increasingly used in nematology research (e.g. Waite et al 2003, Cook et al 2005, Donn et al 2008, Chen et al 2010). Amongst the sequence-based approaches, DNA barcoding has gained increased importance in the field of nematology with a primary focus on elucidating biodiversity of nematodes across different trophic levels.

DNA barcoding relies on the use of a standardized fragment of the genome as a tag or barcode for rapid and accurate species identification (Hebert and Gregory 2005). In 2003, Hebert et al first proposed the use of a 648 bp region from the 5'end of the mitochondrial cytochrome *c* oxidase subunit 1 (COI) gene as a DNA barcode for identifying all metazoan species. The goal of DNA barcoding primarily is to identify all eukaryotic species (Miller 2007). The DNA barcoding has been tested across several phyla and so far have shown promising results (e.g. Hebert et al 2004, reviews by Bucklin et al 2010, Radulovici et al 2010). However, for DNA barcoding studies of nematodes, alternate markers within the nuclear genome namely small and large subunit ribosomal RNA (18S rRNA and 28S rRNA) have been tested (e.g. Floyd et al 2002, De Ley et al 2005, Bhadury et al 2006) partly because of the problem associated with the amplification of mitochondrial COI gene (e.g. De Ley et al 2005). New high-throughput sequencing techniques are accelerating biodiversity research based on DNA barcoding (e.g. Rayé et al 2010, Hajibabaei et al 2011).

High-throughput DNA sequencing technology is a new revolution that can increase our understanding of nematode diversity and its role in ecosystem processes across different environments. Next-generation sequencing, also known as "massively parallel" sequencing or 454 sequencing, was introduced only a few years back (Roche GS FLX; Margulies et al. 2005), with capacities to process millions of sequences in parallel. The next generation sequencing platform sharply differs from the conventional 96-capillary capacity Sanger sequencing and therefore, placed at the forefront of modern biological research. Individual read lengths are usually limited in most of the next-generation platforms but the depth of coverage per base pair and advanced sequence assembly software allow sequencing of 0.5–60 giga base pair (Gbp) in a single run. Massively parallel sequencing has been applied in several fields of biology including microbial diversity (e.g Sogin et al 2006, Ishak et al 2011), phylogenetics (e.g. Roeding et al 2009) and evolution (e.g. Qi et al 2009).

In this review we will discuss the role of DNA barcoding and massively parallel sequencing approaches for elucidating nematode diversity from soil and marine environments.

Soil Nematode Diversity Studies

Soil nematode species diversity is usually high and in a single soil-core the species counts could vary from 10-100 in a given ecosystem (e.g. Freckman and Virginia 1997, Bloemers et al 1997). Because of the tremendous effort required, records on diversity at the species level are so scarce that a valid ecosystem comparison of nematode species diversity is yet impossible. Nevertheless, increasingly molecular techniques are being used to get an idea about the nematode diversity from terrestrial environments. One of the earliest studies involving the application of DNA barcoding was undertaken by Floyd et al (2002). Using a molecular barcode, derived from single-specimen polymerase chain reaction (PCR) and sequencing of the 5' segment of the small subunit ribosomal RNA (SSU) a molecular operational taxonomic unit (MOTU) scheme has been developed for soil nematodes. Based on the MOTU they undertook rapid assessment of nematode taxon diversity in Scottish grassland and found that many MOTU could be assigned to classical, morphologically defined taxonomic units using a database of SSU sequences from named nematode species. In a different study Powers et al (2009) attempted to estimate nematode diversity in a lowland tropical rainforest of La Selva Biological Research Station in Costa Rica based on 18S rRNA barcoding approach. Based on the amplification, sequencing and phylogenetic analyses of the barcodes generated from 360 individual nematodes it was evident that most of the major terrestrial nematode lineages were represented in the samples. The estimated 'species richness ranged from 464 to 502 within the four 40 × 40 m plots. The authors also showed that directed sampling of insects and their associated nematodes produced a second set of barcodes that were not recovered by habitat sampling, although they may constitute a major class of tropical nematode diversity. However, novel nematode barcodes generated as part of their study were hard to link up with existing taxonomic databases. Additionally their study also highlighted the difficulties associated with studying poorly understood organisms in an understudied ecosystem using a destructive (i.e. barcode) sampling method.

Application of molecular methods including high throughput massively parallel sequencing methodology using the 454 platform can increase our understanding about the diversity soil nematodes across different ecosystems. In a recent study by Porazinska et al (2009) the suitability of massively parallel sequencing for analysis of soil nematode diversity from metagenomic samples was assessed. The authors set four artificial metagenomic samples involving 41 diverse reference nematodes in known abundances and 'known' SSU and LSU diagnostic regions to evaluate the potential of the massively parallel sequencing methodology. The total number of reads ranged from 4159 to 14771 per metagenomic sample and 82% of the reads were > 199 bp in length. Among the reads > 199 bp, 86% matched the referenced species with less than three nucleotide differences from a reference sequence. Their study emphasized the importance of using high-throughput sequencing including a high degree of repeatability of the qualitative analysis of a nematode community and a potentially small error for underestimating species richness.

In another study Porazinska et al (2010) evaluated the prediction that nematode diversity is lower in tropical regions than in temperate broadleaf forests, temperate grasslands or even in cultivated soils (Giller 1996, Boag and Yeates 1998) using high-throughput massively parallel sequencing approaches. They conducted a survey of nematode diversity within the soil, litter and canopy habitats of a humid lowland tropical rainforest of Costa Rica based on

12 metagenetic samples sequenced on a 454 platform. The 12 metagenetic samples generated a total of 239776 SSU reads from which 171861 (72%) were of high quality. Within all high quality reads, 40% were of nematode origin and 26% were identified as other than nematode eukaryotic taxa representing 8 phyla. Based on the analyses of the high-throughput data they predicted that the total number of nematode species in the sampled area could range between 390-497. They also found that none of the observed nematode species were present within all 12 metagenetic samples, and 42% of the species were observed only within one sample. Not only that their study highlighted the fact that only 8 of 214 putative species identified matched with the available nematode SSU sequences in GenBank/EMBL/DDBJ databases. The study clearly indicated that nematode diversity is still underestimated and unreported and more so at the molecular level. Overall their study has shown that both richness and diversity of nematode communities in the tropical rainforests of Costa Rica are high.

Marine Nematode Diversity Studies

In the marine realm the diversity of nematodes is often high and there are no definitive species estimates. However, for example in Europe more than 2000 marine nematode species has been documented (SMEBD, 2011). High numerical dominance along with cryptic species complexes in addition to taxonomic paucity has resulted in the decline of marine nematode studies during biodiversity surveys. Molecular techniques including high throughput sequencing method are increasingly used to understand and estimate marine nematode species diversity and concept of nematode biogeography including cosmopolitanism across different oceanic environments (e.g. Derycke et al 2005, 2007; Bhadury et al 2008)

The study by De Ley et al (2005) is one of the foremost where the applicability of DNA barcoding was evaluated based on the amplification and sequencing of 28S rDNA. In their study 37 nematodes were barcoded based on the D2D3 region of the large subunit ribosomal RNA (28S rDNA) and these represented at least 32 species, none of which matched with available 28S rDNA sequences in public databases. The study highlighted that data-rich surveys and phylogenetic tools for analysis of barcode sequences are an essential component for biodiversity discovery. In another study the applicability of barcoding approach for rapid identification of marine nematodes from estuarine and coastal sedimentary environments of British Isles was evaluated based on the 18S rRNA marker (Bhadury et al 2006). The reported success rate was over 97% as part of their study (Bhadury et al 2006). Pereira et al (2010) used morphological and molecular approaches based on DNA barcoding to evaluate the diversity of free-living marine nematodes (order Enoplida) at four coastal sites in the Gulf of California and three on the Pacific coast of Baja California, Mexico. Their results indicated that 28S sequences (D2D3 domain) may be better suited for DNA barcoding of marine nematodes than those from the 18S rRNA, particularly for differentiating closely related or cryptic species. Incidentally as part of the barcoding studies it was observed that the nematode 18S rRNA primers were also co-amplifying 18S fragments from other eukaryotic phyla and this led to the development of more stringent nematode 18S rRNA primers that could be used for barcoding studies (Bhadury and Austen 2010). Based on the barcoding methodology studies have shown that cryptic species complexes are wide-spread among nematodes in the marine realm and also there are cosmopolitan nematode species (e.g. Bhadury et al 2008)

In a very recent study Derycke et al (2010) have explored the possibility of use of cytochrome oxidase c subunit 1 (COI) for barcoding marine nematodes. In their study they evaluated and successfully amplified COI region from 41 species of marine nematodes and found that the variability pattern observed in the COI could be used to increase the identification of marine nematode species, provided there is a good reference sequence database.

High throughput sequencing technologies have been recently applied to assess marine nematode diversity from coastal environments of United Kingdom. Creer et al (2010) analyzed sediment samples collected from intertidal zone at Littlehampton on the south coast of England, UK as well as soil, litter and understory habitats sampled at La Selva Biological Station, Costa Rica. Community PCR and 454 sequencing based on 18S rDNA approaches yielded a total of 29756 high-quality sequences over 200 bases from the marine samples while for the tropical rain forest samples yielded a total of 40334 high-quality sequences of at least 200 bases. In the terrestrial data set, the soil habitat had fewer nematode operational cluster taxonomic units (OCTUs) (35) than either the marine littoral habitat (149) or canopy (97). The plant-parasitic nematodes were more diverse and abundant in the soil environment, with bacterial and fungal-feeding nematodes predominating in the litter and canopy. This study evaluated the effectiveness of 454 sequencing in nematode diversity studies from varied environments including marine realms.

Fonseca et al (2010) applied a metagenetic approach using next generation sequencing of the 18S nuclear small subunit (nSSU) ribosomal RNA (rRNA) to assessthe relative levels of richness and patterns of diversity of multiple metazoan phyla from eight benthic samples collected from the low-tide zone of an estuarine beach near Prestwick on the West coast of Scotland, and from one sample from a beach in Littlehampton in the South of England. The amplicons were sequenced on a Roche 454 FLX platform generating a total of 353,896 sequences which were subsequently filtered to 305,702 for downstream analysis. Based on the annotation, 374 OCTUs were assigned to phylum level. Of the metazoan OCTUs, 182 were from Nematoda, at least three times more than from any other individual meiofaunal taxon. Platyhelminthes (61 OCTUs) was the second richest phylum, followed by Arthropoda (29 OCTUs including Copepoda, Ostracoda and Malacostraca), Mollusca (22 OCTUs), Gastrotricha (7 OCTUs), Annelida (6 OCTUs) and five less-rich phyla (for example, Bryozoa, Echinodermata, Cercozoa, Rotifera and Alveolata with between 1 and 3 OCTUs each). The comparisons of the OCTU sequences with the NCBI databases revealed that the majority of the Nematoda OCTUs (95%) were not known to science. For other phyla only a small fraction of the OCTUs showed 100% identity with previously sequenced specimens. Only fifteen of the OCTUs with <90% identity (300 sequences) were robustly placed within identified phyla (mainly Nematoda). For the Prestwick samples the authors detected 182 Nematoda OCTUs, compared with 450 species of free-living marine nematodes that have been described previously from around the entire British Isles based on morphological characters. In all, 70% of Nematoda OCTUs were unique to Prestwick and 58–100% of the OCTUs for the other phyla were only present in Prestwick. This study clearly shows the applicability of high throughput sequencing methodology to unravel metazoan diversity including nematodes from coastal environments.

From some of these studies it is clearly evident that nematode species richness across soil and marine realms is largely under-estimated. Techniques like DNA barcoding could be integrated with high throughput sequencing like the 454 platform to track nematode diversity across different biogeographic locations. It would be also prudent to develop dedicated databases for nematodes containing sequence information generated using high-throughput approaches.

Future Directions

The massively parallel sequencing methodology is now being used to study the interaction between nematodes and other organisms at the ecological level. For example, Haegeman et al (2011) studied the transcriptome of the nematode *Pratylenchus coffeae* by sequencing expressed sequence tags (ESTs) on a 454 platform and found putative sequences which are involved in the process of parasitism. They identified several plant cell wall modifying and found a homologue to a chorismate mutase, suggesting that this parasitism gene has a wider occurrence in plant-parasitic nematodes than previously assumed. Based on the EST and high throughput sequencing in 454 platform Mayer et al (2011) have shown the horizontal gene transfer of microbial cellulases into nematode genomes as a key adaptation towards parasitism and pathogenicity. There are numerous studies where 454 sequencing platforms are increasingly used to undertake transcriptome based studies of nematodes in order to gain an understanding of pathogenicity.

Environmental transcriptomics approach has been widely used in microbial studies (e.g. Gilbert et al 2008, Poretsky et al 2009), but has not been attempted in nematodes. High throughput sequencing methods involving platforms like 454 can be effectively used to undertake transcriptome based studies of soil and marine nematodes for understanding their role in ecosystem processes. For example, high throughput sequencing of microRNAs (miRNAs) can help us to know the role of miRNA in regulating gene expression and in development, metabolism and behaviour in nematodes which can be subsequently linked with environmental factors. Additionally, identification of miRNAs and other small RNAs could help us understand the molecular regulation mechanisms in soil or marine nematodes. Environmental transcriptomics of nematodes particularly, from soil and marine realm can help in the discovery of completely novel ecological functions and these can be linked with organismal capabilities, functional diversity, ecosystem processes and evolution.

High throughput sequencing methodologies are undergoing revolution and it is becoming increasingly easier to process large number of samples in a matter of hours and generate huge sequence datasets for interpretation and downstream analyses. The study of nematodes which has primarily relied on taxonomy and therefore, neglected to a certain extent in the past could benefit immensely by integrating high throughput sequencing platforms with DNA barcoding methodology as well as morphological characters based taxonomy. In future large scale international efforts have to be initiated to study the nematodes as part of global biodiversity surveys for understanding and interpreting the ecological role of this abundant metazoan group across different environments.

References

Bhadury, P., Austen, M.C., Bilton, D.T., Lambshead, P.J.D., Rogers, A.D. and Smerdon, G.R. (2006) Development and evaluation of a DNA barcoding approach for the rapid identification of nematodes. *Marine Ecology Progress Series*, 320, 1-9.

Bhadury, P., Austen, M.C., Bilton, D.T., Lambshead, P.J.D., Rogers, A.D. and Smerdon, G.R. (2008) Evaluation of combined morphological and molecular techniques for marine nematode (*Terschellingia* spp.) identification. *Marine Biology* 154: 509-51

Bhadury, P. and Austen, M.C. (2010) Barcoding marine nematodes- an improved set of nematode 18S rRNA primers to overcome eukaryotic co-interference. *Hydrobiologia*, 641, 245-251.

Bloemers, G.F., Hodda, M., Lambshead, P.J.D., Lawton, J.H. and Wanless, F.R. (1997) The effects of forest disturbance on diversity of tropical soil nematodes. *Oecologia*, 111, 575-582.

Boag, B. and Yeates, G.W. (1998) Soil nematode biodiversity in terrestrial ecosystems. *Biodiversity and Conservation*, 7, 617–630.

Bucklin, A., Steinke, D. and Blanco-Bercial, L. (2010) DNA barcoding of marine metazoan. *Annual Review of Marine Science*, 3, 471-508.

Chen, X.Y., Daniell, T.J., Neilson, R., O'Flaherty, V. and Griffiths, B.S. (2010) A comparison of molecular methods for monitoring soil nematodes and their use as biological indicators. *European Journal of Soil Biology*, 46, 319-324.

Cook, A.A., Bhadury, P., Debenham, N.J., Meldal, B.H.M., Blaxter, M.L., Smerdon, G.R., Smerdon, G.R., Austen, M.C., Lambshead, P.J.D. and Rogers, A.D. (2005) Denaturing gradient gel electrophoresis as a tool for the identification of marine nematodes. *Marine Ecology Progress Series*, 291, 103-113.

Creer, S., Fonseca, V.G., Porazinska, D.L., Giblin-Davis, R.M., Sung, W., Power, D.M., Packer, M., Carvalho, G.R., Blaxter, M.L., Lambshead, P.J.D. and Thomas, W.K. (2010) Ultrasequencing of the meiofaunal biosphere: practice, pitfalls and promises. *Molecular Ecology,* 19, 4-20.

De Ley, P., De Ley, I.T., Morris, K., Abebe, E. and 8 others (2005) An integrated approach to fast and informative morphological vouchering of nematodes for applications in molecular barcoding. *Philosophical Transactions of the Royal Society B: Biological Sciences*, 360, 1945–1958.

Derycke, S., Remerie, T., Vierstraete, A., Backeljau, T., VanXeteren, J., Vincx, M. and Moens, T. (2005) Mitochondrial DNA variation and cryptic speciation within the free-living marine nematode *Pellioditis marina*. *Marine Ecology Progress Series*, 300, 91–103.

Derycke, S., Backeljau, T., Vlaeminck, C., Vierstraete, A., VanXeteren, J., Vincx, M. And Moens, T. (2007) Spatiotemporal analysis of population genetic structure in *Geomonhystera disjuncta* (Nematoda, Monhysteridae) reveals high levels of molecular diversity. *Marine Biology*, 15, 1799–1812.

Derycke, S., Vanaverbeke, J., Rigaux, A., Backeljau, T. and Moens, T. (2010) Exploring the use of cytochrome oxidase c Subunit 1(COI) for DNA barcoding of free-living marine nematodes. *PLoS One*, 5, e13716.

Donn, S., Griffiths, B.S., Neilson, R. and Daniell, T.J. (2008) DNA extraction from soil nematodes for multi-sample community studies. *Applied Soil Ecology*, 38, 20-26.

Freckman, D.W. and Virginia, R.A. (1997) Low diversity Antarctic soil nematode communities: Distribution and response to disturbance. *Ecology*, 78, 363-369.

Floyd, R., Abebe, E., Papert, A. and Blaxter, M.L. (2002) Molecular barcodes for soil nematode identification. *Molecular Ecology*, 11, 839–850.

Fonseca, V.G., Carvalho, G.R., Sung, W., Johnson, H.F., Power, D.M., Neill, S.P., Packer, M., Blaxter, M.L., Lambshead, P.J.D., Thomas, W.K., and Creer, S. (2010) Second-generation environmental sequencing unmasks marine metazoan biodiversity. *Nature Communications*, DOI: 10.1038/ncomms1095.

Gilbert, J.A., Field, D., Huang, Y., Edwards, R., Li, W., Gilna, P. and Joint, I (2008) Detection of large numbers of novel sequences in the metatranscriptomes of complex marine microbial communities. *PLoS One*, 3, e3042.

Giller, P.S. (1996) The diversity of soil communities, the ''poor man's tropical rainforest''. *Biodiversity and Conservation*, 5, 135–168.

Haegeman, A., Joseph, S. and Gheysen, G. (2011) Analysis of the transcriptome of the root lesion nematode *Pratylenchus coffeae* generated by 454 sequencing technology. *Molecular and Biochemical Parasitology*, 178, 7-14.

Hajibabaei, M., Shokralla, S., Zhou, X., Singer, G.A. and Baird, D.J. (2011) Environmental barcoding: a next-generation sequencing approach for biomonitoring applications using river benthos. *PLoS One*, 6, e17497.

Hallan, J. (2007). Synopsis of the Described Nematoda of the World. http://insects.tamu.edu/research/collection/hallan/Nematoda/Family/NematodaIndex0.htm

Hebert, P.D.N., Cywinska, A., Ball, S.L. and deWaard, J.R. (2003) Biological identifications through DNA barcodes. *Proceedings of the Royal Society B: Biological Sciences, 270, 313-321.*

Hebert, P.D., Stoeckle, M.Y., Zemlak, T.S. and Francis, C.M. (2004) Identification of birds through DNA barcodes. *PLoS Biol*ogy, 2, e312.

Hebert, P.D.N. and Gregory, T.R. (2005) The promise of DNA barcoding for taxonomy. *Systematic Biology*, 54, 852–859.

Ishak, H.D., Plowes, R., Sen, R., Kellner, K., Meyer, E., Estrada, D.A., Dowd, S.E. and Mueller, U.G. (2011) Bacterial diversity in *Solenopsis invicta* and *Solenopsis geminata* ant colonies characterized by 16S amplicon 454 pyrosequencing. *Microbial Ecology*, 61, 821-831.

Lambshead, P.J.D. (2004) Marine nematode biodiversity. In: Chen, Z.X., Chen, S.Y. and Dickson, D.W. (eds*). Nematology: Advances and Perspectives Vol 1: Nematode Morphology, Physiology and Ecology*. CABI Publishing pp. 436-467.

Margulies, M., Egholm, M., Altman, W.E., Attiya, S. and others (2005) Genome sequencing in microfabricated high-density picolitre reactors. *Nature*, 437, 376-380.Mayer, W.E., Schuster, L.N., Bartelmes, G., Dieterich, C. and Sommer, R.J. (2011) Horizontal gene transfer of microbial cellulases into nematode genomes is associated with functional assimilation and gene turnover. *BMC Evolutionary Biology,* doi:10.1186/1471-2148-11-13.

Miller, S.E. (2007) DNA barcoding and the renaissance of taxonomy. *Proceedings of the National Academy of Sciences USA*, 104, 4775-4784.

Pereira, T.J., Fonseca, G., Mundo-Ocampo, M., Guilherme, B.C. and Rocha-Olivares, A. (2010) Diversity of free-living marine nematodes (Enoplida) from Baja California assessed by integrative taxonomy. *Marine Biology*, 157, 1665-1678.

Porazinska, D.L., Giblin-Davis, R.M., Faller, L., Farmerie, W., Kanzaki, N., Morris, K., Powers, T.O., Tucker, A.E., Sung, W. and Thomas, W.K. (2009) Evaluating high-throughput sequencing as a method for metagenomic analysis of nematode diversity. *Molecular Ecology Resources*, 9, 1439-1450.

Porazinska, D.L., Giblin-Davis, R.M., Esquivel, A., Powers, T.O., Sung, W. and Thomas, W.K. (2010) Ecometagenetics confirms high tropical rainforest nematode diversity. *Molecular Ecology*, 19, 5521-5530.

Poretsky, R.S., Gifford, S., Rinta-Kanto, J., Vila-Costa, M. and Moran, M.A. (2009) Analyzing gene expression from marine microbial communities using environmental transcriptomics. *Journal of Visualised Experiments*, pii: 1086. doi: 10.3791/1086.

Powers, T.O., Neher, D.A., Mullin, P., Esquivel, A., Giblin-Davis, R.M., Kanzaki, N., Stock, S.P., Mora, M.M. and Uribe-Lorio, L. (2009) Tropical nematode diversity: vertical stratification of nematode communities in a Costa Rican humid lowland rainforest. Molecular Ecology, 18, 985-996.

Qi, W., Käser, M., Röltgen, K., Yeboah-Manu, D. and Pluschke, G. (2009) Genomic diversity and evolution of *Mycobacterium ulcerans* revealed by next-generation sequencing. *PLoS Pathogen,* 5, e1000580.

Radulovici, A.E., Archambault, P. and Dufresne, F. (2010) DNA barcodes for marine biodiversity: moving fast forward? Diversity, 2, 450-472.

Rayé, G., Miquel, C., Coissac, E., Redjadj, C., Loison, A. and Taberlet, P. (2010) New insights on diet variability revealed by DNA barcoding and high-throughput pyrosequencing: chamois diet in autumn as a case study. *Ecological Research*, 26, 265-276.

Roeding, F., Janus, B., Kube, M., Klages, S., Reinhardt, R. And Burmester, T. (2009) A 454 sequencing approach for large scale phylogenomic analysis of the common emperor scorpion (Pandinus imperator). *Molecular Phylogenetics and Evolution*, 53, 826-834.

SMEBD. World Register of Marine Species. Accessed at *http://www.marinespecies.orgon* [20-6-2011] (2011).

Sogin, M.L., Morrison, H.G., Huber, J.A., Welch, D.M., Huse, S.M., Neal, P.R., Arrieta, J.M. and Herndl, G.H. (2006) Microbial diversity in the deep sea and the underexplored "rare biosphere". *Proceedings of the National Academy of Sciences USA,* 103, 12115-12120.

Waite, I.S., O'Donnell, A.G., Harrison, A., Davies, J.T., Colvan, S.R., Ekschmitt, K., Dogan, H., Wolters, V., Bongers, T., Bongers, M., Bakonyi, G., Nagy, P., Papatheodorou, E.M., Stamou, G.P. and Boström, S. (2003) Design and evaluation of nematode 18S rDNA primers for PCR and denaturing gradient gel electrophoresis (DGGE) of soil community DNA. *Soil Biology and Biochemistry*, 35, 1165-1173.

Index

B

C

D

E

F

G

H

I

J

K

L

M

N

O

P

Q

R

S

T

U

V

W

X

Y

Z